MW00648254

Electric Energy

An Introduction

POWER ELECTRONICS AND APPLICATIONS SERIES

Muhammad H. Rashid, Series Editor
University of West Florida

PUBLISHED TITLES

Complex Behavior of Switching Power Converters
Chi Kong Tse

DSP-Based Electromechanical Motion Control
Hamid A. Toliyat and Steven Campbell

Advanced DC/DC Converters
Fang Lin Luo and Hong Ye

Renewable Energy Systems: Design and Analysis with Induction Generators
M. Godoy Simões and Felix A. Farret

Uninterruptible Power Supplies and Active Filters
Ali Emadi, Abdolhosein Nasiri, and Stoyan B. Bekiarov

Electric Energy: An Introduction
Mohamed El-Sharkawi

Electric Energy

An Introduction

Mohamed A. El-Sharkawi
University of Washington
Seattle

CRC PRESS
Boca Raton London New York Washington, D.C.

Library of Congress Cataloging-in-Publication Data

El-Sharkawi, Mohamed A.
Electric energy / by Mohamed A. El-Sharkawi.
p. cm.
ISBN 0-8493-3078-5 (alk.paper)
1. Power electronics. I. Title.

TK7881.15.E525 2005
621.3–dc22

2004057048

International Standard Book Number 0-8493-3078-5
Library of Congress Card Number 2004057048
Printed in the United States of America 1 2 3 4 5 6 7 8 9 0
Printed on acid-free paper

This textbook is dedicated to my wife Fatma and my sons Adam and Tamer.

The book is also dedicated to all engineers, without whom we would still be living in caves.

A special dedication goes to the fathers of our electric power systems: Nikola Tesla and Thomas Edison.

Preface

The first course on electric energy in engineering schools is traditionally taught as an energy conversion course. This was justified during most of last century as power was the main area of electrical engineering. Nowadays, the field of electrical engineering includes a number of new specializations, such as digital systems, computer engineering, communications, imaging, and networks. With the field being so widespread, energy conversion turned into a topic for students specializing in power only. Consequently, a large number of schools have decided to move their energy conversion classes from the core curricula to the elective curricula. Other schools with limited resources have dropped the energy conversion courses all together.

In recent years, there has emerged a renewed interest in electric energy because of several reasons. Among them are the ongoing search for renewable energy, the societal impact of blackouts, the environmental impact of generating electricity, and the lack of knowledge of most electrical engineers in fundamental subjects such as electric safety and generation of electricity. In addition, the new ABET criterion in the United States encourages the development of curricula that underline a broad education in engineering, contemporary engineering, and the impact of engineering solutions in a global and societal context. All these requirements can be met by restructuring the first electric energy course and making it relevant to all electrical engineering and most mechanical engineering students. This is the main objective of this textbook.

The textbook is authored to assist many schools looking at establishing a course with a wider view to the key aspects of electric energy while maintaining high levels of modeling and analysis. Most of the topics in this textbook are related to issues encountered daily and, therefore, should be of great interest to all engineering students. Most of the chapters in the textbook are structured to be stand-alone topics, so instructors can pick and choose the chapters they want to teach. The instructors can also select the sequence in which they prefer to teach the chapters. Most of the examples in the textbook are from real systems, with real data to make the course relevant to all students. Among the topics covered in the textbook are the energy resources, renewable energy, power plants, environmental impact of generation, electric safety, power market, and blackouts. In addition, the topics of electromechanical conversion, transformers, power electronics, and three-phase systems are included in the textbook to address the needs of some schools for teaching these important topics. However, these traditional topics are written to be more relevant to all students. For example, the sections on electric motors include linear and levitated motors as well as stepper motors.

The first chapter in the textbook is dedicated to the history of the development of the electric energy system. The milestones in the innovation trip start with the Greek philosopher Thales of Miletus around 600 B.C., passing through William Gilbert (1544–1603), Alessandro Giuseppe Antonio Anastasio Volta (1745–1827), André-Marie Ampère (1775–1836), Georg Simon Ohm (1789–1854), Michael Faraday (1791–1867), Hippolyte Pixii (1808–1835), Antonio Pacinotti (1841–1912), Thomas Alva Edison (1847–1931), and Nikola Tesla (1856–1943). In addition to history, the chapter presents current innovations and future possibilities in the field of electric energy.

In Chapter 2, modern power networks are described by three distinct systems: generation, transmission and distribution. Each one is identified and discussed in terms of the overall system operation. At the consumer voltage levels, the chapter addresses the frequency and voltage standards worldwide and gives reasons for the difference among the various standards.

Chapter 3 deals with the energy resources which are often divided into three categories: fossil fuel, nuclear fuel, and renewable resources. The fossil fuel includes oil, coal, and natural gas. The renewable energy resources include hydroelectric, wind, solar, hydrogen, biomass, tidal, and geothermal energy resources. All these resources are also classified as primary and secondary resources. The primary resources include fossil fuel, nuclear fuel, and hydroelectric energy. The secondary resources include all renewable energy minus hydroelectricity. The chapter discusses the known reserves as well as the world consumptions of the primary resources.

In Chapter 4, primary resources power plants are studied. This includes the description of the various designs, the main components of the power plants, the theory of their operation, and their modeling and analyses. The chapter has several examples of key calculations for the hydroelectric, fossil fuel, and nuclear plants.

Chapter 5 deals with electric energy from the viewpoint of the environment. Each of the primary resources is associated with some air, water and land pollutions. Although there is no pollution-free method for generating electricity, the primary resources are often accused of being responsible for most of the negative environmental impacts. In this chapter, the various pollution problems associated with the generation of electricity are presented, discussed, and put into perspective.

In Chapter 6, the renewable energy methods are presented. The phrase "renewable energy" is loosely used to describe energy generated in any manner from resources other than fossil or nuclear fuels. These include hydroelectric, wind, solar, wave and tidal, geothermal, and hydrogen. The chapter describes each of these technologies, the main components of each system, the merits and demerits of each technology, the mathematical modeling, and the energy calculations of key systems.

Chapter 7 is a review of the AC circuit analysis that is required for subsequent chapters. Usually, the topics in this chapter are covered in earlier courses. However, for completion, they are also included in this textbook.

Chapter 8 deals with electric safety, a topic often overlooked. The students will learn what constitutes a harmful levels of currents and their biological effects. They will also learn the primary and secondary shocks and how they are affected by various variables such as currents, voltages, frequencies, and pathways. Body and ground resistance calculations are studied as well as their effects on the step and touch potentials. The impact of ground resistance on the level of electric shock is explained by various common safety scenarios. Various world safety standards for neutral and ground are discussed and justifications for the three-prong plugs and outlets are made. Furthermore, the safety equipment for household applications such as the ground fault interrupter is presented. In addition, the concept of equipotential zones to protect people working adjacent to energized systems is discussed.

Chapter 9 discusses the three-phase systems. In additional to the mathematical analysis, the chapter discusses the reasons for using the three-phase systems and the various connections in power systems.

Chapter 10 deals with the various power electronic devices and circuits. Because of the power electronic revolution, the boundary between direct current and alternating current is becoming invisible. Besides the main solid state devices such as the SCR, IGBT, and MOSFET, the chapter analyzes the four types of converters: ac/dc, dc/dc, dc/ac, and ac/ac. For each of these converters, at least one circuit is discussed and analyzed in detail using waveforms and mathematical models.

Chapter 11 is an introduction to transformers. The chapter includes the single- and three-phase transformers as well as the autotransformer and transformer banks. Analyses for the ideal and actual transformers are given in details.

Chapter 12 discusses the essential electric machines such as the induction motor, the synchronous machines, and the stepper motor. All these machines are addressed in the context of their applications to make machine theory appear more relevant to all students. For example, linear and magnetically levitated trains are used as applications of the induction machine, and the hard disk drive and printer are associated with stepper motors. Besides the basic machine modeling, the students can analyze application scenarios such as a train powered by a linear induction motor traveling under certain road and wind conditions.

Chapter 13 deals with the operation of the power system. The chapter describes the secure and insecure operations as well as the stable and unstable conditions of the power system. The difference between radial and network structures as related to the power system reliability is analyzed. The effect of power imbalance on the power system stability is studied. The concepts of power angle, system capacity, and transmission limits are introduced.

The energy demand and the energy trade between utilities are addressed as an economical necessity as well as a reliability obligation. The concept of spinning reserve and its impact on the system reliability is given. The various conditions that could lead to brownouts and blackouts are addressed, and three major blackout scenarios in the United States are described.

Contents

1

History and Future of Power Systems

A large number of great scientists created the wonderful innovations that led to electric power as we know it today. However, it was not until the end of the 19th century that we envisioned having electric energy on demand by flipping a switch at every home, school, office, or factory. Today, dependence on electric power is so entrenched in our society that we cannot imagine our life without electricity. We indeed take electricity for granted, so when we experience an outage, we realize how much our life depends on electricity. This dependence creates a formidable challenge to engineers to make the power system the most reliable and efficient complex system ever built by man.

The route to creating an electric power system began when the Greek philosopher Thales of Miletus discussed electric charge, around 600 BC. The Greeks observed that when rubbing fur on amber, electric charge would build up on the amber. The charge would allow the amber to attract light objects such as hair. However, the first scientific study of electric and magnetic phenomena was done by the English physician William Gilbert (1544–1603). He was the first to use the term *electric*, which is a derivation from the Greek word for amber ($\eta\lambda\varepsilon\kappa\tau\rho o\nu$). The word amber itself was derived from the Arabic word *Anbar*.

Indeed, several volumes are needed to justifiably credit the geniuses behind the creation of our marvelous power system. However, for the lack of space, we shall only highlight the milestone developments. A good starting spot in history would be the middle of the 18th century when the Italian scientist Alessandro Guiseppe Antonio Anastasio Volta (1745–1827) showed that galvanism occurred whenever a moist substance was placed between two different metals. This discovery eventually led to the first battery in 1800.

Figure 1.1 shows a model of Volta's battery, which is displayed in the U.S. Smithsonian Museum in Washington, D.C. The battery was the source of energy used in subsequent developments to create magnetic fields and electric currents. Today we use the unit *volt* for the electric potential in honor of this great Italian inventor.

The next scientist is the French mathematician and physicist André-Marie Ampère (1775–1836). He was the first to explain the ambiguous link between magnetism and electric current. His work eventually led to the development of electromagnetic devices such as motors, generators, and transformers.

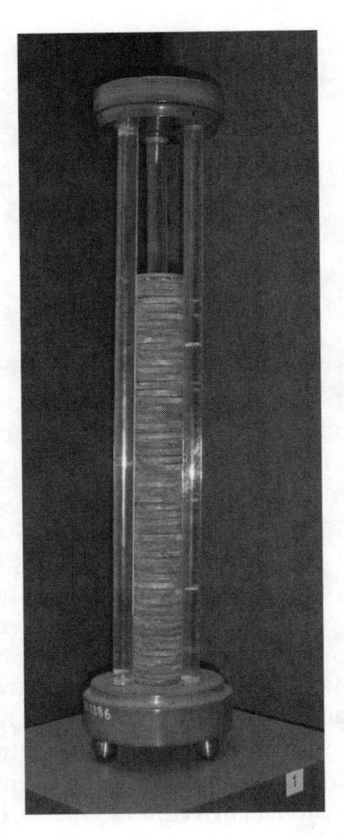

FIGURE 1.1
Model of Volta's battery (courtesy of the Smithsonian Museum).

Today we use *ampere* as a unit for electric current in honor of this French scientist.

The German scientist Georg Simon Ohm (1789–1854) was the first to relate the electric current to the electromotive force. His theory is known as Ohm's law. Although simple, the theory opened the door for circuit analysis and designs. We use *ohm* as the unit for resistance (or impedance) in his honor.

The next genius is the English chemist and physicist Michael Faraday (1791–1867). He was a superb matriculate experimentalist who laid the foundations for all electromechanical theories that we know today. In 1821, Faraday built a device that produced a continuous circular motion, which was the basis for the first alternating current motor. In his memory, the unit of capacitance (*farad*) is named after him.

The next inventor is probably the least known, but had enormous impact on power systems: Hippolyte Pixii (1808–1835) was a French instrument maker who built the first generator (or dynamo). His machine consisted of a magnet rotated by a hand crank, and surrounded by a coil. Pixii found that the rotation of the magnet induced voltage across the coil. Pixii's machine, which

FIGURE 1.2
Pixii's generator (courtesy of the Smithsonian Museum).

was named by him *magnetoelectric*, was later developed into the electrical generator. Figure 1.2 shows one of the early models of Pixii's generators.

The Italian inventor Antonio Pacinotti (1841–1912) invented a device that had two sets of windings wrapped around a common core. The windings were electrically isolated from the core and from one another. When an alternating voltage was applied across one winding, the second winding had induced voltage of similar shape, but different magnitude. This was the transformer we use today. Westinghouse further developed the transformer and had several early models; among them are the Gaulard and Gibbs transformer developed in 1883, shown in Figure 1.3, and the Stanley transformer developed in 1886.

The above marvelous innovations inspired two great scientists who are considered by many to be the fathers of electric power systems: Thomas Alva Edison and Nicholai Tesla. Before we discuss their work, let us first summarize their biographies.

FIGURE 1.3
Gaulard and Gibbs transformer (courtesy of the Smithsonian Museum).

1.1 Thomas A. Edison (1847–1931)

Thomas Alva Edison was born on February 11, 1847, in Milan, Ohio, and died in 1931 at the age of 84. Early in his life, Edison showed brilliance in math and science, and was a very resourceful inventor with a wealth of ideas and knowledge. In addition to his engineering brilliance, Edison was also a businessman, although not highly successful by today's standards. During his career, he had a large number of patents issued under his name, a total of 1093 patents. His first patent was granted at the age of 21 in 1868, and his last one was at the age of 83. This makes him one of the most productive inventors ever, with an average of about 1.5 patents per month. Of course, most of his patents were due to the teamwork of his research assistants.

His first patent was on an electric voting machine that tallied voting results fairly quickly. He tried to convince the Massachusetts Legislatures to adapt his machine instead of the tedious manual process used at that time, but did not succeed. The politicians did not like his invention because the slow process at that time gave the politicians a chance to influence, and perhaps change, the voting results.

During the period from 1862 to 1868, Edison worked as a telegrapher in several places including New England, which was the Silicon Valley of that time; all new and interesting innovations were made in New England. During this period, he invented a telegraphic repeater, duplex telegraph, and message printer. However, Edison could not financially gain from these inventions, so he decided to move to New York City in 1869.

While in New York, the Wall Street stock trade was disrupted by a failure in the stock ticker machine. Coincidentally, Edison was in the area and managed to fix the machine. This landed him a high-paying job that allowed him to continue with his innovations during his off-working hours.

In 1874 at the age of 27, Edison opened his first research and development laboratory in Newark, NJ. In 1876, he moved the facilities to a bigger laboratory at Menlo Park, NJ. These facilities were among the finest research and development laboratories in the world. Great scientists worked in his labs such as Tesla, Lewis Howard Latimer (an African American who worked on the carbon filaments), Jonas Aylsworth (who pioneered plastics), Reginald Fessenden (a radio pioneer), William K.L. Dickson (Movie developments), Walter Miller (Sound recording), John Kruesi (phonograph), and Francis R. Upton (dynamo). At that time, the innovations and productivities of Edison's labs were unmatched. By today's jargon, these labs were the core of the Silicon Valley of the 19th century. Because of his success with various innovations, Edison expanded his research and development laboratory in 1887, and moved it to West Orange, NJ. Then, he created the Edison General Electric Company, which in 1892 became the famous General Electric Corporation.

The large number of Edison's inventions includes the silent movies, alkaline storage battery, electric generator, ore separator, printer for stock ticker, printer for telegraph, electric locomotive, electric pen, motion picture camera, loudspeakers, telephone, microphone, mimeograph, and the telegraph repeater.

Edison was a highly competitive scientist who was willing and able to compete with great scientists in several areas at the same time. Among his innovations is the carbon transmitter, which ultimately was developed into the telephone. However, Alexander Graham Bell was also working on the same invention, and he managed to beat Edison to the telephone patent in 1876. This specific event upset Edison tremendously, but did not slow him down. He worked on the phonograph, which was one of Bell's projects. This time Edison patented the phonograph in 1878 before Bell could apply for his own patent.

One of Edison's most important inventions was the electric light bulb. This simple device is one of his everlasting inventions. The first incandescent light bulb was tested in Edison's lab in 1878 (see Figure 1.4). Although the filament

FIGURE 1.4
Edison's light bulbs (courtesy of the Smithsonian Museum).

of the bulb burned up quickly, Edison was very esthetic about the commercial prospects of his invention. Several developments were later made to produce filaments with longer lifetime.

Before the invention of the electric light bulb, New York streets were illuminated by oil lamps. So when Edison invented his bulb, the technology was resisted by Rockefeller, who was an oil tycoon. But soon after realizing that oil would be used to generate electricity, Rockefeller accepted this new technology and was among its promoters.

Edison received the *U.S. Congressional Gold Medal for Career Achievements* in 1928. When Edison died in 1931, people worldwide dimmed their lights in honor of his achievements. Initially it was suggested that the proper action would be to turn off the electricity completely for a few minutes, but the suggestion was quickly ruled out because of impracticality of the suggestion; the power systems worldwide would have had to shut down and restart, which was considered to be inconceivable task.

Today, a number of museums around the world have a section dedicated to Edison and his inventions. Among the best in the United States are the *Edison National Historic Sites* in West Orange, and at Menlo Park in Edison, NJ. Also, the *U.S. National Museum of American History* has a section dedicated to Edison.

1.2 Nikola Tesla (1856–1943)

Nikola Tesla was born in Smiljan, Croatia on July 9, 1856. Tesla was a gifted person with an extraordinary memory and a highly analytical mind. He had his formal education at the Polytechnic School of Gratz, Austria, and the University of Prague in the areas of math, physics, and mechanics. In addition to his strength in science and math, Tesla spoke six languages.

During his career, Tesla had over 800 patents. Although lesser in number than Edison's record, his patents were the direct result of his own work. Furthermore, unlike Edison, Tesla was broke most of his life and could not afford the cost of the patent applications. Actually, his most resourceful era was his last 30 years, but because of his poor financial condition, he managed to apply for a very few patents.

Tesla moved from Europe to the United States in 1884 and worked for Thomas Edison in his lab as a research assistant. This was during the time Edison patented the light bulb and was looking for a way to bring electricity to homes, offices, and factories. He needed a reliable system to distribute electricity and Tesla was hired to help in this area.

Among Tesla's inventions are the alternating current motor shown in Figure 1.5, the hydroelectric generator, the radio, x-rays, vacuum tubes, fluorescent lights, microwaves, radar, Tesla coil, the automobile ignition system, the speedometer, and the electronic microscope.

In the opinion of many scientists, historians, and engineers, Tesla invented the modern world. This is because he developed the concept of the power

FIGURE 1.5
Tesla's alternating current motor (courtesy of the Smithsonian Museum).

system we use today. By today's business standards, Tesla should have been among the richest in the world. But because of his unfortunate business failures, Tesla was broke when he died at the age of 86 on July 7, 1943.

Among the honors given to Tesla was the IEEE *Edison Medal* in 1917, the most coveted prize in electrical engineering in the United States. Tesla was inducted into the *Inventor's Hall of Fame* in 1975. He received honorary degrees from institutions of higher learning including Columbia University, Yale University, University of Belgrade, and the University of Zagreb. He also received the Elliott Cresson Medal of the Franklin Institute. In 1956, the term *tesla* was adopted as the unit of magnetic flux density in the MKSA system in his honor. In 1975, the IEEE Power Engineering Society established the Nikola Tesla Award for the outstanding contribution in power in his honor.

Among the best museums for Tesla are the *Nikola Tesla Museum* in Belgrade, and *Nikola Tesla Museum of Science and Industry* in Colorado Springs, Colorado.

1.3 The Battle of ac versus dc

Edison continued to develop the light bulb and made it reliable enough for commercial use by the end of the 19th century. Edison powered his bulbs by *direct current* (dc) sources that were commonly used at that time. The voltage and current of a dc source are unchanging with time; an example is shown in Figure 1.6. The battery is an excellent example of a direct current source.

The people in the northeast of the United States were very eager to use this marvelous electrical bulb in their homes because, unlike oil lamps, electrical bulbs did not emit horrible fumes. Edison, who was also a businessman, decided to build a dc power system and sell electricity to these eager customers. He erected poles and extended wires (conductors) above the city streets. In September 1882, his Pearl Street plant in lower Manhattan started operation as the world's first commercial electric lighting power station. One of the most significant events in technological history took place in 1883 when Edison electrified the city of Brockton, MA. The center of the city was the first place on earth to be fully electrified by a central power system.

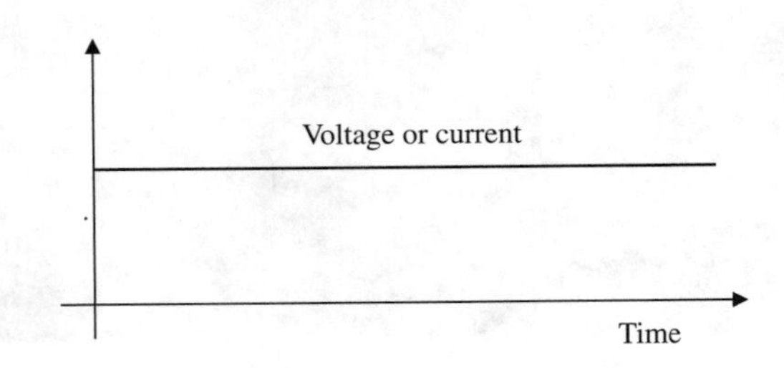

FIGURE 1.6
Direct current waveform.

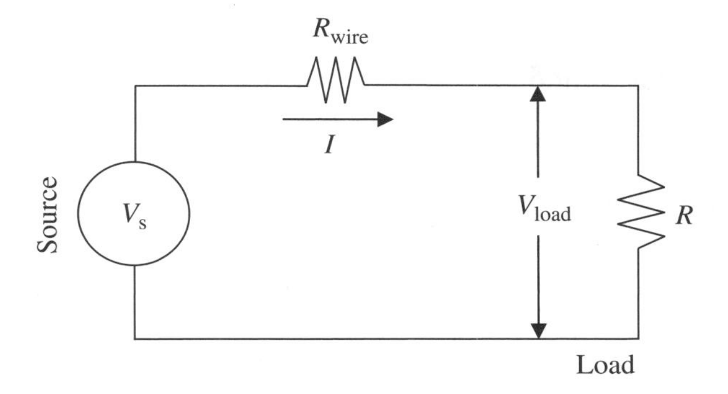

FIGURE 1.7
Simple representation of Edison's direct current system.

Edison dc generators were low voltage (100 V) machines. As a result of the increasing demand for electricity, the wires carried larger currents causing large voltage drop across the wires, thus reducing the voltage at the customers' sites. To understand these relationships, consider the simple representation of Edison's system shown in Figure 1.7. V_s is the voltage of the dc source (100 V), R is the load resistance (large number of bulbs for large number of homes), and R_{wire} is the resistance of the wire (conductor) connecting the source to the load. The voltage across the load V_{load} can be expressed by Equation (1.1).

$$V_{load} = V_s - IR_{wire}$$
$$V_{load} = IR = \frac{V_s}{R + R_{wire}} R = \frac{V_s}{1 + (R_{wire}/R)} \tag{1.1}$$

Because more customers means more load resistances are added in parallel, the total load resistance R decreases when more customers are added to the system. As you can see from Equation (1.1), higher loads (i.e., more customers and smaller R) lead to lower voltage at the customers' end of the line V_{load}.

EXAMPLE 1.1 Assume $V_s = 100$ V, $R = 1\ \Omega$, and $R_{wire} = 0.5\ \Omega$. Compute the following:

1. The voltage at the load side.
2. The percentage of the load voltage with respect to the source voltage.
3. The energy consumed by the load during a 10-hour period.
4. The maximum load (minimum resistance) if the load voltage cannot be reduced by less than 10% of the source voltage.
5. The energy consumed by the new load during a 10-hour period.

Solution

1. By direct substitution in Equation (1.1) we can compute the voltage at the load side.

$$V_{\text{load}} = \frac{V_{\text{s}}}{1 + (R_{\text{wire}}/R)} = \frac{100}{1 + (0.5/1)} = 66.67 \text{ V}$$

This low voltage at the load side may not be high enough to shine the light bulbs.

2. $\dfrac{V_{\text{load}}}{V_{\text{s}}} = 66.67\%$
3. The energy is the power multiplied by time. Hence, the energy consumed by the load is

$$E = Pt = \frac{V_{\text{load}}^2}{R} t = \frac{66.67^2}{1} 10 = 44.444 \text{ kWh}$$

4. If the minimum load voltage is 90%, the load resistance can be computed from Equation (1.1) as

$$\frac{V_{\text{load}}}{V_{\text{s}}} = \frac{1}{1 + (R_{\text{wire}}/R)}$$

$$0.9 = \frac{1}{1 + (0.5/R)}$$

$$R = 4.5 \ \Omega$$

Note that the new load resistance is 4.5 times the original load resistance. This means that less customers are connected, and less energy is consumed by the load.

$$E = Pt = \frac{V_{\text{load}}^2}{R} t = \frac{(0.9 \times 100)^2}{4.5} 10 = 18.0 \text{ kWh}$$

This is less energy than the one computed in step 3.

As seen in Example 1.1, the load voltage V_{load} is reduced when more customers are added. The main reason is because the conductor resistance R_{wire} and the load current I, cause a voltage drop across the conductor, thus reducing the voltage across the load resistance as given in Equation (1.1). The wire resistance increases when the length of the wire increases, or the cross-section of the wire decreases, as given in the following equation.

$$R_{\text{wire}} = \rho \frac{l}{A} \tag{1.2}$$

where ρ is the resistivity of the conductor in Ωm which is a function of the conductor's material. A is the cross-section of the wire and l is the length of the wire.

EXAMPLE 1.2 Assume $V_s = 100$ V and the load resistance $R = 1\ \Omega$. Compute the length of the wire that would not result in load voltage reduction by more than 10%. Assume the wire is made of copper and the diameter of its cross-section is 3 cm.

Solution

From Equation (1.1), we can compute the wire resistance.

$$\frac{V_{\text{load}}}{V_s} = \frac{1}{1 + (R_{\text{wire}}/R)}$$

$$0.9 = \frac{1}{1 + (R_{\text{wire}}/1)}$$

$$R_{\text{wire}} = 0.111\ \Omega$$

From the tables of units in the appendix, the resistivity of copper is $1.673 \times 10^{-8}\ \Omega$m. Hence, the length of the wire can be computed by using Equation (1.2).

$$l = \frac{AR_{\text{wire}}}{\rho} = \frac{(\pi r^2)R_{\text{wire}}}{\rho} = \frac{(\pi (0.015)^2)\,0.111}{1.673 \times 10^{-8}} = 4.69 \text{ km}$$

A wire with length longer than 4.69 km will result in lower voltage across the load.

As seen in Equation (1.2) and Example 1.2, long wires cause low voltage across the load. Because of this reason, the limit of Edison's system was about 3 miles, beyond which the wire resistance would be high enough to render the light bulbs useless. To address this problem, Edison considered the following three solutions:

1. To reduce the wire resistance by increasing its cross-section. This solution was expensive because:
   a. conductors with a bigger cross-section are more expensive;
   b. bigger conductors are heavier and would require bigger poles to be placed at shorter spans.
2. To have several wires feeding areas with high demands. This was also an expensive solution as it would require adding more wires for long miles.
3. To place electrical generators at every neighborhood. This was also an impractical and expensive solution.

Edison was faced with the first real technical challenge in power system planning and operation. He looked at all possible solutions from anyone who could help. That was probably why he hired Tesla in his lab and promised him lucrative bonuses if he could solve this problem.

Tesla knew the problem was related to the low voltage (100 V) Edison was using in his dc system. The viable alternative was to increase the supply voltage so the current could be reduced, and consequently the voltage drop across the wire could be reduced. Since the power consumed by the load is the multiplication of voltage by current [as given in Equation (1.3)], an increase in the source voltage results in a decrease in current. If the current decreases, the voltage drop across the wire IR_{wire} is reduced, and the load voltage $V_{load} = V_s - IR_{wire}$ becomes a higher percentage of the source voltage. Although logical, adjusting the voltage of the dc systems was unknown technology at that time.

$$P = VI \tag{1.3}$$

Instead of the dc system, Tesla promoted a totally different concept. Earlier in his life, Tesla conceived the idea of the *alternating current* (ac) whose waveform is shown in Figure 1.8. The ac is a sinusoidal waveform changing from positive to negative and back to positive in one cycle. Tesla wanted to use the ac to produce a rotating magnetic field that can spin motors (see Chapter 12). While working at the Continental Edison Company in Paris, Tesla actually built the first motor running on ac. The machine was later named the *induction motor* (see Chapter 12).

Tesla was also aware of the device invented by Pacinotti in 1860 that could adjust the ac voltages. This device was further developed by Tesla and became the transformer we use today (see Chapter 11). The transformer requires alternating magnetic fields to couple its two separate windings. Therefore, it is designed for ac waveforms and is ineffective in dc systems.

Tesla molded these various technologies into a new design for the power system based on ac as shown in Figure 1.9. The voltage source of the proposed system was ac at generally low voltage. *Transformer 1* was used to step up (increase) the voltage at the wire side of the transformer to a high level. The

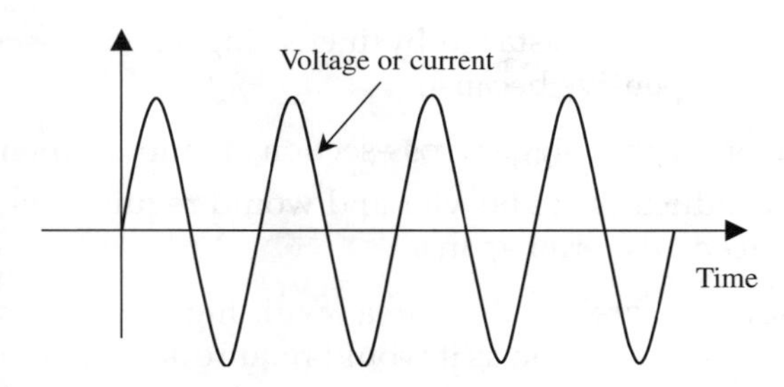

FIGURE 1.8
Alternating current waveform.

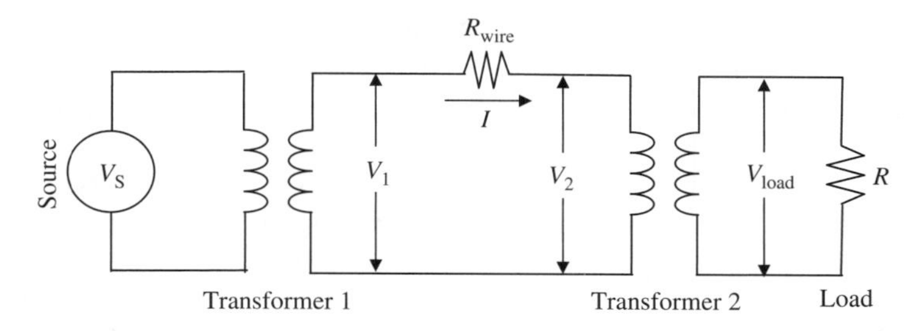

FIGURE 1.9
Tesla's proposed system.

high voltage of the wire resulted in lower current inside the wire. Thus, the voltage drop across the wire IR_{wire} was reduced. At the customers' sites, the high voltage of the wire was stepped down (reduced) by *transformer 2* to a safe level for home use.

Tesla stated three main advantages for his ac system:

1. The ac power can be transmitted over long distances with little voltage drop across its wires.
2. There is no need to install generating plants at residential areas as suggested by Edison for his dc system.
3. The ac can create rotating magnetic fields for electric motors.

Edison was not impressed by Tesla's ac system because of its unsafe high voltage wires that would pass through residential areas. However, most historians believe that Edison's rejection of the ac system was because he had too much money invested in the dc infrastructure. This disagreement created an irreparable rift between these two great scientists. Tesla and Edison continued to disagree on various other issues, and Tesla left Edison's laboratory and eventually worked for Westinghouse.

The dc versus ac battle started between Edison and General Electric on one side, and Tesla and Westinghouse on the other. Edison used unconventional methods to convince the public that Tesla's high-voltage ac system was too dangerous. He made live demonstrations where he deliberately electrocuted animals such as puppies, cats, horses, and even elephants. Edison went so far as to convince the state of New York to use an electric chair powered by a high-voltage ac system to execute condemned inmates on the death row. Most historians believe that his real motive was to further tarnish the safety of the ac system. Sure enough, New York City executed its condemned inmates by the proposed electric chair, and Edison captured the occasion by referring to the execution process as "Westinghoused," a clear reference to the ac system promoted by Tesla and Westinghouse. When Tesla signed a contract with Westinghouse to erect ac wires, posters were put on the power poles to warn residents from the danger of being "Westinghoused."

In addition to the negative publicity, Westinghouse ran into financial troubles and Tesla's contract with Westinghouse was terminated before the ac system was fully implemented. Instead of becoming the world's richest man, Tesla died broke.

Despite all the negative publicity created by Edison, the advantages of the ac systems were overwhelming. In fact, Edison later admitted that ac is superior to dc for power distribution, and was quick to exploit the economical advantages of the ac system through several inventions. One time when Edison met with George Stanley, the son of William Stanley who developed the transformer for Westinghouse, he told him, "Tell your father I was wrong."

Tesla was considered by many historians to be the inventor of the new world. This is probably true, since social scientists use energy consumption as a measure of the advancements of modern societies.

1.4 Today's Power Systems

The great scientists mentioned in this chapter, and many more, have provided us with the power system we know today. The power engineers made use of the various innovations mentioned in this chapter to deliver electricity safely and reliably to each home, commercial building, and factory. Indeed, it is hard to imagine our life without electricity.

In 2002, the world was capable of generating about 16.1×10^{12} kWh of electrical energy. This awesome amount of energy is the main reason for the highly developed societies we enjoy today. To deliver this amount of power to the consumers, power engineers constructed several millions of transmission lines, power plants, and transformers. It is overwhelming to imagine how such a massive system is controlled and operated on a continuous basis.

NASA often publish fascinating panoramic photos of the earth at night; one of them is shown in Figure 1.10. The photo shows the earth's man-made lights and is composed of hundreds of pictures made by the orbiting Defense Meteorological Satellites Program (DMSP). As you see in the photo, man-made lights make it quite possible to identify the borders of the cities and countries. Note that the developed or populated areas of the Earth, such as Europe, Eastern United States, and Japan, are quite bright reflecting their high consumption of electric energy. Also, for countries with high populations around rivers, such as Egypt, you can trace the path of the rivers by following the lights' trail. The Dark areas, such as in Africa, Australia, Asia, and South America, reflect low consumption of electricity or unpopulated regions. In North Africa, you can identify the great Sahara desert by the dark and expanded region. Similarly, you can identify the unpopulated outback region in Australia.

FIGURE 1.10
Earth at night (courtesy of NASA).

1.5 Future Power Systems

Thomas Edison and Nikola Tesla invented the early power generation system. Their systems, however, have gone through tremendous developments and improvements over the past century. Today's power system is probably the most complex system known to man. It is immense in capacity, huge in size, complex in operation, and fast in response. Despite the various blackouts we are experiencing every few years, the reliability of the power system is extremely high considering the complexity and capacity of the system.

The question often asked is what the future power system would look like. Is it going to be the same system we know today, but more expanded and more complex? Will we use the same energy resources? Will the energy resources last for future generations? Will we keep locating our generation near cities? Will we continue polluting the environment? Will we be able to provide electricity to remote places on earth?

These are tough questions that may require a crystal ball to answer them. Nevertheless, researchers are working hard to develop what they perceive as the power system of the future. The following are some of the ideas and concepts that are being researched. Some of them are in the conceptual phases and others are currently being implemented.

1.5.1 Less Polluting Power Plants

Polluting the environment is no longer acceptable by the public and most governments. The health effects and environmental damages due to industrial wastes are well monitored and documented in most of the world. Although regulations and severe penalties are imposed on industrial polluters, the effective solutions to the problem may include the following:

- The development of new industrial processes that produce less pollutions.
- The use of less-polluting material. For example, natural gas can be used instead of coal.
- Preprocessing the material to remove the polluting agents. For instance, reducing the sulfur in coal.
- Develop effective filters at every stage of the production.
- Develop creative ideas to reduce the negative impact on the environment; for example, hydropower plants with fish ladders facilitate fish migration.

It has been wrongly perceived that implementing rigid environmental enhancement measures may lead to higher costs and fewer jobs. In fact, studies have shown that the extra cost added to the products is offset by the reduction in medical expenses. Indeed, numerous studies show that health was improved in areas where the pollution was reduced. Furthermore, reducing the pollution is an industry by itself. It requires the skills of engineers, chemists, physicists, mathematicians, biologists, meteorologists, hydrologists, geologists, medical doctors, several government agencies, and many more. This field is wide open to new ideas and methods.

1.5.2 Alternative Resources

Because of the environmental problems associated with the existing forms of energy resources (fossil, hydro, and nuclear), researchers are looking for other resources that are less polluting and have less impact on the environment. Among the alternative resources are the energy of the sun, wind, tide, geothermal, biomass, hydrogen, and others. Some of these technologies are in the early stages of developments, and much work is needed to make them viable for serious consideration. Others are further along in their development, but more work is needed to increase their efficiencies and reliabilities, and make them cost effective.

1.5.3 Distributed Generation

The power plants we know today are located in areas where energy resources exist, or can be easily accessed. This necessitates the construction of large regional generating plants, which is probably the most cost-effective arrangement. To transmit the power to the users, power lines hundreds of miles long are needed. The reliability of this power system is highly dependent on the transmission systems; if the line is tripped off, the service is interrupted.

An alternative method is to install distributed generation systems where the power plants are located in neighborhoods or even at homes. Since natural gas can reach a large number of homes, it can be used to generate electricity at homes by using small-size thermal power plants the size of regular water heaters.

Another idea is to use fuel cells to generate electricity for home use. These devices use hydrogen, or natural gas, to produce electricity without introducing pollution to the immediate environment.

The concept of distributed generation should make the electricity more reliable and more available. With this technology, blackouts that last for days will no longer affect large areas.

1.5.4 Power Electronics

The electrical apparatus and equipment are often designed to provide their optimum performances at certain loading conditions. When these devices operate at different loading conditions, their efficiencies are normally reduced. To improve the efficiency of most devices, the voltage and frequency must be adjusted continuously. This can be done effectively by using semiconductor power electronic devices. Indeed, most of the appliances we use today are much more efficient than their predecessors. Every year, newer power electronic devices and circuits are introduced that enhance the operation of various equipments and improve their overall efficiencies.

In addition to the enhancement of efficiencies, power electronic devices make possible the development of technologies such as fuel cells, photovoltaic, superconducting magnetic energy storage systems, and levitated systems. Power electronic devices allow us to use motors and actuators in applications that were never envisioned before, such as angio-surgeries, microtechnology, and mechatronics.

For high-power applications, power electronic devices are used to control the bulk power flow in transmission lines through devices such as the Flexible ac Transmission System (FACTS) and the voltage compensators.

1.5.5 Enhanced Reliability

Although the power systems are highly reliable, we tend to disagree with this notion when a blackout occurs. This is because we have become so dependent on electric power that blackouts are often very costly and highly annoying. To totally eliminate blackouts is, probably, an unachievable task. However, reducing the frequency and impact of blackouts should be possible to achieve.

At any given moment, the power system is moving enormous amounts of power through its various components. For stable operation, a delicate balance must always be maintained between demand and generation. If this balance is altered, blackouts could occur unless the balance is rapidly restored. The speed with which the energy balanced is restored depends on the availability of fast and synchronized sensing, fast acting relays and breakers, quick topology restoration, and effective system control. Although the research in these areas is continually advancing, predicting all possible scenarios for blackouts is a very challenging task indeed. Furthermore, to be able to analyze and control the system online requires much more powerful computers and better knowledge of the complex system than what we have attained so far.

The reliability of the power system also depends on the redundancy of the transmission systems. However, due to public concern, fewer and fewer transmission systems are allowed to be constructed. The public resistance to the constructions of new lines is based on ecstatic consideration and concerns regarding the magnetic fields. Newer designs and technologies are needed to address these two issues.

1.5.6 Monitoring and Control of Power Systems

One of the most critical aspects of controlling our massive power system is to sense and synchronize various signals measured at locations thousands of miles apart. Regular terrestrial communication networks have shortcomings with the exact synchronization of high-frequency signals. In recent ongoing research, engineers are looking at using *geosynchronous* (e.g., GPS) and *low-earth orbit* (LEO) satellites for better monitoring and control of the power systems. This is one of the intriguing ideas that will eventually lead to more effective and more reliable power system operation.

1.5.7 Space Power Plants

Solar energy at the outer space is much higher than at the surface of Earth. The U.S. government is pioneering the conceptual idea of having massive solar power plants orbiting the Earth. The envisioned system consists of solar array of the size of a football field, that could generate large amounts of power comparable to large terrestrial power plants. The generated electric power can be used to support spacecrafts and long space missions. The power may even be beamed back to the Earth. The concept of space solar power was considered improbable and silly in the 1970s. However, NASA and NSF have revived the idea in 2000 and several research programs are working on the development of technologies that may lead to this goal.

A space power plant is compatible with the environment and it could lead to less dependency on the conventional energy resources. The plant could potentially provide electric energy to remote and inaccessible areas all over the world. However, the idea of having satellites beaming their energy to earth has remarkable technological challenges. For example, the solar cell technology must be improved substantially to make it possible to place football field sized solar arrays in space. The solar cells must be lightweight, thin, flexible, highly efficient, and able to withstand the harsh environment of space.

Today's solar cell technology is not adequate for space power plants because they are bulky, heavy, of low efficiency, and brittle. But nanotechnology offers a potential solution to this problem.

Beaming the energy to Earth may require the use of microwaves and a highly accurate positioning system. The health risks associated with the exposure to electric and magnetic fields generated by beaming the power to Earth is significant public concern.

Furthermore, space power stations located miles apart in orbits pose maintenance challenges. Among the ideas proposed is to use swarms of autonomous robots as maintenance crew. The research in this area is fascinating and highly challenging.

1.5.8 Intelligent Operation, Maintenance, and Training

The human control in hazardous environments, such as the nuclear power plant, will eventually be replaced by intelligent autonomous robots. A swarm of these robots could provide the maintenance and operation inside the reactors where humans cannot be located. In addition, the power system will eventually benefit from the virtual reality technology as it will make it possible for experts located miles away to perform maintenance on high-voltage equipment while it is energized, must like the remote surgery approach being developed in the medical field.

Virtual reality could also provide an excellent training system that could lead to more experienced operators. This is because various scenarios can be imposed on the system and the operators learn how to steer the system to a secure and stable region. This is much like the flight training simulators for the aircraft pilots.

Exercise

1.1. In your opinion, identify 10 of the most important innovations in electrical engineering, and name the inventors of these innovations.

1.2. Thomas Edison has several innovative inventions to his credit; select one of them and write an essay on the history of its development and the impact of the invention.

1.3. Nikola Tesla has made several innovative inventions. Select one of them and write an essay on the history of its development and the impact of the invention.

1.4. The transformer is one of the major inventions in power systems. Why can we not use it in dc systems?

1.5. State the advantages and disadvantages of using low-voltage transmission lines.

1.6. State the advantages and disadvantages of using high-voltage transmission lines.

1.7. In your opinion, what are the major developments to foresee in future power systems?

1.8. A simple power system consists of a dc generator connected to a load center via a transmission line. The load resistance is 10 Ω. The transmission line is 50 km copper wire of 3 cm in diameter. If the voltage at the generator terminals is 400 V, compute the following:

a. The voltage across the load.

b. The voltage drop across the line.

c. The line losses.

d. The system efficiency.

1.9. A simple power system consists of a dc generator connected to a load center via a transmission line. The load power is 100 kW. The transmission line is 100 km copper wire of 3 cm diameter. If the voltage at the load side is 400 V, compute the following:

a. The voltage drop across the line V_{line}.

b. The voltage at the source side V_{source}.

c. The percentage of the voltage drop $V_{\text{line}}/V_{\text{source}}$.

d. The line losses.

e. The power delivered by the source.

f. The system efficiency.

1.10. Repeat the previous problem assuming that the transmission line voltage at the load side is 10 kV.

2

Basic Components of Power Systems

Modern power networks are made up of three distinct systems: generation, transmission, and distribution. Figure 2.1 shows a sketch of a typical power system. The generation system includes the main parts of the power plants such as the turbines and generators. Most of the energy resources are combustible, nuclear, or hydro. The burning of fossil fuels or the nuclear reaction generates heat that is converted into mechanical motion by the thermal turbines. In hydro systems, the flow of water through the turbines converts the kinetic energy of the water into rotating mechanical energy. These turbines rotate the electromechanical generators that convert the mechanical energy into electrical energy.

The generated electricity is transmitted to all customers by a complex network of transmission systems composed mainly of transmission lines, transformers, and protective equipment. The transmission lines are the links between power plants and load centers. The transformers are used to increase (step up) or decrease (step down) the voltage. At the power plant, a transmission substation with step-up transformers increases the voltage of the transmission lines to very high values (220 to 1200 kV). This is done to reduce

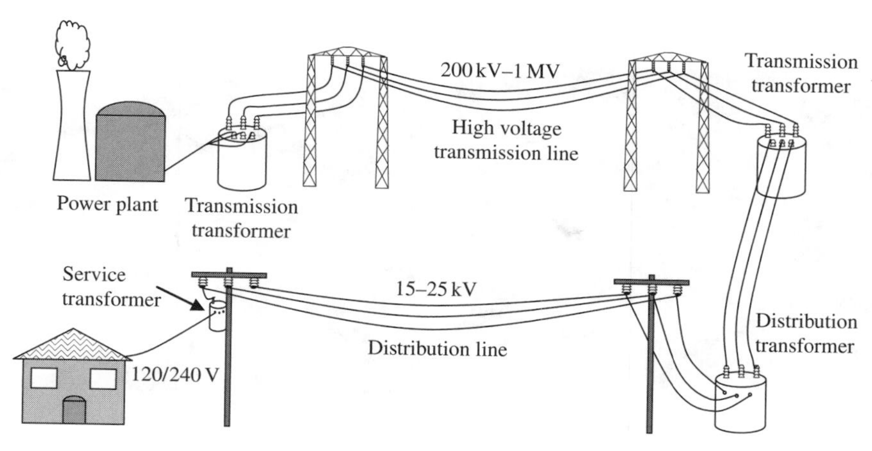

FIGURE 2.1
Main components of power systems.

the current through the transmission lines, thus reducing the cross-section of the transmission wires as well as reducing the overall cost of the transmission system. At the load centers, the voltage of the transmission lines is reduced by step-down transformers to lower values (15 to 25 kV) for distribution within the city limits. At the consumer sites, the voltage is further reduced to a value from 100 to 240 V for household use worldwide.

The power system is extensively monitored and controlled. It has several levels of protection to minimize the effect of any damaged component on the system ability to provide safe and reliable electricity to all customers. A number of the key devices and equipment used in power systems are covered in detail in this book, and a summary of their functions and operations is given in the following.

2.1 Power Plants

At the power plant, the energy resources such as coal, oil, gas, hydro, and nuclear are converted into electricity. The main parts of the power plant are the burner (in fossil plants), the reactor (in nuclear power plants), the dam (in hydro plants), the turbine, and the generator.

The power plants can be huge in size and capacity. For example, the ITAIPU Hydroelectric Power Plant, Figure 2.2, located at the Brazilian–Paraguaian border produces 75 TWh annually. China is building the largest nuclear power plant in the southern province of Guangdong with a capacity of 6 GW.

Although they are enormous in mass, the power plants are delicately controlled. A slight imbalance in power may cause blackouts unless rapidly

Byron Nuclear Power Plant

ITAIPU Hydropower Plant

FIGURE 2.2
(see color insert following Page 208) Power plants. (Images courtesy of the U.S. Department of Energy.)

corrected. The control of the massive amount of water or steam inside the plant creates an enormous challenge to the mechanical and structural engineers. The control of the system frequency and the system protection are immense challenges to the electrical engineers.

2.1.1 Turbines

The function of the turbine is to rotate the electrical generator by converting the thermal energy of the steam or the kinetic energy of water into rotating mechanical energy. The first type of turbines is called thermal turbine and the second is called hydro turbine.

In thermal power plants, fossil fuels or nuclear reactions are used to produce steam at high temperatures and pressures. The steam is passed through the blades of the turbine and causes the turbine to rotate. Since the generator is mounted on the shaft of the turbine, the generator rotates and electricity is generated. The steam flow is controlled by several valves at critical locations to ensure that the turbine is rotating at a precise speed.

A typical hydropower plant consists of a dam that holds the water behind the turbines at high elevations. The difference in height between the water surface behind the dam and the turbine blades is called the *head*. The larger the head, the more potential energy is stored in the water behind the dam. When electricity is needed, the water is allowed to pass to the turbine blades through pipes called *penstocks*. The turbine then rotates and so does the generator. The valves of the penstock are used to regulate the flow of water and thus control the speed of the generator.

2.1.2 Generators

The generator used in all power plants is the *synchronous* machine. The invention of the synchronous generator goes back to Hippolyte Pixii (1808–1835) who was the first to build a dynamo. The synchronous machine has a magnetic field circuit mounted on its *rotor* and is firmly connected to the turbine. The stationary part of the generator, called the *stator*, has windings wrapped around the core of the stator. When the turbine rotates, the magnetic field moves inside the machine in a circular motion. As explained by Faraday, the relative speed between the stator windings and the magnetic field induces voltage across the stator windings. When a load is connected across the stator windings, the load is energized, as explained by Ohm's law.

The output voltage of the generator (5 to 22 kV) is not high enough for transmission lines. Higher voltage generators are not practical to build as they require more insulation, making the generator unrealistically large in size. Thus, the output voltage of the generators is increased by using step-up transformers.

2.2 Transformers

The main function of the transformer is to increase (step up) or decrease (step down) the voltage. As explained in Chapter 1, the voltage of the transmission line must be high enough to reduce the current of the transmission line. When the electric power is delivered to the load centers, the voltage is stepped down for safer distribution over city streets as seen in Figure 2.1. When the power reaches the customers' homes, the voltage is further stepped down to the household level of 100 to 240 V depending on the various standards worldwide.

2.3 High-Voltage Transmission Lines

The transmission lines deliver the electrical power from the generating plant to the customers as shown in Figure 2.1. The bulk power of the generating plant is transmitted to the load centers over long distance lines called *high-voltage transmission lines*. The transmission lines that distribute the power within a city limit are called *medium-voltage distribution lines*. There are several other categories such as *sub-transmission* and *high-voltage distribution lines*, but these are fine distinctions that we should not worry about at this stage.

The transmission lines are high-voltage wires (220 to 1200 kV) mounted on tall towers to prevent the wires from touching the ground, humans, animals, buildings, or equipment. High-voltage towers are normally 25 to 45 m in height; one of these towers is shown in Figure 2.3. The higher the voltage of the wire, the taller is the tower.

High-voltage towers are normally made of galvanized steel to achieve the strength and durability needed in harsh environments. Since the steel is electrically conductive, the high-voltage wires cannot be attached directly to the steel tower. Instead, insulators made of nonconductive material mounted on the tower are used to hold the conductors away from the tower structure. The insulators withstand the static and dynamic forces exerted on the conductor during wind storms, freezing rain, and earth movements. Figure 2.4(a) shows an insulator which consists of a central rod with several conical disks. The top end of the central rod is bonded to the tower and the lower end is attached to the conductor. The cone-shaped disks have two functions:

1. Increase the flashover distance between the tower and the conductor.
2. When rain falls, the insulators are automatically cleaned.

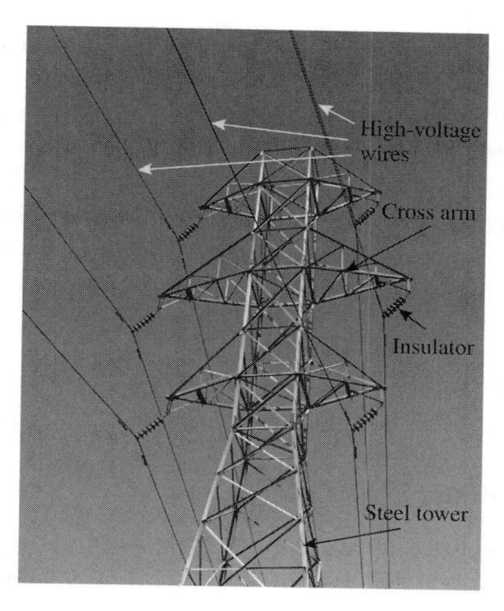

FIGURE 2.3
High-voltage transmission tower.

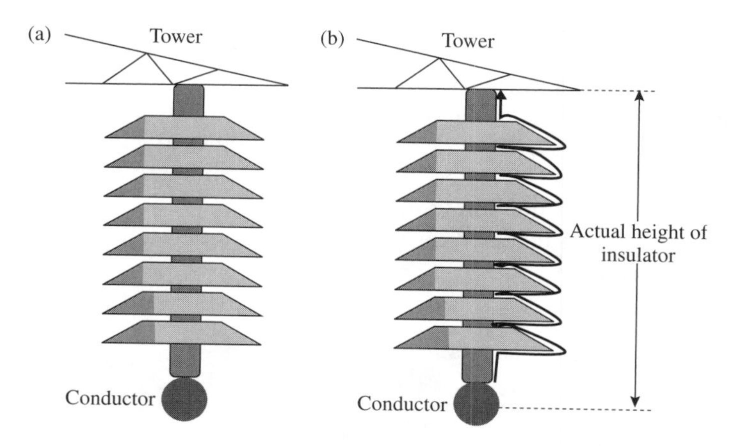

FIGURE 2.4
(a) Insulator; (b) flashover path.

An electrical discharge between any two metals having different potentials depends on the potential difference and the separation between the two metals. If an insulator is placed between two metals, a flashover may occur along the surface of the insulator if the potential difference between the two metals is high enough, or the thickness of the insulator is small. To protect against the flashover in high-voltage transmission lines, the insulator must be unrealistically long. However, if a disk-type insulator is used, the distance between the tower and the conductor over the surface of the insulator is more than the actual length of the insulator. This is explained in Figure 2.4(b), where the jagged arrow shows the flashover path. This path is longer than the actual height of the insulator. But why do the conical disks have upper radii smaller than the lower radii? The simple reason is that the configuration in Figure 2.4 makes it easy to clean the insulators. When rain falls, the deposed salt and dust on the insulator are washed away. If it is mounted upside down, there will be pockets where dust is trapped, thus reducing the insulation capability of the insulators.

2.4 Medium-Voltage Distribution Lines

The conductors of the medium-voltage distribution lines are either buried underground or mounted on poles and towers. In most cities, the distribution network is mainly underground for aesthetic and safety reasons. However, it is also common to have overhead distribution lines; one of these is shown in Figure 2.5.

Since the voltage of the distribution lines is much lower than that of the transmission lines, the distribution towers are shorter and their insulators are smaller. The distribution towers are often made of steel, wood, concrete, or composite material. Most of the commercial and industrial plants have direct access to the distribution network, and they use their own transformers to step down the voltage to the level needed by their equipment. In residential areas, utilities install transformers in vaults or on towers to reduce the distribution line voltage to any value between 100 and 240 V depending on the standard practice in the country.

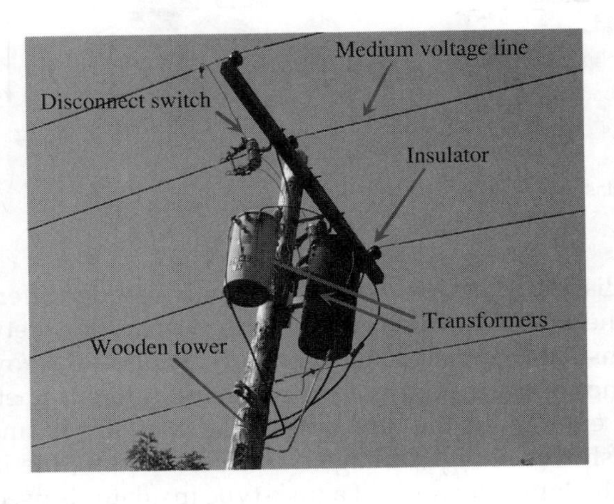

FIGURE 2.5
Medium-voltage distribution tower.

2.5 Worldwide Standards for Household Voltage and Frequency

The magnitude of the household voltage or frequency is not the same worldwide. This is mostly because of three reasons:

- Due to the competition between the United States and Europe in the late 1800s and early 1900s, the manufacturing companies in both continents have independently developed their power system equipment without coordinating their efforts.
- Safety concerns, being different in Europe and the United States, have led to various voltage standards.
- Wiring and equipment costs have led nations to select the most economical voltage standards.

Table 2.1 shows the voltage and frequency standards in selected countries. The voltage standard is for single-pole outlets used in most household appliances. In the United States, 120 V is used everywhere in the house, except for the heater, dryer, and oven where the receptacles are double-pole at 240 V. The various voltage and frequency levels have created confusion among manufacturers and consumers who wish to use equipment manufactured based on one standard, in a country with a different standard. For example, refrigerator compressors designed for 60 Hz standard, slow down by 17% when used in

TABLE 2.1

Standards for Voltage and Frequency in Several Nations

Country	Voltage (V)	Frequency (Hz)
Australia	230	50
Brazil	110/220	60
Canada	120	60
China	220	50
Cyprus	240	50
Egypt	220	50
Guyana	240	60
Korea, South	220	60
Mexico	127	60
Japan	100	50 and 60
Oman	240	50
Russian Federation	220	50
Spain	230	50
Taiwan	110	60
United Kingdom	230	50
United States	120	60

a 50 Hz system. Also, electrical equipment designed for 100 V standard, will be damaged if it is used in 240 V systems.

For small appliances, modern power electronics circuits have addressed this problem very effectively. Almost all power supplies for travel equipment (electric razors, portable computers, digital cameras, audio equipment, etc.) are designed to operate at all voltage and frequency levels worldwide.

2.5.1 Voltage Standard

It appears that the standard of 120 V was chosen somewhat arbitrarily in the United States. Actually, Thomas Edison came up with a high-resistance lamp filament that operated well at 120 V. Since then, the 120 V was selected in the United States. The standard for voltage worldwide varies widely from 100 V in Japan to 240 V in Cyprus. Generally, a wire is less expensive when the voltage is high; the cross-section of the copper wires is smaller for higher voltages. However, from the safety point of view, lower voltage circuits are safer than the high-voltage ones; 100 V is perceived to be less harmful than 240 V.

2.5.2 Frequency Standard

Only two frequencies are used worldwide: 60 Hz and 50 Hz. The standard frequency in North America, Central America, most of South America, and some Asian countries is 60 Hz. Everywhere else, 50 Hz is the standard frequency.

In Europe, major manufacturing firms such as Siemens and AEG have established 50 Hz as the standard frequency for their power grids. Most of Asia, parts of South America, all of Africa, and the Middle East have adopted the same 50 Hz standard.

In the United States, Westinghouse adopted the 60 Hz standard. Nikola Tesla actually wanted to adopt a higher frequency to reduce the size of the rotating machines, but 60 Hz was eventually selected. Among the reasons mentioned for the selection of 60 Hz are:

- It is high enough frequency to eliminate light flickers in certain types of incandescent lamps.
- It is conveniently synchronized with time.
- Machines designed for 60 Hz can have less iron and smaller magnetic circuits than the ones designed for 50 Hz.

In Japan, both frequencies are used; in Eastern Japan (Tokyo, Kawasaki, Sapporo, Yokohoma, and Sendai), the grid frequency is 50 Hz; and in Western Japan (Osaka, Kyoto, Nagoya, and Hiroshima), the frequency is 60 Hz. The first generator in Japan was imported from Germany for the Kanta area during the Meiji Era, and the frequency was 50 Hz. Subsequent acquisitions of power equipment were mainly from Europe and were installed in the east side of Japan, making the frequency of the eastern grid 50 Hz. More recently, all 60 Hz equipment was imported from the United States for Western Japan; the first was for the Kanzai area. The two-frequency standards created two separate grids. The dividing line between the two grids is from the Fuji River in Shizuoka upwards to the Itoi River in Niigata. These two grids are not directly connected by an ac line, but are connected by direct current lines.

It is interesting to know that for stand-alone systems, such as aircraft power systems, the frequency is 400 Hz. This is selected to reduce the size and weight of the rotating machines and transformers aboard the aircraft.

2.5.2.1 Frequency of a Generating Plant

The frequency of the system voltage is directly proportional to the speed of the generators of the power plants. The relationship, which is explained in detail in Chapter 12, is given in Equation (2.1):

$$f = \frac{P}{120} n \tag{2.1}$$

where n is the speed of the generator in revolutions per minute (rpm), P is the number of magnetic poles of the field circuit of the generator, and f is the frequency of the generated current.

One of the earliest ac generators was a 10-pole machine running at 200 rpm. The frequency of this generator, as explained by Equation (2.1), was $16\frac{2}{3}$ Hz. This anomalous frequency was low enough to allow the series-wound dc motor to operate from ac supply without any modification. This was a justifiable reason, because the series-wound motor was used extensively in locomotive tractions. However, this low frequency created noticeable flickers in incandescent lamps, and was therefore rejected as a standard.

When the Niagara Falls was built, the engineers used 12-pole generators running at 250 rpm, and the frequency was 25 Hz. The power plant was built to produce compressed air, and the low frequency was not a major concern. In following developments, the frequency was raised to 40 Hz, then 60 Hz to reduce the flickers.

2.5.2.2 Frequency of Power Grids

In early days, the generators operated independently without any connection between them. The typical system was composed of a single generator, feeders (transmission lines), and several loads. In such a system, the frequency can drift without a major impact on the stability of the system. In the 1940s, it became economically important to interconnect the generators by a system of transmission lines, and the power grid was born. The interconnections demanded that the frequency of the grid be fixed to a single value. If the frequencies of all generators are not exactly equal, the power system would collapse as explained in Chapter 13.

Exercise

2.1. What is the function of a power plant turbine?

2.2. What is the function of a power plant generator?

2.3. What is the function of the hydro dam?

2.4. Why are transformers used with transmission lines?

2.5. A 2-pole generator is to be connected to a 60 Hz power grid. Compute the speed of its turbine.

2.6. A 2-pole generator is to be connected to a 50 Hz grid. Compute the speed of its turbine.

2.7. Why are insulators used on power line towers?

2.8. Why are the tower insulators built as disk shapes?

2.9. Why is the frequency of the airplane power system 400 Hz?

2.10. Why are there different voltage and frequency standards worldwide?

2.11. Why are the transmission line towers higher than the distribution line towers?

2.12. What are the various forces that an insulator must withstand?

3

Energy Resources

As shown in Figure 3.1, the energy resources are often divided into three, loosely defined categories:

1. Fossil fuel.
2. Nuclear fuel.
3. Renewable resources.

The fossil fuel includes oil, coal, and natural gas. The renewable energy resources include hydro, wind, solar, hydrogen, biomass, tidal, and geothermal. All these resources can also be classified as primary and secondary resources. The primary resources are:

1. Fossil fuel.
2. Nuclear fuel.
3. Hydro energy.

The secondary resources include all renewable energy minus hydro. Over 99% of all electric energy worldwide is generated from the primary resources.

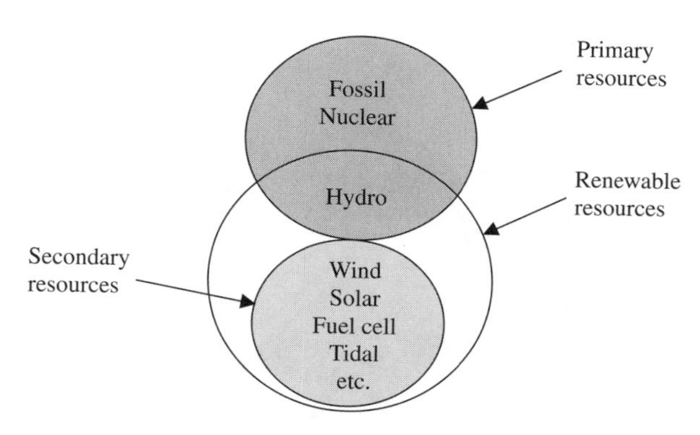

FIGURE 3.1
Energy resources.

The secondary resources, although increasing rapidly, have not yet achieved a level comparable to the primary resources.

In the subsequent statistics three terminologies are used: *generated electricity*, *consumed electricity*, and *generation capacity*. The generated electricity is the amount of electrical energy produced at the power plant. This is equal to the consumed energy plus the losses in various power system equipment such as transformers and transmission lines. The generation capacity is the amount of energy that could be generated if all generators operate at their full capacity at all times. The generation capacity is always more than the generated energy.

The distribution of electricity generated worldwide by primary resources is shown in Figure 3.2. As you can see, most of the electrical energy is generated by oil, coal, and natural gas. Hydroelectric energy is limited to about 6% of the world electrical energy because of the limited water resources suitable for generating electricity. Nuclear energy is only 7% of the total electrical energy because of public resistance to building new nuclear facilities.

In the United States, most of the electrical energy is generated by fossil fuel as depicted in Figure 3.3. Because of its abundance, coal is by far the prevailing source of energy in the United States. Actually, the United States has the largest coal reserve among all the other countries. After coal, nuclear energy accounts for about 20% of the generated electricity. Unfortunately, the share of the secondary resources is only about 2% of the total generated electricity. Some attributed this low percentage to the low cost of electricity in the United States.

The generation capacity worldwide in 2002 is shown in Figure 3.4. The total world generation capacity was about 16,100 Terawatt hour (16.1×10^{12} kWh) in 2002. About 30% of this capacity was in North America; the United States generation capacity was about 4027 Terawatt hour.

In 2002, the estimated worldwide consumption of electrical energy was about 14×10^{12} kWh; the distribution is shown in Figure 3.5. The United States' share of this energy was about 3.6×10^{12} kWh, which is about 25% of the world's energy consumption. In second place is China, with 9% of the

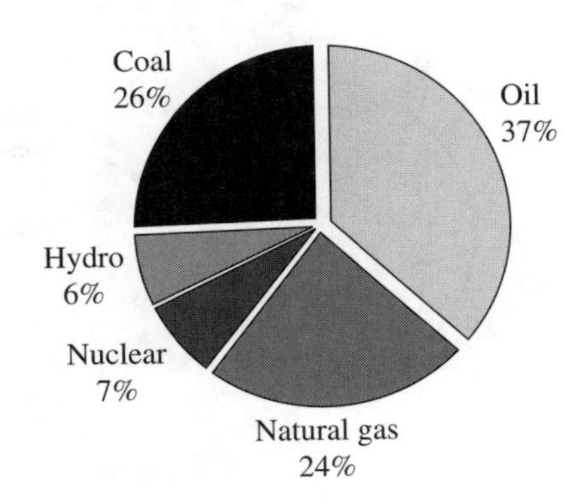

FIGURE 3.2
Primary fuel used to generate electricity worldwide (2002) (*Source*: British Petroleum Statistical Review of World Energy, June 2003).

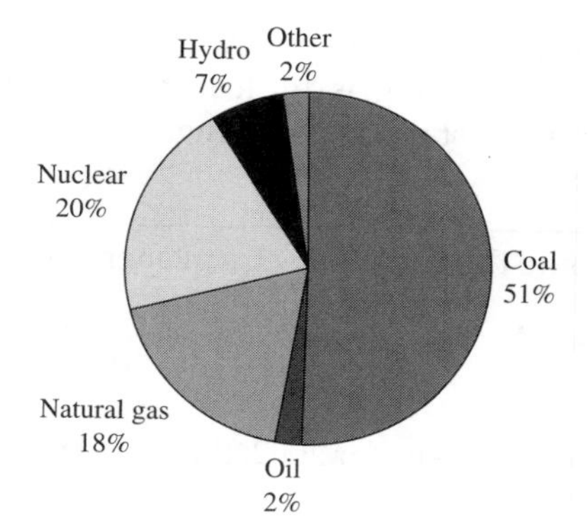

FIGURE 3.3
Fuel used to generate electricity in the United States (2002) (*Source*: U.S. Department of Energy, Annual Energy Review 2003).

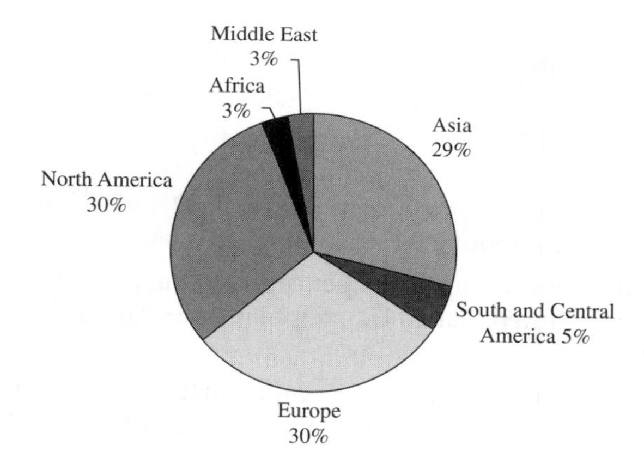

FIGURE 3.4
Generation capacity worldwide as of 2002 (*Source*: British Petroleum Statistical Review of World Energy, June 2003).

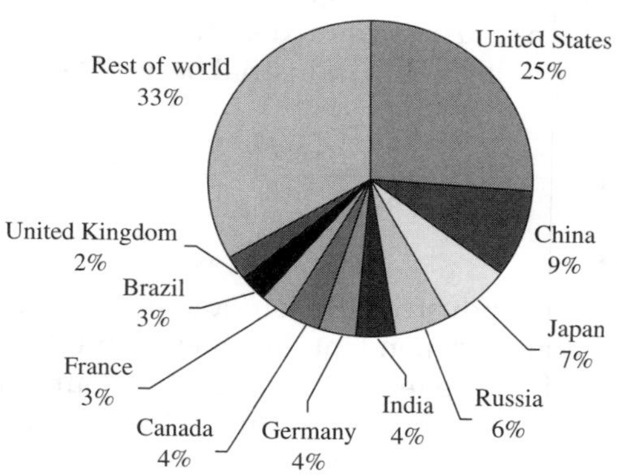

FIGURE 3.5
Consumed electric energy worldwide in 2002 (*Source*: British Petroleum Statistical Review of World Energy, June 2003).

world's electric energy consumption, followed by Japan. The high demand in China, which is rapidly increasing, is due to its recent industrial developments and the enhancement in its standard of living.

EXAMPLE 3.1 Compute the annual electrical energy consumption per capita worldwide in 2002.

Solution

According to the world census for 2002, the world population by the end of 2002 was about 6.3×10^9 people.

Annual electric energy consumed per capita worldwide = Total world consumption/world population $= 14 \times 10^{12}/6.3 \times 10^9 = 2.22$ MWh.

EXAMPLE 3.2 Compute the annual electric energy consumed per capita in the United States in 2002. Compare the result with the worldwide average.

Solution

According to the United States' estimate for 2002, the U.S. population by the end of 2002 was about 2.92×10^8 people.

Annual electric energy consumed per capita in the United States = Total consumption in United States/U.S. population $= 3.6 \times 10^{12}/2.92 \times 10^8 =$ 12.33 MWh.

Annual electric energy consumed per capita in the United States/Annual electric energy consumed per capita worldwide $= 12.33/2.22 = 5.55$.

In 2002, the average person in the United States consumed more than five times the electrical energy consumed by the average person worldwide. This high consumption is not just attributed to the luxury living in the United States, but is also due to its highly industrial base.

3.1 Fossil Fuel

Since the start of the Industrial Revolution in Europe in the 19th century, the world became dependent on energy produced by fossil fuel, which was realized as an effective and reliable source for energy. Fossil fuel is formed from fossils buried for millions of years. It is composed of high carbon and hydrogen elements such as oil, natural gas, and coal. Because the formation of fossil fuel takes millions of years, it is considered nonrenewable.

The use of fossil fuel to generate electricity is always a hot subject for debates. Among the main issues associated with fossil fuel are:

1. Fossil fuel has been an indispensable source of energy until other forms of energy resources became viable.
2. Burning fossil fuel causes a wide range of pollutions that include the release of carbon dioxides, sulfur oxides, and the formation of nitrogen oxides. These are harmful gases that cause some health and environmental problems, as discussed in Chapter 5.
3. The availability and prices of fossil fuel are vulnerable to world politics. Oil production and distribution are often interrupted during war times and due to political tensions between nations.

Most of the fossil fuel is used in transportation and industrial processes, and a relatively small percentage is used to generate electricity. Almost 70% of the world's oil is consumed by the transportation sector, and only 2% is used to generate electricity.

3.1.1 Oil

Oil is the most widely used fossil fuel worldwide. It is also called *petroleum*, which is composed of two words, "petro" and "oleum." The first is a Greek word for rock and the second is a latin word for oil. The petroleum is extracted from fields with layers of porous rocks filled with oil.

The Chinese discovered oil as early as AD 300. They used bamboo tubes to extract it and use it as a medical substance. In the 12th century, the famous traveler Marco Polo, while in Persia near the Caspian Sea, observed a geyser gushing a black substance, used by the inhabitants to cure skin problems or burned for heat and light. It was not until the 15th century that oil was first commercialized in Poland and used to light street lamps.

In North America, the first oil wells were drilled in Ontario, Canada, and Pennsylvania, U.S. in the middle of the 18th century. Around 1870, John D. Rockefeller formed the Standard Oil of Ohio which eventually controlled about 90% of the U.S. refineries. In the early 19th century, oil was discovered in Persia, then in the rest of the Middle East, and Central and South America.

Today, oil is used mainly in the transportation and industrial sectors as seen in Figure 3.6. In the United States, the generation of electrical power uses only 2% of the total oil consumed annually, mainly because of the abundance of coal in the United States.

The estimated oil reserve worldwide, as of 2002, is about 1.05×10^{12} Barrel, mainly in the Middle East region as seen in Figure 3.7. The world consumption of oil in 2002 was about 2.76×10^{10} Barrel, distributed as seen in the chart of Figure 3.8. This high rate of consumption is troubling, and new fields must always be discovered before the available supply is dried out.

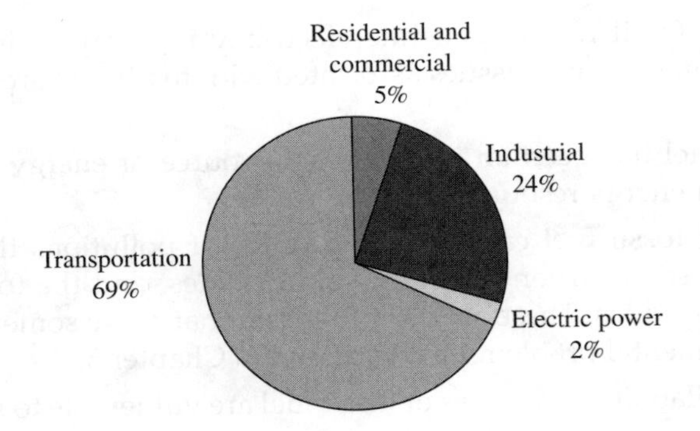

FIGURE 3.6
Consumption of oil in the United States by sectors (*Source*: British Petroleum Statistical Review of World Energy, June 2003).

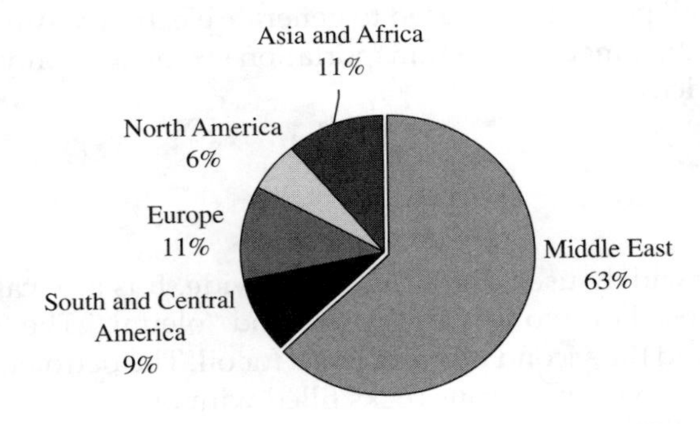

FIGURE 3.7
Distribution of known oil reserve (*Source*: British Petroleum Statistical Review of World Energy, June 2003).

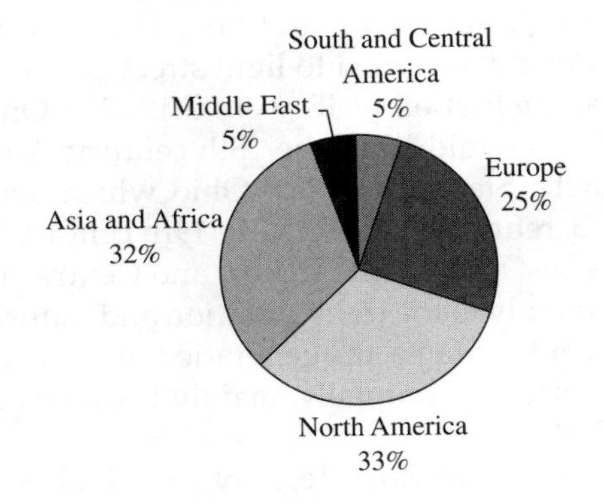

FIGURE 3.8
Consumption of oil worldwide (*Source*: British Petroleum Statistical Review of World Energy, June 2003).

EXAMPLE 3.3 Assume no new oil field is discovered. Then, how long can we maintain the consumption of oil at the 2002 rate?

Solution

The world known reserve is 1.05×10^{12} Barrel, and the annual world consumption is 2.76×10^{10} Barrel. Hence, the world reserve will last for $1.05 \times 10^{12}/2.76 \times 10^{10} = 38.04$ years. This is a disturbing short period.

3.1.2 Natural Gas

Although natural gas was discovered in Pennsylvania, U.S., as early as 1859 by Edwin Drake, it was not utilized commercially, because the process of liquefying natural gas at low temperatures was unknown until the 19th century. Natural gas was considered a by-product of the crude oil extraction, and was often burned out in the field. After the discovery of large reservoirs of natural gas in Wyoming in 1915, the United States developed cryogenic liquefaction methods and reliable pipeline systems for the storage and transportation of natural gas. In the 1920s, Frank Phillips (the founder of Phillips Petroleum) commercialized natural gas in the form of propane and butane.

The distribution of world reserve of natural gas is shown in Figure 3.9. The world reserve, as of 2002, is about 1.55×10^{14} m^3. The majority of this reserve is in Russia and the Middle East. The 2002 estimated reserve in the United States was 5.19×10^{12} m^3, which is about 3.35% of total world reserve.

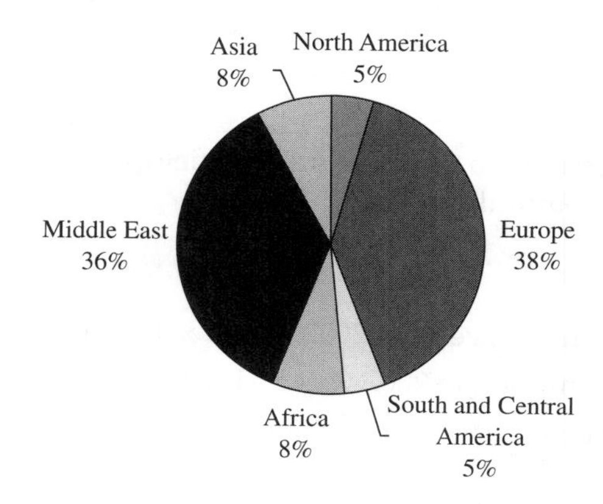

FIGURE 3.9
Known natural gas reserve as of 2002 (*Source*: British Petroleum Statistical Review of World Energy, June 2003).

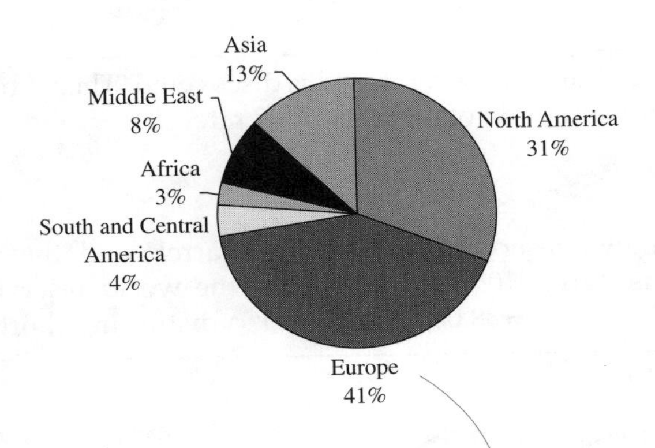

FIGURE 3.10
World consumption of natural gas (*Source*: British Petroleum Statistical Review of World Energy, June 2003).

In 2002, the world consumed about 2.54×10^{12} m^3 of natural gas according to the distribution shown in Figure 3.10. The majority of the consumed natural gas was in North America and Europe. The United States alone consumed about 6.68×10^{11} m^3.

EXAMPLE 3.4 Assume no new natural gas field is discovered in the United States. Then, how long can it be self-sustained at the 2002 consumption rate?

Solution

The known reserve of natural gas in the United States as of 2002 is 5.19×10^{12} m^3.

The U.S. consumption in 2002 was 6.68×10^{11} m^3.

Hence the reserve in the United States can sustain the 2002 consumption rate for $5.19 \times 10^{12}/6.68 \times 10^{11} = 7.8$ years.

EXAMPLE 3.5 Assume no new natural gas field is discovered worldwide. How long will the natural gas last?

Solution

The world known reserve as of 2002 is 1.55×10^{14} m^3.

The world consumption in 2002 was 2.54×10^{12} m^3.

Hence the world reserve can sustain the 2002 consumption rate for $1.55 \times 10^{14}/2.54 \times 10^{12} = 61$ years.

3.1.3 Coal

Charcoal, the black porous carbon substance produced by burning wood, is one of the earliest sources of heat and light known to man. Five thousand years ago, the Ancient Egyptians used charcoal for cooking, heating, baking pottery, and melting metals such as gold. Coal is another form of the carbonized vegetations formed in the carboniferous period millions of years ago. The heat, pressure, geological activity, and millions of years are needed to form the coal.

After coal was discovered in England, a large number of people have suffocated because of the carbon monoxide released from burning coal in closed quarters without adequate ventilation. This led King Edward I in the 12th century to impose the death penalty upon anyone caught burning coal. The ban lasted for two centuries.

In the 17th century, English scientists discovered that coal burned cleaner and produced more heat (2 to 4 times) than wood charcoal. This discovery opened the door wide open for the exploration of coal worldwide to provide Europe with the stable energy source needed to power its industrial revolution. Because of coal, the Scottish James Watt invented the steam engine that propelled ships, drove trains, and powered industrial machines. Coal was later used in the 1880s to generate electricity.

In the United States, the French explorers Louis Joliet and Jacques Marquette discovered coal in Illinois in 1673. This discovery was followed by more discoveries in Kentucky, Wyoming, Pennsylvania, West Virginia, and Texas. The largest producers of coal in the United States are Wyoming and West Virginia.

The world's reserve of coal as of 2002 is estimated at about 9.8×10^{11} tons, distributed as shown in Figure 3.11. North America's reserve is 27% of the world's reserve. The United States' share of coal reserve is about 2.5×10^{11} tons.

The world consumption of coal in 2002 is shown in the chart of Figure 3.12. As you see, although Asia has 30% of world's coal reserve, it accounts for almost half of the world's consumption. In North America and Europe,

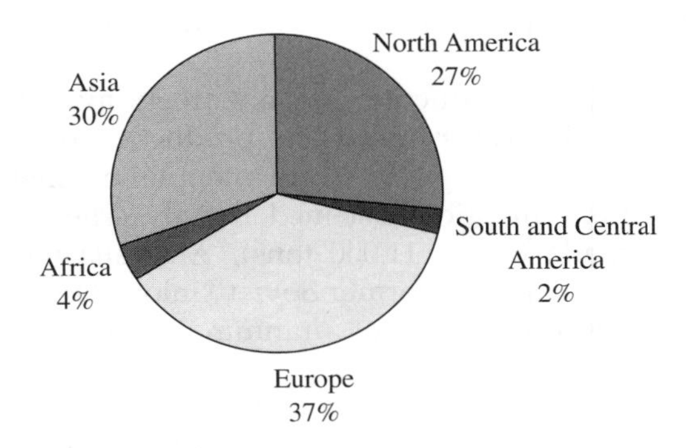

FIGURE 3.11
World's coal reserve in 2002 (*Source*: British Petroleum Statistical Review of World Energy, June 2003).

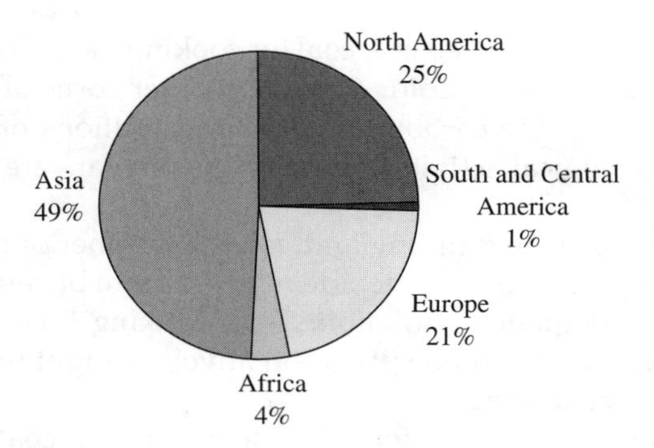

FIGURE 3.12
World consumption of coal in 2002 (*Source*: British Petroleum Statistical Review of World Energy, June 2003).

the consumption of coal is very high, but the dependency on coal for the production of electricity is diminishing because of the wider use of natural gas.

3.2 Nuclear Fuel

Nuclear fuel is heavy nuclei material that releases energy when its atoms are forced to split and, in the process, some of its mass is lost. The nuclear fuel used to generate electricity is mostly uranium, but plutonium is also used. The uranium is found in nature and contains several isotopes. Natural uranium is almost entirely a mixture of three isotopes, uranium-234 (U^{234}), U^{235}, and U^{238}; the subscript indicates the atomic mass of the isotope. The concentration of these isotopes in natural uranium is 99.2% for U^{238} and 0.7% for U^{235}. However, only U^{235} can fission in nuclear reactors. Since the concentration of U^{235} in uranium ores is very low (0.7%), an enrichment process is used to increase its concentration in the nuclear fuel. For a nuclear power plant, U^{235} concentration is about 3 to 5%, and for nuclear weapons it is over 90%.

The data for uranium production are not freely available for political and security reasons. However, the world production of uranium in 2002 was estimated at 34,000 tons. It is worth mentioning that since the end of the cold war, uranium prices declined rapidly. The largest producers of uranium are Canada (about 11,000 tons), Australia (about 5000 tons), Niger, Namibia, Russia and the former Soviet Union states, and the United States. United States production of uranium is estimated at 1900 tons annually.

Plutonium (Pu) is the other nuclear fuel used to generate electricity, but is less common than uranium. Plutonium is mainly a man-made element,

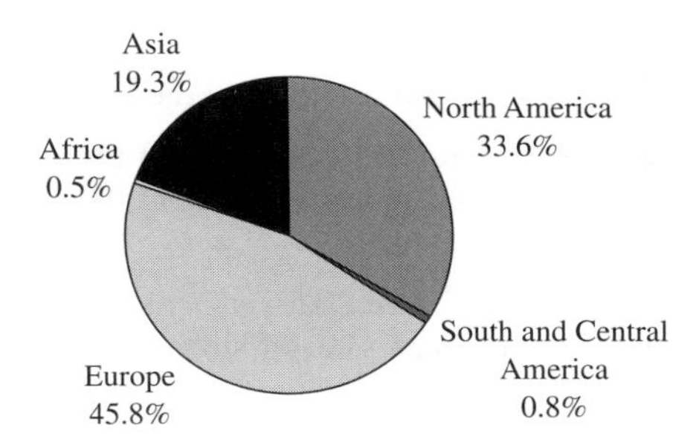

FIGURE 3.13
Nuclear fuel consumed worldwide in 2002 (*Source*: British Petroleum Statistical Review of World Energy, June 2003).

discovered by a group of scientists from the University of Berkeley in 1941. They created plutonium when U^{238} absorbed a neutron to become U^{239} and ultimately decayed to Pu^{239}. Minute amounts of Pu can also be found naturally. Plutonium is produced in the breeder reactors and it has three common isotopes: Pu^{238}, Pu^{239}, and Pu^{240}. The other Pu isotopes are created by different combinations of uranium and neutron. Pu^{239} is used in nuclear weapons and Pu^{238} is used in nuclear power plants. In some cases, Pu^{238} is mixed with uranium to form a mixed-oxide fuel that increases the power plant output.

Another use of Pu is in stand-alone power systems, such as the ones installed in satellites. Also, a minute amount of plutonium can provide long lasting power for medical equipments such as heart pacemakers.

The consumption of nuclear fuel worldwide is distributed as shown in Figure 3.13. The world consumption of nuclear fuel in 2002 was about 610×10^9 tons oil equivalent. Europe and the United States led the world in the amount of nuclear fuel used to generate electricity. The United States consumed about 186×10^9 tons oil equivalent of nuclear fuel in 2002.

Exercise

3.1. Exclude the United States and compute the generation capacity of the rest of the world. Find the per capita capacity.

3.2. Compute the generation capacity per capita in the United States. Compare the result with the world average.

3.3. Exclude the United States and compute the annual world consumption of electric energy per capita. Compare the result with the U.S. consumption per capita.

3.4. Find the ratio of the electric energy capacity to the electric energy demand (consumption) in the United States. Identify the amount of surplus or deficit.

3.5. Exclude the United States and find the ratio of the electric energy capacity to the electric energy demand worldwide. Identify the amount of surplus or deficit.

3.6. Assume the demand in the United States is increasing at a rate of 5% annually. For how long can the United States be self-sustained with respect to electricity?

4

Power Plants

The vast majority of electricity generated worldwide (about 99%) is generated from power plants using primary energy resources: hydroelectric, fossil fuel, and nuclear fuel. The descriptions of the power plants that use primary resources are covered in this chapter, and the methods to generate electricity from secondary resources (solar, wind, geothermal, etc.) are discussed in Chapter 6.

The geological and hydrological characteristics of the area where the power plant is to be erected determine, to a large extent, the type of power plant. For example, fossil fuel power plants in the United States are concentrated mainly in the East and Midwest regions, where coal is abundant. Similarly, hydroelectric power plants are concentrated in the Northwest region where water and water storage are available. Nuclear power plants, however, are distributed in all regions since their requirement of natural resources is limited to the availability of cooling water.

The percentage of the energy produced worldwide by the primary resources is given in Chapter 3. The data in Chapter 3 shows that fossil fuel is the main source of electric energy (over 80%), and about 63% of the electric energy is produced by coal- and oil-fired power plants. In the United States, coal counts for about 50% of the fuel used to generate electricity.

4.1 Hydroelectric Power Plants

Hydro is a Greek word meaning water. The hydroelectric power plant harnesses the energy of the hydrologic cycle. Water from oceans and lakes absorbs solar energy and evaporates into air, forming clouds. When the clouds become heavy, rains and snows occur. The rain and melted snow travel through streams and eventually end up in oceans. The motion of water toward oceans is due to its kinetic energy, which can be harnessed by the hydroelectric power plant that converts it into electrical energy. If the water is stored at high elevations, it possesses potential energy proportional to the elevation. When this water is allowed to flow from the high elevation to a lower one, potential energy is transformed into kinetic energy that is converted into electrical energy by the hydroelectric power plants.

TABLE 4.1
World's Largest Hydroelectric Power Plants

Name of Dam	Location	Capacity (MW)	Year of Operation
Three-Gorges	China	18,000	—
Itaipu	Brazil/Paraguay	14,000	1983
Guri	Venezuela	10,000	1986
Tucurui	Brazil	8,370	1984
Grand Coulee	Washington	6,494	1942
Sayano-Shushensk	Russia	6,400	1989
Krasnoyarsk	Russia	6,000	1968
Churchill Falls	Canada	5,428	1971
La Grande 2	Canada	5,328	1979
Bratsk	Russia	4,500	1961

The world's first hydroelectric power plant was constructed across the Fox River in Appleton, WI, U.S.A., and began its operation on September 30, 1882. The plant generated only 12.5 kW, which was enough to power two paper mills and the private home of the mill's owner. The latest and largest hydroelectric power plant, so far, is the one being built in China's Three-Gorges, which has a capacity of 18 GW. Some of the world's largest hydroelectric plants are given in Table 4.1.

4.1.1 Types of Hydroelectric Power Plants

The common types of hydroelectric power plants are the impoundment hydropower, diversion hydropower, and pumped storage hydropower.

1. *Impoundment hydropower*: It is the most common type of hydropower and is suitable for water bodies with high heads. The dam in these power plants creates a reservoir at high elevation behind the dam. A good example is the Grand Coulee Dam in Figure 4.1.
2. *Diversion hydropower*: An example of this type of hydropower plant is shown in Figure 4.2. It is suitable for low heads and is based on diverting some water of a river with strong current through the turbines. This hydropower plant does not require a water reservoir at high elevation, so its generating capacity is less than that for the impoundment hydropower.
3. *Pumped storage hydropower*: A pumped storage hydropower plant operates as a dual action water flow system. When the power demand is low, the generated electricity is used to pump water from the lower water level in front of the dam to the higher level behind the dam. This increases the potential energy behind the dam for later use.

FIGURE 4.1
(see color insert following Page 208) The Grand Coulee Dam and Franklin D. Roosevelt Lake. (Image courtesy of the U.S. Army Corps of Engineers.)

FIGURE 4.2
(see color insert following Page 208) Fox River diversion hydropower plant, Wisconsin. (Image courtesy of the U.S. Army Corps of Engineers.)

4.1.2 Impoundment Hydroelectric Power Plants

A typical impoundment hydroelectric power system has six key components: the dam, the reservoir, the penstock, the turbine, the generator, and the governor. A schematic of a hydropower plant is shown in Figure 4.3.

1. *The hydroelectric dam*: It is a barrier that prevents water from flowing downstream, thus creating a lake behind the dam. The potential

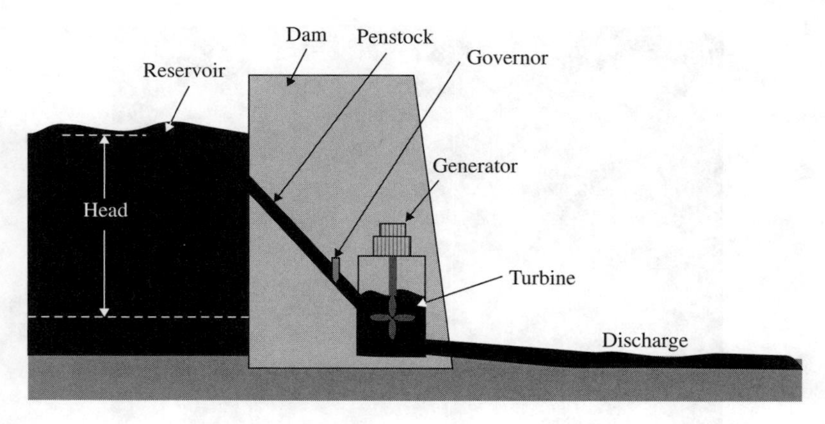

FIGURE 4.3
Simple schematic of hydroelectric power plant.

energy of the water behind the dam is directly proportional to the volume and height of the lake. The hydroelectric dam can be enormous in size; the Grand Coulee Dam in Washington State, U.S., is 170 m in height, 1.5 km in length, and its crest is 9 m wide. Its base is 150 m wide, which makes the base about four times as large as the base of the Great Pyramid of Egypt. The volume of the concrete used to build the dam is almost 3 km^3, enough to build 10,000 km of two-lane highway. Although massive, the Grand Coulee is not the biggest dam in the world. The Three-Gorges Dam in China is the biggest dam ever built so far, followed by the ITAIPU Dam in Brazil.

2. *The reservoir*: The dam creates a lake behind its structure called reservoir and often covers a wide area of land. Grand Coulee Dam created the Franklin D. Roosevelt artificial lake shown in Figure 4.1, which is about 250 km long, and has over 800 km of shore line. Its surface area is about 320 km^2, its lake ranges from 5 to 120 m deep.
3. *The penstock*: It is a large pipeline that channels water from the reservoir to the turbine; Figure 4.4 shows the penstock of the Grand Coulee Dam during its construction. The water flow in the penstock is controlled by a valve called the governor.
4. *The turbine*: A turbine is an advanced water wheel. The high-pressure water coming from the penstocks pushes against the blades of the turbine causing the turbine shaft to rotate. The electrical generator is mounted directly on the same turbine's shaft, and thus rotates at the turbine's speed. A photo of a turbine–generator units is shown in Figure 4.5.
5. *The generator*: It is an electromechanical converter that converts the mechanical energy of the turbine into electrical energy. The generators used in all power plants are the synchronous machine

FIGURE 4.4
Penstock of Grand Coulee Dam (Image courtesy: U.S. Bureau of Reclamation).

FIGURE 4.5
(see color insert following Page 208) Hydroturbine–generator units at the Lower Granite power plant, Walla Walla, WA, U.S.A. (Image courtesy of the U.S. Army Corps of Engineers.)

type. The generator is equipped with various control mechanisms such as the excitation control and stabilizers to control the output voltage and enhance the stability of the generator's operation.

6. *The governor*: It is the valve that regulates the flow of water in the penstock. When it is fully open the kinetic energy of the water is at its maximum. If the system is to shut down, the valve is closed fully.

4.1.3 Analysis of Impoundment Hydroelectric Power Plant

The amount of the generated electric energy of the impoundment hydroelectric power plant depends on several parameters. The most important ones are:

- The water head behind the dam.
- The reservoir capacity.
- The flow rate of the water inside the penstock.
- The efficiencies of the penstock, turbine, and generator.

4.1.3.1 Reservoir

The water behind the dam forms a reservoir (lake). The potential energy of the water in the reservoir PE_r is a linear function of the water mass and head.

$$PE_r = MgH \tag{4.1}$$

where M is the water mass in kilograms, g is the acceleration of gravity in meters per second, and H is the water head (average elevation) behind the dam in meters. The unit of PE_r is joules (watts). The mass of the water is a function of the water volume and water density.

$$M = \text{vol}\ \rho \tag{4.2}$$

where vol is the volume of water in cubic meters, ρ is the water density in kilograms per cubic meter. At temperatures up to 20°C, ρ is 1000 kg/m^3.

EXAMPLE 4.1 A hydroelectric dam forms a reservoir of 20 km^3. The reservoir average head is 100 m. Compute the potential energy of the reservoir's water.

Solution

$$PE_r = \text{vol}\ \rho g H = 20 \times 10^9 \times 1000 \times 9.81 \times 100 = 1.962 \times 10^7 \text{ GJ}$$

This immense energy is the potential energy of the entire reservoir. Keep in mind that seasonal variations in rainfall or snow spills do change the volume of the reservoir; thus the potential energy of the reservoir is accordingly varied.

4.1.3.2 Penstock

The potential energy of the water entering the penstock, PE, is

$$\text{PE} = mgH \tag{4.3}$$

where m is the mass of water entering the penstock. This potential energy is converted into kinetic energy as the water moves inside the penstock. The kinetic energy, KE, of the water leaving the penstock is

$$\text{KE} = \tfrac{1}{2}mv^2 \tag{4.4}$$

where v is the velocity of water exiting the penstock in meters per second. The PE and KE of the penstock are not equal unless the penstock is vertical and the water friction is ignored. In reality, the penstock is inclined; therefore, its length is longer than the water head, which results in some energy losses. Hence, the penstock efficiency η_{p} is defined as the ratio of its output energy KE to its input energy PE.

$$\eta_{\text{p}} = \frac{\text{KE}}{\text{PE}} = \frac{v^2}{2gH} \tag{4.5}$$

Since the power is energy divided by time, the mechanical power of the water exiting the penstock P_{w} is

$$P_{\text{w}} = \frac{\text{KE}}{t} = \frac{1}{2}\frac{m}{t}v^2 = \frac{1}{2}fv^2 \tag{4.6}$$

where f is the flow of water inside the penstock in kilogram per second and is defined as

$$f \equiv \frac{m}{t} = \frac{\text{vol}\ \rho}{t} \tag{4.7}$$

EXAMPLE 4.2 The penstock of a hydroelectric dam allows $800\,\text{m}^3/\text{s}$ of water to flow at a speed of $30\,\text{m/s}$. Compute the mechanical power of the water exiting the penstock.

Solution

$$P_{\text{w}} = \frac{1}{2}\frac{\text{vol}}{t}\rho v^2 = \frac{1}{2}800 \times 1000 \times 30^2 = 360\,\text{MW}$$

This power is not completely converted into electrical power due to the losses incurred in the turbine and the generator.

The volume of water passing through the penstock during an interval of time t is

$$\text{vol} = Avt \tag{4.8}$$

where A is the area of cross-section of the penstock, and t is the time interval. Hence, the mechanical power of the water exiting the penstock can be rewritten as

$$P_{\text{w}} = \frac{1}{2}\frac{\text{vol}}{t}\rho v^2 = \frac{1}{2}A \times \rho \times v^3 \tag{4.9}$$

EXAMPLE 4.3 The diameter of the penstock of a hydroelectric dam is 4 m. The water velocity inside the penstock is 40 m/s. Compute the power of the water exiting the penstock.

Solution

$$P_{\text{w}} = \frac{1}{2}A\rho v^3 = \frac{1}{2}(\pi \times 2^2)1000 \times 40^3 = 402\,\text{MW}$$

4.1.3.3 Turbine

Hydroelectric turbines are specially designed water wheels that come in three main types; two of them are shown in Figure 4.6. The Kaplan turbines, named after Viktor Kaplan, are used mainly in diversion power plants with small

(a)

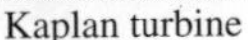

Kaplan turbine

(b)

Francis turbine: Grand Coulee

FIGURE 4.6
Two different types of hydroelectric turbines. (Images are courtesy of the U.S. Army Corps of Engineers.)

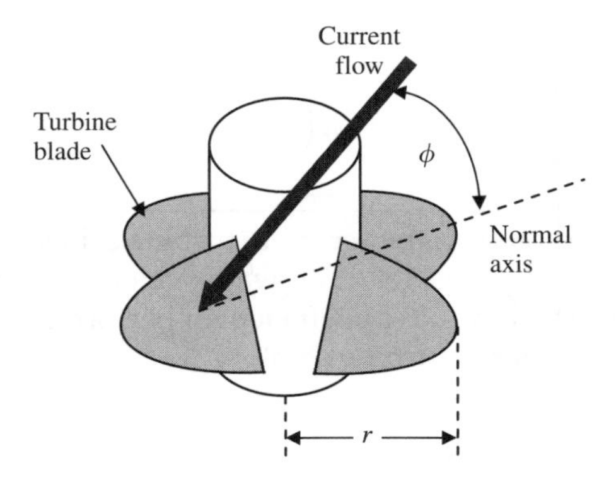

FIGURE 4.7
Simple hydroelectric turbine.

heads. Pelton turbines, invented by Lester Pelton, are used in high head impoundment power plants. Francis turbines, invented by James Francis are used in either type of power plant.

A schematic of a simple turbine is shown in Figure 4.7. Its main parts are the shaft and the blades. The shaft of the turbine is directly connected to the shaft of the generator, and its blades are designed to rotate the shaft similar to the water wheels. The kinetic energy captured by the turbine is a function of the sweep area A_s of the blades.

$$A_s = \pi r^2 \tag{4.10}$$

where r is the radius of the sweep area. If the water enters the turbine at an incident angle ϕ from the normal axis of the blade, the sweep area can be modified as follows:

$$A_s = \pi r^2 \cos\phi \tag{4.11}$$

The mechanical power of the water hitting the turbine P_t is

$$P_t = \tfrac{1}{2} A_s \rho v_t^3 \tag{4.12}$$

where v_t is the velocity of water when it hits the blades of the turbine; it is slightly lower than the speed of water v exiting the penstock in Equation (4.9).

Because of the various mechanical losses of the turbine, the power P_t cannot entirely be converted into the mechanical power P_m entering the generator. The ratio of P_m to P_t is known as the coefficient of performance C_p of the turbine, or the hydroelectric conversion efficiency.

$$C_p = \frac{P_m}{P_t} \tag{4.13}$$

Hence,

$$P_m = C_p\left(\tfrac{1}{2}A_s\rho v^3\right) \tag{4.14}$$

EXAMPLE 4.4 The sweep diameter of a turbine's blades is 3 m and the incident angle of the water is 10°. The water velocity at the surface of the blades is 30 m/s. The turbine has a coefficient of performance of 0.5. Compute the mechanical power of the turbine shaft.

Solution

$$P_m = C_p\left(\tfrac{1}{2}A_s\rho v^3\right) = 0.5\left(\tfrac{1}{2}\left(\pi \times 1.5^2 \cos 10\right)1000 \times 30^3\right) = 46.99\,\text{MW}$$

EXAMPLE 4.5 The Grand Coulee Dam has six 40-ft diameter penstocks for its third powerhouse. Each penstock passes 250,000 gal/s of water when the average head of the water behind the dam is 380 ft.

a. Compute the volume of the discharged water per second in metric units.
b. Assume the penstock efficiency is 90%. Compute the water power exiting the penstock.
c. Compute the speed of the water inside the penstock.
d. Assume that the coefficient of performance of the turbine is 0.5 and the generator efficiency is 95%. Estimate the generated electrical power for each penstock.
e. Compute the overall system efficiency.

Solution

The first step is to convert all data into metric values. Use the conversion table in Appendix A.

The diameter of the penstock = $40 \times 0.3048 = 12.2$ m.

The penstock flow = $250{,}000 \times 3.7854 = 9.46 \times 10^5$ kg/s.

The water head = $380 \times 0.3048 = 115.82$ m

a. $\dfrac{vol}{t} = \dfrac{m/t}{\rho} = \dfrac{9.46 \times 10^5}{10^3} = 946\ \text{m}^3/\text{s}$

   The total volume of water discharged by the six penstocks = $946 \times 6 = 5676\ \text{m}^3/\text{s}$.

b. The input power to the penstock = $P_{in} = PE/t = (m/t)gH = 9.46 \times 10^5 \times 9.81 \times 115.82 = 1.075$ GW.

The output power of penstock $P_w = \eta_p P_{in} = 0.9 \times 1.075 = 967.5\,\text{MW}$.

c. The speed of water inside the penstock can be computed from the following equation (4.6):

$$P_w = 0.5 f v^2 = 0.5 \frac{m}{t} v^2 = 0.5 \times 9.46 \times 10^5 v^2$$

$$v = \sqrt{\frac{967.5 \times 10^6}{4.73 \times 10^5}} = 45.23\,\text{m/s}$$

d. The generator's output power P_g is

$$P_g = P_w C_p \eta_g$$

where η_g is the efficiency of the generator.

$$P_g = 967.5 \times 0.5 \times 0.95 = 459.56\,\text{MW}$$

e. Overall efficiency $\eta_{total} = \eta_p C_p \eta_g = 0.9 \times 0.5 \times 0.95 = 0.4275 = 42.75\%$.

4.2 Fossil Fuel Power Plants

Fossil power plants (coal, oil, or natural gas) utilize the thermal cycle described by the laws of thermodynamics to convert heat energy into mechanical energy. This conversion, however, is highly inefficient as described by the second law of thermodynamics, where a large amount of the heat energy must be wasted in order to convert the rest into mechanical energy. The description of the conversion process is depicted in Figure 4.8. Assume the energy source in the figure produces heat energy Q_1 at temperature T_1. Since heat flows only from high temperature to low temperature, a heat sink of temperature $T_2 < T_1$ is needed to facilitate the flow of heat. The second law states that the ideal efficiency η_{ideal} of a heat engine (turbine, internal combustion engine, etc.) is

$$\eta_{ideal} = \frac{T_1 - T_2}{T_1} \tag{4.15}$$

The equation shows that the engine efficiency is increased when T_2 is decreased. Keep in mind that this efficiency does not include the friction and other mechanical losses, heat leakages, etc. Therefore, the real efficiency is less than η_{ideal}. In modern thermal power plants, the temperature of the heat source T_1 is about 500–600°C, while the temperature of the heat sink T_2 is about 30–70°C.

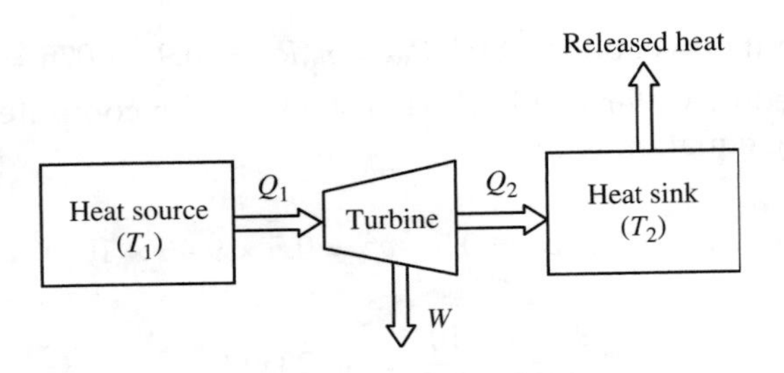

FIGURE 4.8
Second law of thermodynamics.

As seen in Figure 4.8, the turbine of the power plant is installed between the heat source and the heat sink. The turbine is a thermomechanical device (heat engine) that converts the heat energy into mechanical energy. It extracts some of the thermal energy of Q_1 and converts it into mechanical energy W. The rest is dissipated in the heat sink, without which no heat travels through the turbine.

The mechanical energy W is the difference between the source energy Q_1 and the energy dissipated in the heat sink Q_2.

$$W = Q_1 - Q_2 \tag{4.16}$$

The ideal efficiency of the turbine η_{ideal} can be written in terms of heat energy as

$$\eta_{\text{ideal}} = \frac{W}{Q_1} = \frac{Q_1 - Q_2}{Q_1} \tag{4.17}$$

Note that if $T_2 = T_1$, the heat sink does not dissipate any heat energy, that is, $Q_2 = Q_1$. In this case, no mechanical energy is produced by the turbine, and the turbine efficiency is zero.

4.2.1 Thermal Energy Constant

The *Thermal Energy Constant* (TEC) is defined as the amount of thermal energy produced per 1 kg of burned fuel. The unit of the TEC is called the *British Thermal Unit* (BTU), where one BTU is equivalent to 252 calories, or 1.0544 kJ. Table 4.2 shows typical TEC values for various fossil fuels. The table shows that oil and natural gas produce the highest BTU among all fossil fuels.

EXAMPLE 4.6 A coal-fired power plant has a condenser (heat sink) that extracts 18,000 BTU/kg of burned coal. Compute the mechanical energy of the turbine and the overall system efficiency.

Solution

According to Table 4.2, coal has a thermal energy constant of 27,000 BTU/kg or 28.469 kJ/kg.

The condenser extracts 18,000 BTU/kg or 18.979 kJ/kg.

The mechanical energy of the turbine is

$$W = Q_1 - Q_2 = 28.469 - 18.979 = 9.49\,\text{kJ/kg}$$

The ideal efficiency of the thermal turbine η_{ideal} is

$$\eta_{\text{ideal}} = \frac{\text{Ouput mechanical Energy}}{\text{Input thermal Energy}} = \frac{W}{Q_1} = \frac{9.49}{28.469} = 33.3\%$$

It is normal that the efficiency of the thermal cycle is below 50%, since the condenser dissipates (wastes) a large amount of the thermal energy to complete the thermal cycle.

TABLE 4.2

Thermal Energy Constant for Various Fossil Fuels

Fuel Type	Thermal Energy Constant (BTU/kg)
Petroleum	45,000
Natural gas	48,000
Coal	27,000
Wood (oven dry)	19,000

4.2.2 Description of a Thermal Power Plant

Generally, most fossil fuel power plants have similar designs. The main differences among them are the designs of their burners, fuel feeders, and stack filters. Nevertheless, these differences are not essential for the description of the operation of any thermal plant, and therefore, we shall discuss the coal-fired type only.

As explained in Chapter 3, coal is a combustible black rock that is abundant in the Midwest and the East of the United States. Coal comes from organic matter that built up in the swamp millions of years ago during the Carboniferous period (about 30 million years ago) and became fossilized.

A typical view of the powerhouse of a thermal power plant is shown in Figure 4.9. It consists mainly of a turbine and a generator. The turbine consists of blades mounted on a shaft. The angles and contours of the blades are designed to capture the maximum thermal energy from the steam. Figure 4.10 shows a large thermal turbine under construction.

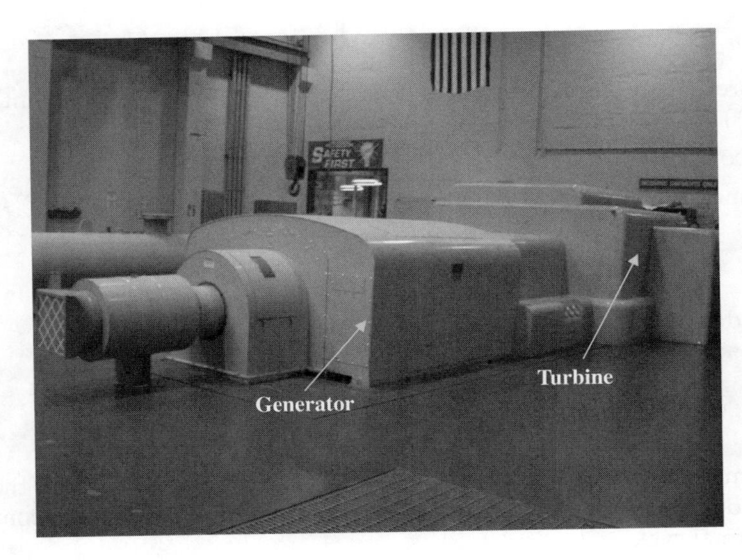

FIGURE 4.9
(see color insert following Page 208) Inside a thermal power plant.

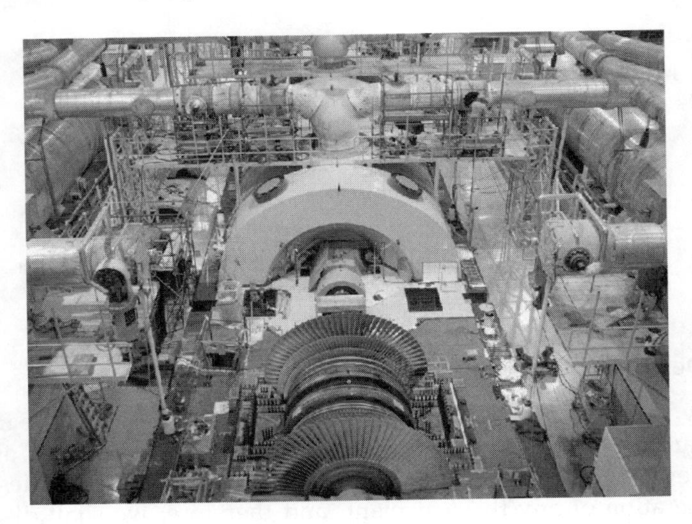

FIGURE 4.10
(see color insert following Page 208) Thermal turbine. (Image courtesy of the Oregon Department of Energy.)

The schematic of the main components of a coal-fired power plant is shown in Figure 4.11, and a photo of the plant is shown in Figure 4.12. The process starts when coal is delivered to the plant by trucks and railroad trains. The coal is then crushed and delivered to the burner via conveyor belts. The coal is then burned to generate heat that is absorbed by water pipes inside the boiler. The water turns into high-pressure steam at high temperature. The steam leaves

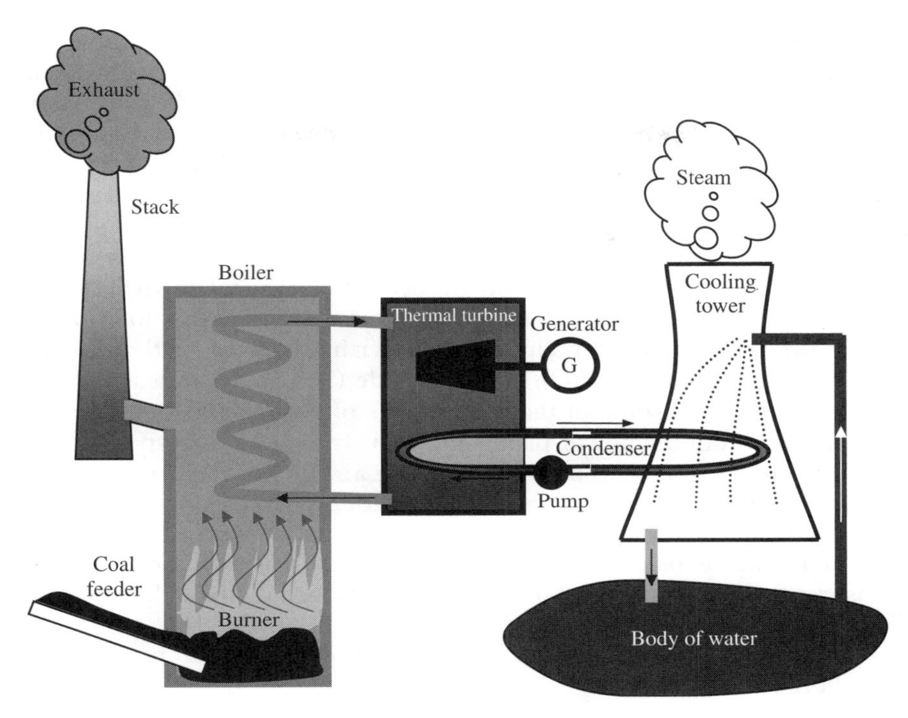

FIGURE 4.11
(see color insert following Page 208) Main components of coal-fired power plant.

FIGURE 4.12
(see color insert following Page 208) Weston coal-fired power plant. (Image courtesy of the U.S. Department of Energy.)

the boiler at a temperature higher than 500°C and enters the turbine at a velocity greater than 1600 km/h. The high-speed steam hits the blades of the turbine and causes the turbine to rotate. The rotating turbine's shaft is connected to the shaft of the generator, thus causing the generator to rotate, and electricity is generated. Because of the presence of the condenser (heat sink), the thermal cycle is completed as described by the second law of thermodynamics. The condenser cools the steam to about 50°C, turns it to a liquid form and goes back to the boiler to complete the thermal cycle. Inside the cooling tower, the condenser uses water from the nearby lake to cool down the steam.

Although coal-fired power plants are simple in design and easy to maintain, they are major producers of pollution, as shown in Chapter 5. Carbon dioxide (CO_2), carbon monoxide (CO), sulfur dioxide (SO_2), nitrous oxide (NO_x), soot, and ashes are some of the by-products of coal combustions. In fact, coal-burning from the power plants and industrial sector is responsible for 30 to 40% of the total carbon dioxide in the air. In older and unregulated plants, most of these pollutants are vented through the stack. However, with newer technology, large amounts of the pollutants are trapped by filters or removed from the coal before it is burned. Examples of the pollution reduction measures that are taken in most coal-fired plants include the following:

- The coal is chemically treated to remove most of its sulfur before it is burned.
- Filters are used to remove the particulate (primarily fly ash) and some of the exhaust gases from the boilers. There are various types of filters; among them are the wet scrubber system and the fabric filter system. With the wet scrubber system, the exhaust gas passes through liquid, which traps flying particulate and sulfur dioxide before the gas is vented through the stacks. The fabric filter system works like a vacuum cleaner where the particulates are trapped in bags.
- Sulfur dioxide is removed by its own scrubber system.
- Nitrous oxide is reduced by upgrading the boilers to low NO_x burners.
- Carbon dioxide removal is too expensive to implement and not all power plants use a carbon dioxide scrubber system.

4.3 Nuclear Power Plants

From the point of view of availability, nuclear fuel is the most abundant source of energy. As given in Chapter 3, the common fuel for nuclear power plants is uranium, where an atom of uranium produces about 10^7 times the energy produced by an atom of coal. In the United States, there are about 109 commercial nuclear power plants in operation generating about 20% of the total electric energy. Worldwide, there are about 400 power plants, generating as much as

70% of the energy demand in nations such as France. However, because of the public concern, few nuclear power plants have been constructed in the United States, and several are expected to be mothballed (retired) in the near future.

Nuclear power plants can generate electricity by one of two methods:

- *Fission,* which is the splitting of a heavy nucleus element such as uranium, plutonium, or thorium into many lighter elements. By this process, mass is converted into energy. Fission power plants have two main designs: *boiling water reactor* (BWR) and *pressurized water reactor* (PWR). About two thirds of the nuclear reactors are pressurized water reactors, and almost all of the commercial nuclear power plants worldwide are fission reactors.
- *Fusion,* which is a process by which two lighter elements are combined into heavier elements. The fusion technique is not yet fully developed for commercial power plants.

4.3.1 Nuclear Fuel

An atom is composed of subatomic particles known as protons, neutrons, and electrons. Atoms that differ in the number of protons they posess form different *elements.* Atoms with the same atomic number (number of protons in the nucleus) but differing in the number of neutrons are called *isotopes.* Different isotopes of the same element can have different masses due to the difference in the number of neutrons they posess. Heavier elements have more subatomic particles and their atomic masses are greater.

Natural uranium is almost entirely a mixture of three isotopes, U^{234}, U^{235}, and U^{238}; the superscript indicates the atomic mass of the isotope. The concentration of these isotopes in natural uranium is 99.2% for U^{238} and 0.7% for U^{235}. Unfortunately, only U^{235} can fission in nuclear reactors. Since the concentration of U^{235} in uranium ores is very low, an enrichment process is used to increase its concentration in nuclear fuel. In nuclear power plants, U^{235} concentration is about 3 to 5%, and for nuclear weapons it is over 90%.

Another way to develop a fissionable nuclear fuel is through the *breeder reactors*: The breeder reactor uses the widely available, nonfissionable uranium isotope U^{238}, together with small amounts of fissionable U^{235}, to produce a fissionable isotope of plutonium, Pu^{239}.

Commercial uranium fuel comes in the form of ceramic pellets. The pellets are made of enriched uranium containing 3 to 5% U^{235} isotopes by mass. Each of these pellets is about one inch in size, and contains the energy equivalent of 1000 kg of coal, 700 liter of oil, or 1500 m^3 of natural gas. The pellets are loaded into zirconium alloy metal tubes. Zirconium alloy is chosen because of its ability to resist radiation and thermal stresses. About 800 tube assemblies are placed inside the reactor.

4.3.2 Fission Process

The fission process is depicted in Figure 4.13. When a nuclear power plant starts up, neutrons are let loose to strike uranium atoms at high speeds, causing it to split (fission). The fission process produces the following:

- *Fission fragments*: These are the leftover materials after the atoms have split. They are cesium-140 and rubidium-93, which are radioactive materials.
- *Released neutrons*: After each fission action, three neutrons are released. If they hit other uranium atoms, they cause more fission to occur. This is known as the *chain reaction*. The neutrons produced by nuclear fission are fast-moving and must be slowed down to initiate further fission. The material used for this purpose is called the *moderator*; the common moderators are graphite and heavy water. Heavy water is chemically similar to regular water (H_2O), but the hydrogen atoms are replaced by an isotope of hydrogen called deuterium (the symbol of the heavy water is D_2O). Deuterium has one neutron more than hydrogen, which makes D_2O heavier than H_2O by about 10%.
- *Energy*: The mass of the original uranium atom is more than the combined mass of the fission fragments and the released neutrons. The lost mass is converted into energy as described by Albert Einstein's formula

$$E = mc^2 \tag{4.18}$$

where E is the released energy, m is the lost mass, and c is the speed of light. Each fission event releases approximately 3.2×10^{-11} J of energy. To put

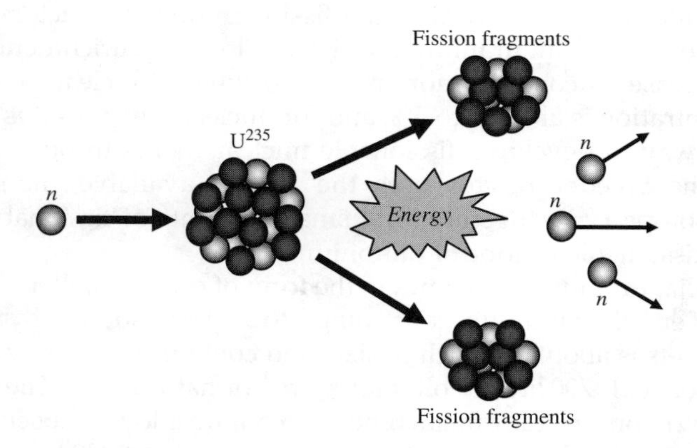

FIGURE 4.13
Fission reaction.

the number into perspective, 1 J of energy requires approximately 31×10^9 fission events. One kilogram of U^{235} can have approximately 25.4×10^{23} fission events.

EXAMPLE 4.7 A nuclear reactor produces an average of 1 GW of thermal power annually. Compute the number of its fission events per year.

Solution

1 J requires 31×10^9 fission events.

1 W requires 31×10^9 fission events per second.

1 GW requires $10^9(31 \times 10^9) = 31 \times 10^{18}$ fission events per second.

1 GW for 1 h requires $3600(31 \times 10^{18}) = 11.16 \times 10^{22}$ fission events.

1 GW for 1 year requires $8760(11.16 \times 10^{22}) = 9.7762 \times 10^{26}$ fission events.

The number of fission events is staggering. However, the mass of U^{235} required to produce 1 GW of thermal energy for the entire year is very small. This is explained in Example 4.8

EXAMPLE 4.8 Compute the mass of U^{235} to produce an average of 1 GW of thermal energy annually. Compare the mass of the nuclear fuel with the equivalent mass of coal.

Solution

In Example 4.7, the number of annual fission events is 9.7762×10^{26}.

Since 1 kg of U^{235} can have 25.4×10^{23} fission events, the fuel needed for the reactor annually is

$$\text{Mass of } U^{235} \text{ annually} = \frac{9.7762 \times 10^{26}}{25.4 \times 10^{23}} = 385\,\text{kg}$$

About 1 kg of coal produces 28.47×10^3 kWh of thermal energy as given in Table 4.2.

To produce 1 GW of thermal power for one hour, we need to burn $10^6/(28.47 \times 10^3) = 35.12$ kg of coal.

To produce 1 GW of thermal power annually, we need to burn $8760 \times 35.12 = 307.65 \times 10^3$ kg of coal.

$$\text{Mass of coal/mass of uranium} = 800$$

This example shows that a small amount of nuclear fuel can unleash a tremendous amount of thermal energy.

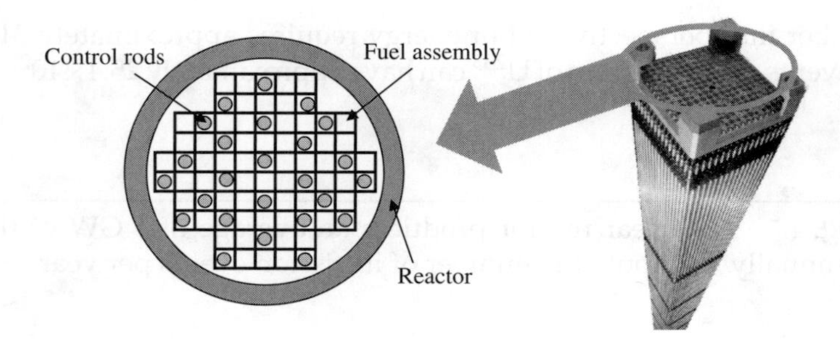

FIGURE 4.14
Fuel assembly and control rods.

4.3.3 Fission Control

The fission process can be sustained indefinitely if the fuel is available and the chain reaction is maintained. To control the amount of the released energy, the chain reaction must be regulated by controlling the number of neutrons available for fission. This is done by using control rods made of material absorbent to neutrons and inserted inside the reactor between the fuel tubes as shown in Figure 4.14. The common materials used for control rods are hafnium, cadmium, or boron. The control rods can be inserted and removed from the reactor by a motion control mechanism. When they are inserted, the rods absorb neutrons so fewer are available for fission. The reactor can be completely shut down by fully inserting the control rods into the core. In emergency situations, the rods are released and the gravity pulls them down the core to their fully inserted positions. This is called the *Safety Control Rod Axe Man* (SCRAM) process; the phrase is developed for the early generation of test reactors when the rods were suspended by rope. In an emergency, the robe was cut by an axe, and the person in charge of it was called the SCRAM.

4.3.4 Boiling Water Reactor (BWR)

The Boiling Water Reactor (BWR) used in commercial power plants was originally designed by General Electric (GE) Company. The first installation was at Humboldt Bay, CA in 1963. The BWR typically boils the water inside the reactor vessel. The operating temperature of the reactor is approximately 300°C, and its steam pressure is about 7.0×10^5 kg/m^2. Current BWRs can generate as much as 1.4 GW with overall efficiency of 33%.

The schematic of a BWR nuclear power plant is shown in Figure 4.15. The plant has the following key components:

- *Containment structure*: The dome structure houses the reactor and has multiple barriers of thick steel and concrete to contain the radiation inside the structure.

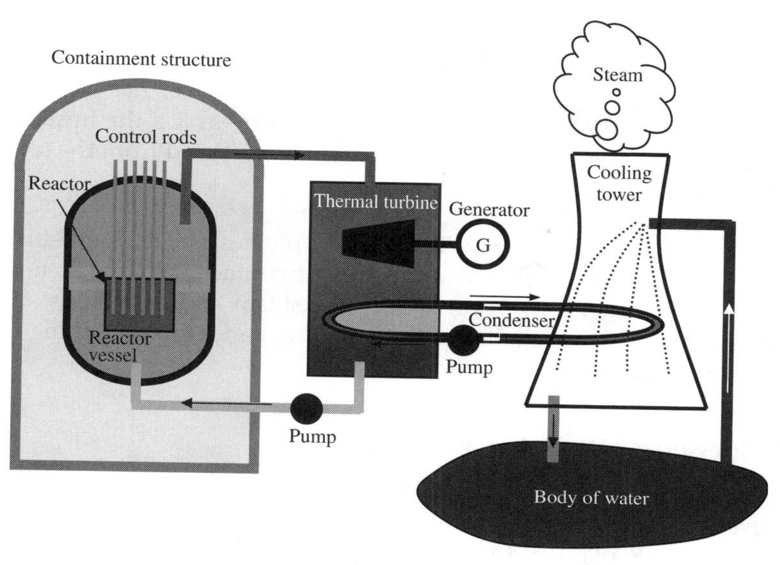

FIGURE 4.15
(see color insert following Page 208) Main components of a BWR.

- *Reactor vessel*: The reactor vessel houses the reactor, nuclear fuel, and fuel rods. The vessel is filled with water that acts as a moderator for the chain reaction, and also extracts the heat energy generated by the nuclear reaction. The vessel walls are made of a thick barrier of steel and concrete to guard against radiation leakage and accidental meltdown.
- *Reactor*: The reactor contains the nuclear fuel assembly and the control rods. The control rods can be inserted or removed from the reactor by an actuation system to control the amount of generated heat. In case of an emergency shut down, the rods are released and dropped to the fully inserted position to halt the chain reaction.
- *Turbine*: The energy generated by the nuclear reaction heats the water inside the reactor vessel, and steam is produced at high temperature and pressure. The steam passes through the turbine, which is a thermo-mechanical converter that converts the thermal energy into mechanical energy.
- *Generator*: The turbine is connected to the generator via a drive-shaft. The generator is an electro-mechanical converter, where the mechanical energy of the turbine is converted into electrical energy.
- *Condenser*: This is the heat exchanger that extracts heat from the steam exiting the turbine. It acts as a thermal link between the turbine steam and the cooling tower.

- *Cooling tower*: The function of the cooling tower is to act as the heat sink of the thermal cycle. It uses cold water from a nearby reservoir, lake, river, or ocean. The cold water is poured over the condenser pipes to cool them down. The cooling water extracts the heat from the condenser and turns into steam, which is vented from the top of the cooling tower.

Among the advantages of the BWR are its simple design and high efficiency. However, because the fuel rods are in direct contact with the steam, any leakage from the fuel rods is carried by the steam and eventually reaches the turbine. This poses a challenge to the maintenance crew working on the turbine or the generator.

4.3.5 Pressurized Water Reactor (PWR)

The Pressurized Water Reactor (PWR) was originally designed by Westinghouse as the power source for navy ships and submarines. The first commercial PWR plant in the United States was the Shippingport Atomic Power Station near Pittsburgh, which operated from 1958 until 1982. The PWR is the most widely used type of nuclear reactor worldwide.

The PWR design is distinctly different from the BWR because, in the PWR, a heat exchanger is placed between the water of the reactor and the steam entering the turbine. This is done to prevent the radioactive water from contaminating the turbine. This heat exchanger is called a *steam generator*. A schematic of the PWR power plant is shown in Figure 4.16 and photos of

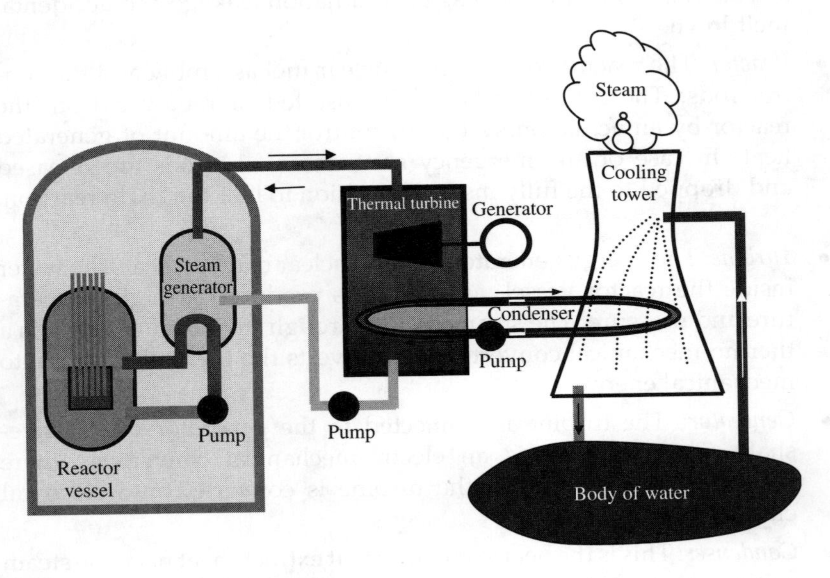

FIGURE 4.16
(see color insert following Page 208) Pressurized water reactor.

FIGURE 4.17
(see color insert following Page 208) PWR nuclear power plants.

PWR power plants are shown in Figure 4.17. The water of the PWR reactor is under enough pressure to remain in liquid form even when the water reaches 300°C or more; thus it is called a pressurized water reactor. Raising the temperature of the water under pressure makes it absorb more energy. The PWR has three separate heat exchange loops, or water loops, but the water in these loops never mix. In the first loop, called primary loop, the pressurized water is pumped through the reactor to extract the thermal energy of the nuclear reaction and then pass through extremely strong pipes that lead to a steam-generator (heat exchanger). The water in this loop is radioactive. The secondary loop includes the heat exchanger and the turbine and generator. The water in the second loop is free from any significant radioactive material. The third loop is the cooling loop that includes the turbine–generator and the condenser.

4.3.6 Safety Features in Nuclear Power Plants

Safety has been an important consideration from the very beginning of the nuclear era. However, in the 1940s, the developments were focused on demonstrating the potential of nuclear power as a reliable and cheap source of energy. Unfortunately, several nations sacrificed safety measures by adopting cheap and substandard designs.

The world enjoyed relatively safe nuclear power from 1940 until late 1979, with no major accident reported. However, on March 28, 1979, a reactor at the Three Mile Island nuclear power facility near Harrisburg, PA suddenly overheated, releasing radioactive gases. Equipment failure and human error were the main reasons for the worst nuclear accident in the history of the United States. Although the plant engineers managed to shut down the reactor without damaging the core, public concern regarding the safety of nuclear power plants greatly intensified.

On April 26, 1986, unit 4 of the Chernobyl nuclear power plant in the Ukraine was destroyed due to a combination of weak design and a series of human errors. The accident killed 31 people almost immediately and about 7 tons of uranium dioxide products escaped and began to spread across the surrounding areas.

These two accidents reaffirmed the need for effective safety measures to prevent future nuclear accidents. The nuclear industry is often faced with a public who is skeptical regarding the integrity and safety of nuclear power. Among the components with safety concerns are the fuel assembly, control rods, and the water of the reactor. Each of these components is discussed in the following paragraphs:

Fuel: Fuel pellets are enriched uranium in the size of jellybeans. Most of the pellets are enriched with 3 to 5% U^{235}. The pellets are placed into a tube made of an alloy of zirconium and niobium. This alloy resists corrosion and has a low neutron absorption property. It is capable of maintaining its property even at high-temperature, high-pressure, and high-irradiation environments. The fuel assembly is often replaced every 3 to 5 years.

Control rods: The control rods are made of materials such as of boron, which absorbs neutrons. The boron, therefore, impedes the chain reaction and controls the rate of the fission reaction. The control rods are inserted between the fuel tubes, or mixed with the fuel rods. During emergency conditions, the control rods would drop between the fuel rods and stop the nuclear reactions completely.

Reactor water: Reactor water removes excess heat from the reactors to prevent any meltdown. Water also slows down the neutrons and increases the probability of fission. Without water, neutrons would become too energized to initiate fission. Therefore, any loss of water would slow down the fission process dramatically.

4.3.7 Disposal of Nuclear Waste

The fission fragments are radioactive with a half-life of thousands of years. The spent fuel rods from a nuclear reactor are the most radioactive of all nuclear wastes. The spent fuel rods must then be stored in special storage facilities that prevent radiation leakages. The storage facilities are of two types: temporary and permanent. The United States lacks permanent storage facilities, and its temporary facilities store the nuclear waste for a very long time. The temporary facilities are composed of two types: wet storage and dry storage.

4.3.7.1 Wet Storage

When the spent fuel rods are removed from the core of the reactor, they are extremely hot and must be cooled down. The spent rods are placed in a pool filled with boric acid. A dilute water solution of boric acid is quite safe and is commonly used as a mild antiseptic and eyewash. The boric acid absorbs some of the radiation of the fission fragments. The spent fuel rods are immersed in the fluid for at least 6 months before they are transferred to permanent storage facilities. As an additional safety measure, control rods are placed among the fuel rods to inhibit any fission action of leftover U^{235}.

4.3.7.2 Dry Storage

After the spent fuel rods are cooled down in the wet storage facilities for a number of years, they can be stored in temporary dry storage made of reinforced casks, or buried in concrete bunkers. The casks are steel cylinders that are welded or bolted closed. Each cylinder is surrounded by additional steel and concrete as a further measure against the radiation leaks. The casks can be used for both storage and transportation.

4.3.7.3 Permanent Storage

Nuclear waste is composed of low- and high-level radiation waste. The low-level waste loses its radioactivity in a few hundred years, and is often buried in shallow sites. The high-level waste, such as the spent fuel rod, is harder to dispose of because it contains fission fragments that are radioactive for thousands of years. So, they are buried in deep geological permanent storage. The deep site must have no or little groundwater to prevent the erosion of the containments (steel cylinders). The site must also be stable geologically, so earthquakes do not damage the containments. In the United States, a permanent storage site has been recently selected at Yucca Mountain, Nevada. The site, which is expected to be ready by 2010, is about 460 m underground and is far from any population center.

Exercise

4.1. Name three types of hydroelectric power plants.

4.2. Why are cooling towers used in thermal power plants?

4.3. What are the two nuclear reactions?

4.4. What is the enrichment process of nuclear fuel?

4.5. How is the chain reaction controlled in nuclear power plants?

4.6. What is the main difference between boiling water reactors and the pressurized water reactors?

4.7. What is the heavy water? Why it is used in nuclear power plants?

4.8. Write an essay on the latest technology used to dispose of the spent fuel rods.

4.9. A man owns a land that includes a low-head waterfall and wants to build a small hydroelectric plant. He used a tube 3 m in length and 1 m in diameter as a penstock. He computed the speed of the water inside the penstock by dropping a small ball at the entrance of the tube and measuring the time it took to reach the other end of the tube. The ball traveled the penstock in 2 s.

a. Compute the mechanical power of the water.

b. If the coefficient of performance of the turbine is 0.5 and the efficiency of the generator is 90%, compute the expected generation of the site.

c. If the cost of building the small hydroelectric system is $20K, compute the payback period if the cost of electricity from the neighboring utilities is $0.2/kWh. Assume that the water flow is always constant.

d. Is building this small hydroelectric system a good investment?

4.10. A hydroelectric dam creates a reservoir of 10 km^3. The average head of the reservoir is 100 m. Compute the potential energy of the reservoir.

4.11. A penstock is used to bring water from behind a dam into a turbine. The penstock is 6 m in diameter and moves water at a rate of 500 m^3/s. Compute the mechanical power entering the turbine.

4.12. The penstock of a hydroelectric power plant is 4 m in diameter. The penstock efficiency is 95% and the water head is 60 m. Compute the mechanical power of the water at the exit of the penstock. Also, compute the water flow inside the penstock.

4.13. A hydroelectric dam has a penstock that discharges 10^5 kg/s of water. The head of the dam is 80 m.

a. Compute the volume of the discharged water per second.

b. Assume the penstock efficiency is 85%. Compute the power of the water entering the turbine.

c. Compute the speed of the water inside the penstock.

d. Assume that the coefficient of performance of the turbine is 0.5 and the generator efficiency is 92%. Estimate the generated electrical power.

e. Compute the overall system efficiency.

4.14. A natural gas power plant has a condenser that extracts 18,000 BTU/kg of natural gas. Compute the mechanical energy of the turbine and the overall system efficiency.

4.15. An oil-fired power plant has a condenser that extracts 18,000 BTU/kg of oil. Compute the mechanical energy of the turbine and the overall system efficiency.

4.16. Compare natural gas to oil in terms of thermal power and efficiency. Use the results of the previous two problems to verify your assessment.

4.17. Why are heat sinks used in thermal power plants?

4.18. Write an essay on the general design of oil-fired power plants.

4.19. Write an essay on the storage of nuclear waste.

4.20. Write an essay on the disposal of the contaminated structure of a nuclear power plant.

4.21. Write an essay on the nuclear accident of the Three Mile Island power plant. Show the sequence of events and comment on the safety measures taken to prevent any subsequent catastrophic failure.

4.22. Write an essay on the Chernobyl nuclear accident. Show why such an accident is unlikely to happen in the United States.

4.23. Estimate the amount of nuclear energy produced by 10 kg of uranium-235.

5

Environmental Impact of Power Plants

Beginning from the Stone Age era when fire was used for heating, humans have continually been involved in activities with harmful effects on the environment and their health. Unfortunately, as we become more industrially advanced, the Earth's atmosphere has become more polluted, the water resources further contaminated, and the Earth's crust more acidic.

A good starting point in the history of pollution is the 11th century, when wood was extensively used in Europe as a major heat source. The pollution from burning wood was relatively limited, but wood became scarce and was replaced by coal in the beginning of the 12th century. The dense smoke created by burning coal was initially viewed as just a minor discomfort. However, when coal was burned inside closed quarters with inadequate ventilation, the carbon monoxide released from the combustion suffocated a large number of people. This led King Edward I in the 12th century to impose the death penalty upon anyone who was caught burning coal. The ban lasted for two centuries.

In the late 17th century, the industrial revolution started in Europe and spread all over the world. During this period, coal propelled the heavy industries as it was the only reliable source of energy. In the early 20th century, oil started to replace coal since it was a cleaner form of fuel. Because of petroleum, the automobile industry flourished along with several other heavy industries such as steel and rubber. This was the period of unchecked assaults on the environment, with the automobile responsible for 60% of all atmospheric pollution.

Until recently, pollution was primarily an urban phenomenon in industrial countries, but now it has spread all over the world with more than 20% of the world population living in communities that do not meet the World Health Organization air quality standards. Only recently, humans have begun to comprehend the severity of the problems created by pollution. In the early 1960s, several public and scientific organizations intensified their efforts to enforce various regulations to maintain the delicate ecological balance of nature. In the United States, President Richard Nixon signed into law the National Environmental Policy Act on January 1, 1970. The law led to the formation of the Environmental Protection Agency (EPA), which is chartered with the protection of human health and the environment.

Because air pollution respects no national boundaries, international cooperation and treaties are established to reduce the flow of pollution across borders. The United Nations Commission on Sustainable Development is an example of international umbrella organizations that promote control measures leading to cleaner air and a healthier environment.

In the electric energy sector, over 99% of the electric energy generated worldwide is produced by the primary resources: fossil fuel, hydroelectric, and Nuclear. In 2002, the estimated world consumption of electric energy from primary resources was about 14×10^{12} kWh. The extensive use of these primary resources is because they are readily available, produce enormous amount of energy, and are cheaper than the alternative methods (solar, wind, etc.). From the viewpoint of the environment, each of these primary resources is associated with some air, water, and land pollution. Although there is no pollution-free method for generating electricity (from primary or secondary resources), the primary resources are often accused of being responsible for most of the negative environmental impacts. In this chapter, the various pollution problems associated with the generation of electricity are presented and discussed.

5.1 Environmental Concerns of Fossil Fuel Power Plants

The Earth's atmosphere is a mixture of gases and particles surrounding the plant. Clean air is generally composed of the following mixture of gases. PPM in the list below stands for *parts per million* of the substance in air by volume, which is the number of molecules of the substance in a million molecules of air.

- Nitrogen (N_2), 78.1%
- Oxygen (O_2), 21%
- Argon (Ar), 0.9%
- Carbon Dioxide (CO_2), 330 PPM
- Neon (Ne), 18 PPM
- Water vapor (H_2O)
- Small amounts of Krypton (Kr), Helium (He), Methane (CH_4), Hydrogen (H), Nitrous Oxide (N_2O), Xenon (Xe), and Ozone (O_3)

This delicate balance must be maintained in order for the air to be healthy for humans, animals, and vegetations. Unfortunately, the combustion of large quantities of fossil fuel could alter this balance, and could introduce other polluting gases and particles in the local areas. Indeed, it is alleged that fossil fuels are probably the most air-polluting energy resources with the worst among them being raw coal. A single raw coal power plant without adequate fitters can pollute an entire city of the size of New York. Although pollution from burning fossil fuels can have a direct effect on the environment, it can

TABLE 5.1

EPA's National Ambient Air Quality Standards (NAAQS)

Pollutant	Primary Standards	Averaging Times	Secondary Standards	Additional Restrictions
Carbon monoxide	9 PPM (10 mg/m^3)	8 h	None	Not to be at or above this level more than once per year.
	35 PPM (40 mg/m^3)	1 h	None	Not to be at or above this level more than once per year.
Lead	1.5 μg/m^3	Quarterly	Same as primary	
Nitrogen dioxide	0.053 PPM (100 μg/m^3)	Annual	Same as primary	
Particulate matter 2.5–10 μm	50 μg/m^3	Annual	Same as primary	
	150 μg/m^3	24 h		Not to be at or above this level for more than 3 days over a 3-year period.
Particulate matter <2.5 μm	15 μg/m^3	Annual	Same as primary	
	65 μg/m^3	24 h		
Ozone	0.08 PPM	8 h	Same as primary	The average of the annual 4th highest daily 8 h maximum over a 3-year period is not to be at or above this level.
	0.12 PPM	1 h	Same as primary	Not to be at or above this level on more than 3 days in a 3-year period.
Sulfur oxides	0.03 PPM	Annual	—	
	0.14 PPM	24 h	—	Not to be at or above this level more than once per year.
	—	3 h	0.5 PPM (1300 μg/m^3)	Not to be at or above this level more than once per year.

also be combined with other gases and particles to create more potent effects. These are known as synergistic effects among pollutants.

The question often asked is, what level of pollution is considered harmful? The U.S. Environmental Protection Agency (EPA) has established National Ambient Air Quality Standards (NAAQS), shown in Table 5.1, for six pollutants: carbon monoxide, lead, nitrogen dioxide, particulate matter, ozone, and sulfur dioxide. The table shows two types of national air quality standards: *primary* and *secondary*. The primary standards set limits to protect the health of the "sensitive" populations such as asthmatics, children, and the elderly.

The secondary standards set limits to protect the general public health as well as to guard against decreased visibility, and damage to animals, crops, vegetation, and buildings.

When a pollutant level in an area causes a violation of a particular standard, the area is classified as *nonattainment* for that pollutant. In this case, the EPA imposes federal regulations on the polluters and gives deadlines by which time the area must satisfy the standard.

5.1.1 Sulfur Oxides

Fossil fuels such as coal, sulfated oil, and diesel are not pure substances, but are often mixed with other minerals, such as sulfur (S) and nitrogen (N); mined coal, in particular, contains more than 6% sulfur. When fossil fuel is burned, the released sulfur is combined with oxygen to form sulfur oxides (sulfur dioxide, SO_2, and sulfur trioxide, SO_3).

$$S + O_2 \Rightarrow SO_2 \tag{5.1}$$

Sulfur dioxide (SO_2) is a corrosive, acidic, and colorless gas with a suffocating odor that can cause severe health problems. When the concentration of the gas reaches 2 PPM, the suffocating odor of the gas can be easily detected. Inhaling large amounts of SO_2 can damage the upper respiratory tract and lung tissues. This problem is more severe for the very young and the very old. In addition, asthmatic patients are more sensitive to SO_2 and their health status can deteriorate quickly. What makes SO_2 particularly dangerous is its quick effect on people, usually within the first few minutes of exposure. Table 5.2 summarizes various epidemiological studies linking SO_2 to respiratory and cardiovascular problems. Note that a small PPM can increase the health risk.

It is hard to estimate the exact amounts of sulfur dioxide released after burning fossil fuels. However, a coal-fired power plant can produce as much as 7 kg/MWh of SO_2, while a natural gas power plant emits just about 5 g/MWh. It is estimated that 20 million tons of SO_2 are released by power plants worldwide.

Our history has several examples of human tragedies due to excessive release of SO_2 from industrial plants other than the power plants. In 1930, in Meuse Valley, Belgium, industrial pollution in the form of sulfur dioxide killed

TABLE 5.2

Health Effect of SO_2

Concentration of SO_2 (PPM)	Exposure Time	Effect
3	3 min	Increases airway resistance
0.2	4 days	Increases cardiorespiratory diseases
0.04	1 year	Increases cardiovascular diseases

63 people. In 1948, in Donora, Pennsylvania, sulfur dioxide emissions from industrial plants caused various respiratory illnesses to about 6000 people, and eventually caused the death of 20 people in a few days. In 1952, London experienced the worst air pollution disaster ever reported from burning coal during a dense foggy day; about 4000 people died mainly because of sulfur dioxides. During the first Gulf war, high concentrations of sulfur dioxides were released when oil fields were set on fire. Soldiers and civilians suffered from severe cardiorespiratory ailments.

To reduce sulfur emissions, most governments impose various regulations (such as the NAAQS in the United States), and are closely monitoring the level released by industrial plants. Penalties are often imposed on those who exceed the government set "quota" of sulfur emissions. Since fossil fuel power plants produce almost half of all sulfur emissions worldwide, the penalties are severe for utilities with inadequate sulfur filtering systems. These regulations have led to a decline in sulfur emissions in the United States, Canada, and many cities in Western Europe. The regulations also encourage owners of coal-fired plants to convert their facilities to natural gas. However, the sulfur emissions are still very high in a number of cities in Eastern Europe, Asia, Africa, and South America.

5.1.2 Nitrogen Oxides

Other harmful gases produced by burning fossil fuel are the nitrogen oxides NO_x (O_x represents various stages of oxidation). Coal or natural gas power plants produce about 2 kg of NO_x for every MWh of electricity.

The direct health effect of nitrogen oxides on humans is minor. However, nitrogen dioxide (NO_2) plays major roles in the formation of smog and acid rain. NO_2 absorbs sunlight resulting in the brownish color of the smog. Further, it is one of the greenhouse gases: nitrous oxide absorbs 270 times more heat per molecule than carbon dioxide.

5.1.3 Ozone

Ozone, O_3, can be found at two different altitudes: the troposphere (up to 10 km altitude) and the stratosphere (10 to 50 km altitude). The stratosphere has a high concentration of ozone of about 3×10^4 PPM.

The *stratosphere ozone* protects the Earth by absorbing the dangerous ultraviolet radiation of the sun. However, because of gases such as the chlorofluorocarbons (CFCs), the ozone in the stratosphere area can be depleted. When the CFC is decomposed by ultraviolet radiation, it releases chlorine atoms that react with the ozone to create chlorine oxide and oxygen. By this process, one chlorine atom can destroy as much as 100,000 ozone molecules. Keep in mind that CFC is not released by any power plant, but it is used as refrigerant and can also be found in aerosol products. In 1978, CFC products were banned in the United States and more recently in most countries.

The *troposphere ozone* is formed when NO_2 is released by industrial plants. This pollutant when excited by solar radiations is converted into nitric oxide (NO), and in the process releases free oxygen (O) that can be combined with oxygen molecules (O_2) to produce ozone at low elevations (up to 10 km).

$$NO_2 + \text{light energy} \Rightarrow NO + O \tag{5.2}$$

$$O + O_2 \Rightarrow O_3 \tag{5.3}$$

Nitric oxide can also be formed during lightning storms by the reaction of nitrogen and oxygen.

Although the stratosphere ozone is beneficial, the troposphere ozone is one of the main ingredients of smog that irritates lungs and causes damage to vegetations. In addition, ozone can make asthmatic patients more sensitive to SO_2.

The troposphere ozone can be recycled back into NO_2 and O_2 when NO is available.

$$NO + O_3 \Rightarrow NO_2 + O_2 \tag{5.4}$$

The chemical equations (5.2) to (5.4) form a delicately balanced process where ozone is formed then destroyed. The reaction in Equation (5.4) is rapid, causing the concentration of O_3 to remain low as long as NO is available. However, when hydrocarbons are present (e.g., from automobile emissions), they react with NO to form organic radicals that cause irritation to sensitive tissues such as eyes. These hydrocarbons compete for NO so less is available to destroy the troposphere ozone. The result is an increase in the concentration of ozone at the troposphere level.

5.1.4 Acid Rain

Sulfur dioxide produced by burning fossil fuels can react with oxygen to form sulfur trioxide.

$$2SO_2 + O_2 \Rightarrow 2SO_3 \tag{5.5}$$

When SO_3 reaches the clouds, it reacts with water to form sulfuric acid (H_2SO_4).

$$SO_3 + H_2O \Rightarrow H_2SO_4 \tag{5.6}$$

Similarly, when nitrogen dioxide NO_2 produced by burning fossil fuels reaches the clouds, it reacts with water and nitric acid (HNO_3) is formed.

$$3NO_2 + H_2O \Rightarrow 2HNO_3 + NO \tag{5.7}$$

The rain from clouds impregnated with these acids (known as the *acid rain*) can be very damaging to crops, agricultural lands, and structures. When it reaches lakes, acid rain increases the acidity of water, which can have a severe effect on fish populations. Acid rain can also damage limestone, precious buildings, and statues.

The acidity in the acid rain depends on the concentration of its hydrogen ions. Scientists have developed a scale called the *potential of Hydrogen* (pH) to quantify the degree of acidity in a solution. It is a negative logarithmic measure of hydrogen-ion concentration in moles per liter of solution. The mole is a chemical unit used to measure the amount of a substance that contains as many elementary entities (atoms, molecules, etc.) as there are atoms in 12 g of the isotope carbon-12 (6.023×10^{23}).

$$pH = -\log H^{+} \tag{5.8}$$

Since pH is a negative logarithmic scale, a change in just one unit from pH 3 to 2 would indicate a tenfold increase in the acidity. The hydrogen ion concentration in pure water is about 1.0×10^{-7} mole, hence its pH is 7. Increasing the concentration of hydrogen ions increases the acidity of the liquid, thus pH < 7 is considered acidic, while a pH > 7 is considered alkaline or basic. Milk is about pH 7, while battery acid is pH 0 to 1. Clean rain usually has a pH of 5.6. Rain measuring less than 5 on the pH scale is considered acid rain. The most acidic rain reported in the United States was about pH 4.3.

In the United States, the areas that suffer the most from acid rain are the northeast and midwest regions, which are known for their coal-based industries and coal-fired power plants.

5.1.5 Carbon Dioxide

The global temperature of the Earth's surface is determined by the difference between the solar energy reaching the Earth, and the energy radiated back to space from the Earth. Because some naturally occurring gases form thermal blankets at various altitudes, not all solar energy is radiated back to space. This phenomenon is loosely known as the *greenhouse effect*. The ratio of the solar energy reaching the Earth to the radiated energy is delicately maintained in nature to keep the temperature on Earth at the level that can sustain life. Some scientists believe that industrial plants can release more of the *greenhouse gases* that could reduce the amount of the radiated energy and therefore increase the temperature on Earth. The greenhouse gases include carbon dioxide CO_2, CFC, methane (CH_4), nitrous oxide (N_2O), and troposphere ozone (O_3). This phenomenon, which is known as *global warming*, could lead to a partial melting of glaciers and ice sheets, which can result in raising the sea level and flooding dry lands. Also, the increase in sea temperature could result in changes in rain and wind patterns. Some scientists believe that the greenhouse gases have been increasing in concentration since the beginning of the industrial revolution in the 1700s. These scientists estimated that the global temperature on Earth has increased by as much as 0.6 to 1°C during the last century. A counter argument made by other scientists, though, is that while the greenhouse gases increase the temperature, other gases such as the stratospheric ozone and sulfate cause the atmosphere to cool down. Therefore, global warming may not be as severe as initially thought, and the debate is not likely to end anytime soon.

Carbon dioxide, which is one of the greenhouse gases, is a colorless, odorless, and slightly acidic gas. Nature recycles CO_2 through water, animals, and plants. Humans exhale carbon dioxide as they breathe, and plants absorb it during photosynthesis.

$$CO_2 + H_2O + \text{light energy} \Rightarrow O_2 + \text{carbohydrates} \tag{5.9}$$

CO_2 is also absorbed in oceans. The process of generating and absorbing CO_2 is precisely sustained in nature, resulting in the right amount of CO_2 in air necessary to keep the Earth's temperature at the current level. However, industrial activities that require burning of material containing carbon (coal, wood, oil) can increase CO_2 concentration in air, thus contributing to the warming of the atmosphere.

Coal-fired power plants produce as much as 1000 kg/MWh of CO_2 and natural gas power plants produce about half of this amount. But keep in mind that CO_2 is only one of the greenhouse gases, with relatively limited effect, since its presence in the atmosphere is for short periods. Other gases with more sinister effects include the CFC (such as Freon) and nitrogen oxides. One molecule of CFC has the same effect as 10,000 molecules of CO_2. What makes these gases much worse than CO_2 is that they remain in the atmosphere for a very long period.

5.1.6 Ashes

Ashes are small particles (0.01 to 50 μm) that are suspended in air. The combustion process in fossil fuel power plants produces large amounts of ash that could stay suspended in air for days before they reach the Earth. About 7 million tons of ash are released each year by electric power plants and industrial smelters. Ash from power plants may include a variety of metals such as iron, titanium, zinc, lead, nickel, arsenic, and silicon.

The ash affects breathing, weakens the immune system, and worsens the condition of patients with cardiovascular disease. Smaller ash (less than 10 μm) can reach the lower respiratory tract and cause severe respiratory problems.

Most power plants install several filtering devices to eliminate, or substantially reduce, the discharged ash. There are various types of filters; among them are the wet scrubber system and the fabric filter system. With the wet scrubber system, the exhaust gas is passed through a liquid that traps flying particulate and sulfur dioxide before the gas is vented through the stacks. The fabric filter system works like a vacuum cleaner where the particulates are trapped in bags.

5.2 Environmental Concerns Related to Hydroelectric Power Plants

Hydroelectric power is the major form of renewable energy systems. Although it does not emit any gas, the hydroelectric plant impacts the environment in other ways as given below:

Flooding: This is the most obvious impact of the impoundment type hydroelectric power plant. The dam is constructed to form a reservoir by flooding the land upstream. The decaying vegetation, submerged by the flooding, produces greenhouse gases and can release dangerous substances embedded in some rocks such as mercury, which accumulates in fish resulting in some health hazards when the fish is consumed. People can also be affected by the dam; when China constructed the Three-Gorges Dam, thousands of people were evicted from areas behind the dam.

Water flow: Hydroelectric dams alter the flow of water downstream, which may change the quality of water in the river.

Silt: Dams and reservoirs can trap silt that would normally be carried downstream to fertilize lands and prevent shore erosions. In addition, trapped silt behind the dam can reduce the amount of water stored in the reservoir, which can reduce the amount of generated electricity.

Oxygen depletion: In deep reservoirs, cold water sinks to the bottom because of its high density. This could reduce the oxygen at the bottom of the reservoir, thus altering the biota and fish population that live at the bottom of the lake.

Nitrogen: When water spills over the dam, air is trapped in the water creating turbulences. This makes the water more dissolvent to nitrogen, which could be toxic to fish.

Fish: Dams alter the downstream flow of the river, which may affect the migration of certain species of fish. The hydroelectric turbines also kill the fish that pass through the penstocks.

Case Study: The Aswan Dam

After Egypt had built its Aswan Dam in the 1960s, inexpensive electricity was generated and the flow of the Nile was controlled all year round. However, several ecological and social problems occurred. For example, the Aswan dam submerged many valuable ancient monuments that are over 5000 years old.

Some of the monuments, such as Ramses II temple Abu Simbel, were raised to higher dry lands. But the monuments in the new locations lost some of their miraculous engineering features such as the exact timing of the annual sunlight projections, and the daily melodies played in various quarters. In addition, the sediment due to the annual flood is trapped behind the dam creating a number of problems such as:

- The sediment is slowly filling the lake, thus spilling the water over its shores. The increase of the surface area of the lake increases the evaporation, which leads to undesired changes in the otherwise dry climate.
- The sediment trapped behind the dam is believed by some scientists to be the main cause for the frequent weak tremors in the area.
- The silt that used to make fertile the agricultural land downstream is virtually eliminated by the dam. Thus, farming in Egypt is negatively impacted.
- The lack of sediment downstream exposes the banks of the Nile to steady erosion.
- Since the sediment that used to reach the Mediterranean shores counteracted the effects of coastal erosion, the shoreline of the Delta after the construction of the dam is continuously receding.
- The Mediterranean saltwater is reaching the upstream of the Nile, thus contaminating farmland and groundwater.
- Because of the lack of sediments in the river, the water is clearer allowing more sunlight to penetrate the water. This increases the population of several biota and plants such as the phytoplankton, which affects the fish population in the river and makes the water more polluted, thus more chlorine is used to treat the drinking water.
- The influx of nutrients associated with the annual silt-rich flood can no longer reach the Mediterranean Sea to feed the local biota and fish. After the dam was built, certain types of fish, such as sardines and shrimps, were virtually eliminated.

5.3 Environmental Concerns Related to Nuclear Power Plants

Recently, nuclear power plants have received most concern from the public. Although some of this concern is justifiable, most of it is based on misinformation or the erroneous association of nuclear power plants with nuclear weapons. Nuclear power plants operated safely for almost 40 years until the Three Mile Island nuclear power accident followed by the Chernobyl nuclear disaster (see Chapter 4). These two accidents intensified public pressure to set stricter safety regulations or to abandon nuclear power all together. The debate is still going strong and is not likely to end anytime soon. Among the debated issues are the following:

Radioactive release during normal operation: In boiling water reactors (BWRs), the fuel rods are in direct contact with the steam. Hence, any leakage from the fuel rods is carried by the steam and eventually reaches the turbine and condenser. Although the BWR has several layers of protection that hold the radioactive steam for several half-life cycles before it is released, accidental release is a concern. In pressurized water reactors (PWRs), this problem does not exist since the fuel rods are separated from the steam cycle.

The release of small amounts of radioactive gases is not necessarily a heath hazard. The radioactive limit on radiation set by the EPA is 170 mREM/year above the background radiation from all natural resources. The REM stands for the *Radiation Equivalent Man*, which is a measure of a biological damage done to tissues by specific amounts of radiation. High doses of REM give rise to immediate radiation sickness such as hypodermal bleeding, hair loss, and sickness. Under 25 REM, no short-term effects are observed, but long-term exposure may lead to greater possibility for cancer and genetic abnormality in offspring.

Loss of coolant: One of the major public concerns is the loss of coolant inside the reactor. If the flow of water is interrupted due to breaks in pipes or pumps, fuel rods could become excessively hot and could melt. This scenario is sensationalized in movies and is given the name "*China Syndrome*," where melted rods at the bottom of the reactor go through the reactor bottom into Earth and all the way to China on the other side of the globe.

Needless to say, this problem is not as severe as it is being depicted in some circles. The water inside the reactor is both a coolant and a moderator. The moderator slows down the neutrons to make fission possible. If the water inside the reactor is lost, the chain reaction is substantially curtailed. In addition, nuclear power plants have emergency core cooling and emergency core shutdown systems that prevent the meltdown of rods.

Reactor explosion: Nuclear material is wrongly associated with nuclear bombs, and nuclear explosion inside the reactor is often stated. This is an unfounded concern since the uranium used in nuclear power plant is about 3% enriched, while that used in nuclear bombs is over 90% enriched. Furthermore, the chain reaction in nuclear power plants is slowed down by the moderator when the temperature of the reactor increases to above the design limits; the moderator density increases thus slowing down the chain reaction.

Disposal of radioactive waste: This is one of the crucial issues facing the nuclear power industry. The spent fuel is a high-level radioactive waste with a half-life of thousands of years. In the United States, about 2000 tons of spent fuel is produced annually from nuclear power plants, and much more is produced by several nuclear defense programs.

Because of its long half-life, the spent fuel can stay hot for hundreds of years. Therefore the storage facilities must prevent the radioactive material from leaking for very long periods. One type of storage vessel used in the industry is composed of a container with multiple layers of steel and concrete barriers. The containers are eventually buried in areas that are stable geologically and surrounded by rocks to prevent any leakage from reaching the water table. Some nations store the container in deep ocean floor. Some scientists have proposed some wild ideas such as shooting the radioactive waste into the sun, or storing the radioactive material in space. This is, obviously, a highly sensitive issue to the public.

A more realistic option is to reduce or eliminate the long-lived radioactive waste by transmutation. Nuclear transmutation is a process by which the long-lived fuel waste is converted into shorter-lived waste that is easier to store. Transmutation occurs by bombarding the radioactive waste with charged material from accelerators.

Exercise

5.1. Name three types of major power plants.

5.2. How is acid rain produced?

5.3. What is the greenhouse effect?

5.4. State four drawbacks of hydroelectric power plants.

5.5. What are the health effects of SO_2?

5.6. Fill in the blanks in the following sentences:

a. Sulfur dioxide emitted from power plants reacts with oxygen to form ________.

b. When sulfur trioxide reaches the clouds, it reacts with water to form ________.

c. When nitrogen dioxide reachs the clouds, it reacts with water and ________ is produced.

5.7. A coal-fired power plant produces an average of 100 MW. Estimate the sulfur dioxide released daily and annually if no filtering system is implemented.

5.8. A Natural gas power plant produce an average of 100 MW. Estimate the sulfur dioxide released daily and annually if no filtering system is implemented.

5.9. If a 100 MW coal-fired power plant is converted into natural gas, compute the annual percentage reduction of SO_2.

5.10. In 2002, the world consumed 14×10^{12} kWh of electricity. Estimate the maximum amount of sulfur dioxide released from the coal-fired power plant worldwide.

5.11. Write an essay on an area with severe acid rain. Discuss the pH values and their effect on the local environment.

5.12. How many moles of hydrogen ions are in a liquid with pH 5?

5.13. Choose a case study for a hydroelectric power plant. Identify the pros and cons of the plant.

5.14. Write an essay on the impact of hydroelectric power plants on the fish populations.

5.15. Identify a few ideas to lessen the effect of hydroelectric power plants on fish migrations.

5.16. Identify a few ideas to increase the silt in the upstream flow of hydroelectric power plants.

5.17. Write an essay on the Three Mile Island accident. Identify the sequence of events that led to the accident. In your opinion, were the steps taken to correct the accident adequate?

5.18. Write an essay on the Chernobyl nuclear disaster. Identify the sequence of events that led to the disaster. In your opinion, can this accident occur in the United States? Why?

6

Renewable Energy

As seen in the previous chapters, the world relies heavily on fossil fuel (coal, oil, and natural gas) for its ever-growing appetite for energy. In the early 1950s, the public concerns regarding the negative environmental impact of burning fossil fuel have encouraged engineers and scientists to seriously develop reliable alternative energy resources. The efforts were accelerated in the 1970s when the oil prices soared. Many countries began investing in renewable energy through various programs that promote the development and testing of reliable renewable energy systems. Tax credits, investments in research and development, subsidies, and developing favorable regulations are some of the various supports by the governments to accelerate the development of the technology. However, unfortunately, the development efforts are still largely dependent on the prices of the crude oil as well as the intensity of the societal pressures.

Renewable energy is a phrase that is loosely used to describe any form of generating electric energy from resources other than fossil or nuclear fuels. Renewable energy resources include hydroelectric, wind, Solar, wave and tidal, geothermal, and hydrogen. Sunrays are the source of all these renewable energies with the exception of geothermal energy, which is the heat stored in the Earth's magma, and the ocean's tides, which are due to the gravitational pull of the Moon and the Sun. These resources produce much less pollutions than burning fossil fuels, and are constantly replenished; thus called renewable.

As early as the 19th century, scientists and engineers worked on developing technologies to generate electric energy using various forms of renewable resources. The Italian Prince Piero Ginori Conti invented the first geothermal power plant in 1904. The plant is located at Larderello in Italy. In 1842, Sir William George Armstrong invented a hydroelectric machine that produced frictional electricity. However, the world's first hydroelectric power plant was constructed across the Fox River in Appleton, WI in 1882.

The developments in the renewable energy field during the past few years have led to more efficient and more reliable systems. Some of these systems, such as solar and fuel cells, are already used in aerospace and transportations. In this chapter, we shall discuss various renewable energy systems, but the reader should be aware that the area is growing fast, and new technologies or systems are invented every year.

6.1 Solar Energy

The Sun is the primary source of energy in the solar system, and the Earth receives 90% of its total energy from the Sun. Other natural forms of energy on the Earth include the geothermal, tidal resources, volcanoes, and earthquakes.

The extraterrestrial power density of the Sun is about $1.366 \pm 4\%$ kW/m^2. This tremendous power is weakened at the Earth's surface due to several reasons, among them are:

- The various gases and water vapor in the Earth's atmosphere absorb some of the solar energy.
- The distance from the Sun, which can be measured by the projection angle of the sunrays (zenith angle).
- The reflections and scattering of the sunrays.

The computation of the solar power density on the Earth (solar irradiance) is complex and requires the knowledge of several hard to find parameters. However, approximate models can be used such as the one developed by *Atwater and Ball* and given in Equation (6.1).

$$\rho = \rho_0 \cos \xi \, (\alpha_{\mathrm{dt}} - \beta_{\mathrm{wa}}) \, \alpha_p \tag{6.1}$$

where ρ is the solar power density on the Earth's surface in kilowatts per meter square, ρ_0 is the extraterrestrial power density (the number often used is 1.353 kW/m^2), ξ is the zenith angle (angle from the outward normal on the Earth's surface to the center of the Sun), α_{dt} is the direct transmittance of gases except for water vapor (the fraction of radiant energy that is not absorbed by gases), α_p is the transmittance of aerosol, and β_{wa} is the water vapor absorptions of radiation. The term *aerosol* refers to atmospheric particles suspended in the Earth's atmosphere (sulfates, volatile organic compounds, black carbon, etc.). The size of these particles normally ranges from 10^{-3} to 10^3 μm. As explained in Chapter 5, some of the aerosols (such as CO_2, N_2O, and troposphere ozone) cause the atmosphere's temperature to rise, and some (such as stratospheric ozone, sulfate, and biomass burning) cause it to cool down.

Due to the reflection, scattering, and absorptions, the solar power at the Earth's surface is a fraction of the extraterrestrial solar power. The ratio of the two solar power densities is known as the *solar efficiency* η_{s}.

$$\eta_s = \frac{\rho}{\rho_0} = \cos \xi \, (\alpha_{\mathrm{dt}} - \beta_{\mathrm{wa}}) \, \alpha_p \tag{6.2}$$

The solar efficiency varies widely from one place to another, and is also a function of the season and the time of the day. It ranges between 5 and 70%. The zenith angle has a major effect on the efficiency; for the same absorption conditions, the maximum efficiency occurs at noon in the equator when $\xi = 0°$.

EXAMPLE 6.1 A person wants to install a solar energy system in Nevada. At a certain time in the afternoon, the zenith angle is 30°, the transmittance of all gases is 70%, the water vapor absorption is 5%, and the transmittance of aerosol is 90%. Compute the power density at that time and the solar efficiency.

Solution

By directly substituting the parameters in Equation (6.1):

$$\rho = \rho_0 \cos \xi \,(\alpha_{dt} - \beta_{wa})\,\alpha_p$$
$$= 1353 \times \cos(30) \times (0.7 - 0.05) \times 0.9 = 685.5 \text{ W/m}^2$$
$$\eta_s = \frac{\rho}{\rho_0} = \frac{685.5}{1353} = 50.7\%$$

Figure 6.1 shows the map of the solar power density in various regions in the United States. In the Mohave Desert and Nevada, the solar power density can be as much as 750 W/m^2. In the Northwest region, it can be as high as 300 W/m^2. The solar power density during a typical day with stable weather follows the bell-shaped curve expressed by the normal distribution function in Equation (6.3).

$$\rho = \rho_{max} e^{-(t-t_0)^2/2\sigma^2} \tag{6.3}$$

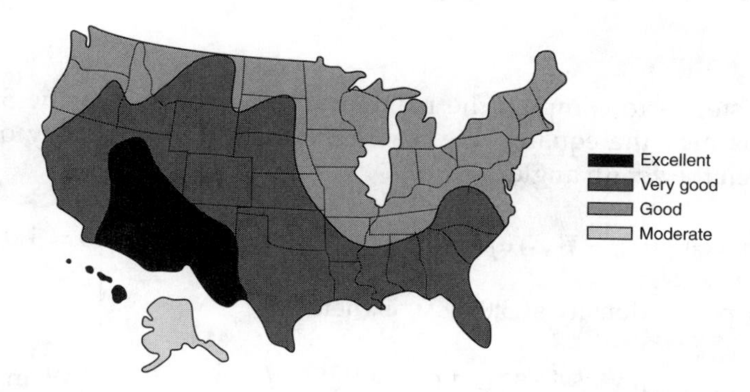

FIGURE 6.1
Solar power density in the United States.

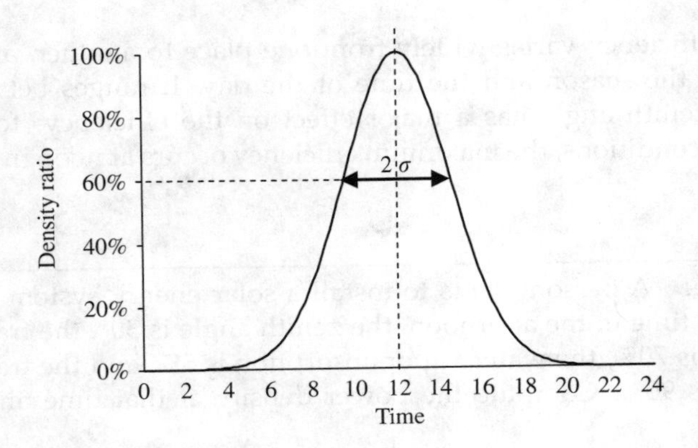

FIGURE 6.2
A typical solar distribution function (solar power density in 24-h period).

where t is the hour of the day using the 24-h clock, ρ_{max} is the maximum solar power density of the day at t_0 (noontime in the equator), and σ is the standard deviation of the normal distribution function. The bell curve in Equation (6.3) is shown in Figure 6.2. The density ratio in the curve is the percentage of the ratio ρ/ρ_{max}. Note that when $t = 12 \pm \sigma, \rho/\rho_{max} = 0.607$.

A large σ means wider areas under the distribution curve; i.e., more solar energy is acquired during the day. In high latitudes, σ is smaller in winter than in summer.

EXAMPLE 6.2 An area located near the equator has the following parameters:

$$\alpha_{dt} = 80\%, \qquad \alpha_p = 95\%, \qquad \beta_{wa} = 2\%$$

Assume that the parameters are unchanged during the day, and the standard deviation of the solar distribution function is 3.5 h. Compute the solar power density at 3:00 P.M. in the afternoon.

Solution

The first step is to compute the maximum solar power density. Since the location is near the equator, the maximum solar power density occurs at noon when the zenith angle is zero.

$$\rho_{max} = \rho_0 \cos\xi\, (\alpha_{dt} - \beta_{wa})\, \alpha_p = 1353 \cos(0) \times (0.8 - 0.02)0.95 = 1.0\,\text{kW/m}^2$$

The solar power density at 3:00 P.M. is then

$$\rho = \rho_{max} e^{-(t-t_0)^2/2\sigma^2} = 1.0 \times e^{-(15-12)^2/2(3.5)^2} = 0.693\,\text{kW/m}^2$$

6.1.1 Types of Solar Energy Systems

Solar energy is harnessed by two methods: passive and active. A passive solar energy system uses the sunrays directly to heat water or gas. The active system converts the Sun's energy into electrical energy by using a photovoltaic semiconductor material called solar cell.

EXAMPLE 6.3 A 2 m^2 panel of solar cells is installed in the Nevada area of Example 6.1. The panel converts the Sun's energy into electrical energy. The conversion efficiency of the solar panel is 10%.

1. Compute the electrical power of the panel.
2. Assume the panel is installed on a geosynchronous satellite. Compute its electrical power output.

Solution

1. In Example 6.1, the computed power density is 685.5 W/m^2. The Sun power of the solar panel P_{sun} is the power density ρ multiplied by the area of the panel A.

$$P_{sun} = \rho A = 685.5 \times 2 = 1.371 \text{ kW}$$

The electrical power output of the panel P_{panel} is the solar power input P_{sun} multiplied by the efficiency of the panel η.

$$P_{panel} = \eta P_{sun} = 0.1 \times 1371 = 137.1 \text{ W}$$

As you can see, because of the low efficiency of the solar panel, the electric power obtained from a 2 m^2 panel in one of the best areas in the United States produces only 137 W. This is enough to power two light bulbs. Higher power can be obtained if the solar panel is larger and the efficiency of the panel is higher.

2. If the same solar panel is mounted on a satellite, the solar power density on the panel is ρ_o. Hence,

$$P_{sun} = \rho_0 A = 1353 \times 2 = 2.706 \text{ kW}$$

The electrical power output P_{panel} of the panel is the solar power input P_{sun} multiplied by the panel efficiency η

$$P_{panel} = \eta P_{sun} = 0.1 \times 2706 = 270.6 \text{ W}$$

As you can see, the electric power of the same panel doubles when it is moved to outer space.

6.1.1.1 *Passive Solar Energy System*

An example of the passive solar system is the thermosiphon hot water appliance shown in Figure 6.3 and Figure 6.4. The system consists of a solar collector, tank, and water tubes. The tank is located above the collector. The collector has an outer lens (or transparent glass) facing the Sun, and houses long zigzagged water tubes. The lens concentrates the sunrays, thus increasing the temperature of the water inside the tubes. The warm water moves naturally upward to the tank (hot water rises above cold water). The water at the top part of the tank is warmer than the water in lower part, hence, the cooler water at the bottom of the tank goes back to the collector. When warm

FIGURE 6.3
(see color insert following Page 208) Passive thermosiphon hot water solar system.

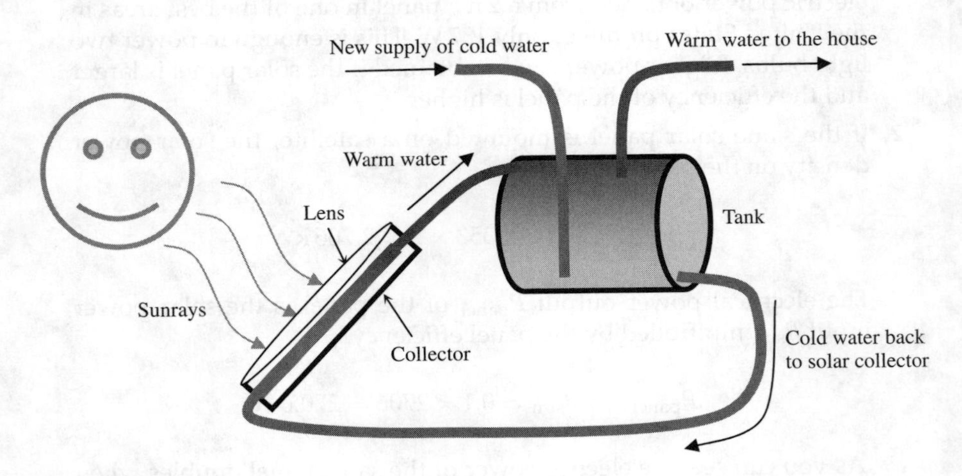

FIGURE 6.4
Thermosiphon hot water system.

water is needed inside the house, it is extracted from the top of the tank. This water is replaced by water from the main water feeder of the house.

The passive solar system is simple, inexpensive, and requires little maintenance. However, it demands enough solar power density to make it viable, and it is most effective during daytime.

6.1.1.2 Active Solar Energy System (Photovoltaic)

Light consists of particles called *photons*, which are the energy by-products of the nuclear reaction in the Sun. Each photon is a packet of energy, but not all photons have the same amount of energy. Photons with shorter wavelengths (higher frequencies) such as the gamma rays (about 10^{20} Hz) have more energy than photons with longer wavelengths such as visible light photons. The frequency of sunlight is about 4.7×10^{14} Hz. According to the photoelectric phenomenon discovered by Heinrich Hertz in 1887, when photons hit certain materials, the electrons in the materials absorb the energy of the photons. If the acquired energy is higher than the binding energy of the electrons, the electrons break away from their atoms. This phenomenon is used to develop the solar cell, wherein the breakaway electrons are collected to form an electric current.

The solar cell, which is also called a *photovoltaic* (PV), consists of two layers of semiconductor material such as crystalline silicon. Silicon is a good insulator unless additives, such as phosphorus, are added to make the silicon compound an n-type with extra free electrons. When boron is added, the silicon has fewer electrons and the silicon compound is p-type with empty locations for electrons called holes. The PV consists of two layers, p-type and n-type, bonded together in a structure that resembles the semiconductor diode.

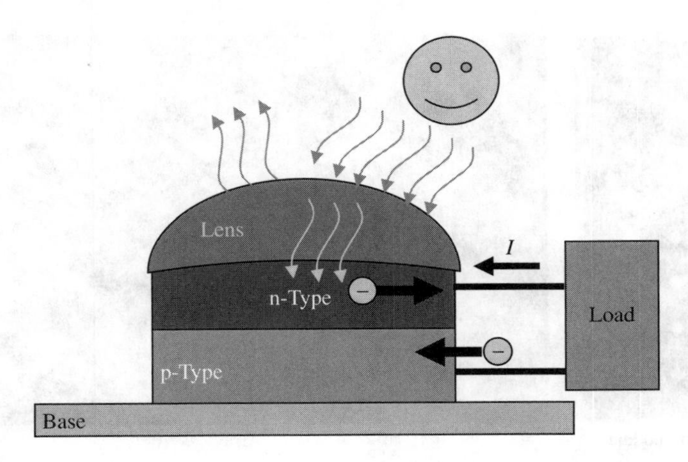

FIGURE 6.5
Concentrating PV cell.

There are two major types of PV cells: concentrating and flat-plate. The structure of the concentrating PV cell is shown in Figure 6.5. The cell consists of a lens mounted on top of an n-type material. The p-type material is at the base of the cell. An electric load is connected between the two layers. When the sunlight reaches the lens, the photons are concentrated and most of them pass through the lens to the n-type material. The energy of the photons is absorbed by the electrons in the n-type material causing them to escape from the n-type layer to the external load and back to the p-type material to close the electrical circuit. The electrons form the electrical current that delivers the power to the load. Note that the PV cell is just a semiconductor diode, so electrons flow only in one direction and the current produced is direct current.

The flat-plate PV cell, as the name implies, is rectangular and flat. This is the most common type of PV array used in commercial applications. Flat-plate cells are often mounted at fixed angles that maximize the exposure to the Sun throughout the year; in the United States, it is the southern direction. In more flexible systems, the angle of the solar panel changes to track the optimal Sun exposure during the day.

During cloudy days, diffused light is produced by sunrays traveling through the clouds. This condition poses no challenge to the flat-plate PV cells. On the other hand, concentrating PV cells generate less power with diffused light; hence, they are limited to the very sunny places.

A PV cell 10 cm in diameter produces about 1 W of power, which is enough to run a calculator. The efficiency of the solar cell ranges from 2 to 20% depending on the material and the structure of the cell. Since the PV cell is a diode, its voltage is about 0.5 V, which is the forward-biased voltage of a diode. To increase the power rating, the PV cells are connected together in parallel and series arrangements. The parallel connection increases the overall current, and the series connection increases the overall voltage. These interconnected PV cells are called the *module* or *panel*; a typical module is shown in Figure 6.6.

Panel or module

Array

FIGURE 6.6
(see color insert following Page 208) PV module and PV array. (Images courtesy of the U.S. Department of Energy.)

For higher power needs, several modules are connected together to form a PV *array* as shown in Figure 6.6. These arrays form a single PV *system*. More sophisticated PV arrays are mounted on tracking devices that follow the Sun throughout the day. The tracking devices tilt the PV arrays to maximize the exposure of the cells to the sunrays, thus increasing the power of the PV system.

EXAMPLE 6.4 Estimate the maximum power, current, and voltage ratings of the panel and array in Figure 6.6. Assume that each PV cell produces a maximum power of 2.5 W at the best solar conditions.

Solution

The panel consists of four columns of nine series PV cells. This is a total of 36 cells. If you assume that the voltage of each cell is 0.5 V, the total voltage of the panel is

$$V_{panel} = 0.5 \times 9 = 4.5\,\text{V}$$

If the power of each cell is 2.5 W, the total power of the panel is

$$P_{panel} = 2.5 \times 36 = 90\,\text{W}$$

The total current of the panel is

$$I_{panel} = \frac{P_{panel}}{V_{panel}} = \frac{90}{4.5} = 20\ \text{A}$$

The array consists of two columns of four series modules. The total voltage of the array is

$$V_{array} = V_{panel} \times 4 = 4.5 \times 4 = 18\,\text{V}$$

The total power of the array is

$$P_{array} = P_{panel} \times 8 = 90 \times 8 = 720\,\text{W}$$

The total current of the panel is

$$I_{array} = \frac{P_{array}}{V_{array}} = \frac{720}{18} = 40\ \text{A}$$

Note that the above ratings are the maximum values based on the best solar conditions.

6.1.2 Daily Power Profile of the Solar Array

The power produced by the PV panel can be approximately modeled by assuming a linear correlation between the solar power density and the output power of the panel. Since the solar power density is a bell-shaped curve as shown in Figure 6.2, the panel power is also a bell-shaped curve that can be expressed by

$$P_{panel} = P_{max} e^{-(t-t_0)^2/2\sigma^2} \tag{6.4}$$

where P_{panel} is the electric power produced by the solar panel at any time t, and P_{max} is the maximum power produced during the day at t_o (noon in the equator). The energy produced by the PV panel E_{panel} in one day is the integral of Equation (6.4).

$$E_{panel} = \int_0^{24} P_{max} e^{-(t-t_0)^2/2\sigma^2} dt \approx P_{max} \sqrt{2\pi}\ \sigma \tag{6.5}$$

EXAMPLE 6.5 The solar power density at a given site is $\rho = \rho_{max} e^{-(t-12)^2/12.5}$. A solar panel is installed at the site and its maximum electric power measured during the noontime is 100 W. Compute the daily energy produced by a PV panel.

Solution

From the equation of the solar power density, $2\sigma^2 = 12.5$.

Hence, $\sigma = \sqrt{\frac{12.5}{2}} = 2.5$.

Using Equation (6.5), we get

$$E_{panel} = P_{max}\sqrt{2\pi}\sigma = 100 \times \sqrt{2\pi} \times 2.5 = 0.627\ \text{kWh}$$

EXAMPLE 6.6 The solar power density at a given site is represented by the following equation:

$$\rho = 0.7 e^{-(t-12)^2/18}\ \text{kW/m}^2$$

A 4 m^2 solar panel is installed at the site. The maximum electric power of the panel is 320 W.

1. Compute the efficiency of the panel.
2. Compute the output power of the panel at 4:00 P.M.

Solution

1. The maximum electric power of the panel per unit area is $320/4 = 80\ \text{W/m}^2$.

The maximum solar power density from the equation given above is 700 W/m^2.

$$\text{Efficiency of solar panel } \eta = \frac{80}{700} = 11.43\%$$

2. The output power of the panel at 4:00 P.M. can be directly computed using Equation (6.4).

$$P_{\text{panel}} = P_{\max} e^{-(t-12)^2/2\sigma^2} = 320 \times e^{-(16-12)^2/18} = 131.56\,\text{W}$$

6.1.3 Photovoltaic System

Generating a dc current at low voltage is not useful for most power equipment and appliances designed for alternating currents at 120/240 V. Therefore, a converter is needed to change the low voltage dc waveform of the PV array to an ac waveform at the frequency and voltage levels required by the equipment. The solar array and the converter are the main components of the PV system. These PV systems are quite popular for household and commercial buildings; some are shown in Figure 6.7.

FIGURE 6.7
(see color insert following Page 208) Various photovoltaic systems. (Images courtesy of the U.S. Department of Energy.)

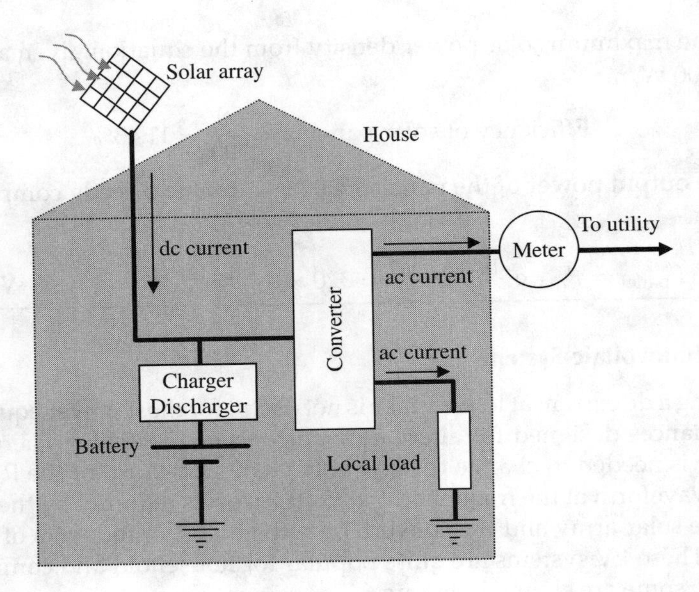

FIGURE 6.8
Storage PV System.

The PV systems typically have two designs: storage and direct systems. The storage PV system shown in Figure 6.8 consists of four main components: solar array, charger/discharger, battery, and converter. The function of the battery is to store the excess energy of the PV system during the day so that it can be recovered during the night. These batteries are 12 V deep-cycle types that can be fully discharged without being damaged. They are more expensive than normal car batteries, and are often used in boats and recreation vehicles. The converter is used to transform the dc power of the PV array, or the battery, into ac for household or commercial use. Several of these converters are discussed in Chapter 10; the commercial ones are in the range of 1000 to 6000 W, which are adequate for most household and small commercial applications. The output of the converter is used to power the household loads, and the extra power is exported to the utility. The utility meter in this application must run backward to allow for *net metering*, which is the difference between the imported and exported powers. After sunset, the energy stored in the battery is used to power the loads through the converter, and any additional power is imported from the utility. A variation of this system without the utility connection is very popular in remote areas without any utility service; it is known as a stand-alone PV system.

The use of storage batteries is the major drawback of the storage PV system. This is because the batteries decrease the overall efficiency of the PV system; about 10% of the energy stored is normally lost, they add to the total expense of the PV system, their life span is about 5 years, and they occupy considerable

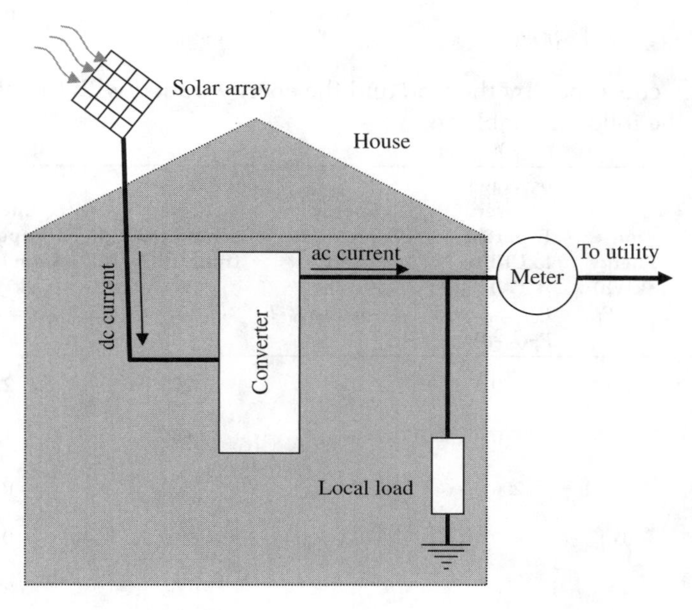

FIGURE 6.9
Direct PV system.

floor space. In addition, the leak from the batteries is very acidic and causes corrosion and damage to surrounding areas.

A less expensive system that requires a direct connection to the utility grid is shown in Figure 6.9. In this system, the storage batteries and their charger are eliminated. The excess power in this solar system is pumped back to the power grid. The meter in the figure measures the net energy (PV power minus the consumed power of the house).

EXAMPLE 6.7 A house is equipped with the solar system in Figure 6.9 The following table shows the daily load P_L and the average output of the PV system P_{PV}. Assume that the daily pattern is repeated throughout one month. Compute the saving in energy cost due to the use of the PV system for one month. Assume that the cost of energy from the utility is $0.20/kWh.

Time	Average Output of PV Array P_{PV} (W)	Average House Load P_L (W)
8.00 P.M.–6:00 A.M.	0	200
6:00–8:00 A.M.	200	800
8:00–11:00 A.M.	300	100
11:00 A.M.–5:00 P.M.	500	100
5:00–8:00 P.M.	200	1200

Solution

The energy consumed by the load and the energy traded with the utility are shown in the following table.

Time T	House Energy (kWh) $E_L = P_L T$	Average Power Exported to Utility (W) $P_{export} = P_{PV} - P_L$	Energy Exported to Utility (kWh) $E_{export} = P_{export} T$	Average Power Imported from Utility (W) $P_{import} = P_L - P_{PV}$	Energy Imported from Utility (kWh) $E_{import} = P_{import} T$
8.00 P.M.–6:00 A.M.	2.0	0	0	200	2.0
6:00–8:00 A.M.	1.6	0	0	600	1.2
8:00–11:00 A.M.	0.3	200	0.6	0	0
11:00–5:00 P.M.	0.6	400	2.4	0	0
5:00–8:00 P.M.	3.6	0	0	1000	3.0
Total daily energy	8.1		3.00		6.2

Without the solar system, the daily cost of the electricity C_1 is

$$C_1 = 8.1 \times 0.2 = \$1.62$$

With the solar system, the daily cost of energy C_2 is

$$C_2 = (6.2 - 3.0) \times 0.2 = \$0.64$$

The saving in one month S is given by

$$S = (C_1 - C_2) \times 30 = \$29.4$$

Keep in mind that this saving does not reflect the cost of the PV system.

6.1.4 Assessment of Photovoltaic Systems

The most common applications of the PV cell are in consumer products such as calculators, watches, battery chargers, light controls, and flashlights. The larger PV systems are extensively used in space applications (such as satellites) where their usage has increased 1000-fold since 1970. In higher power applications, three factors determine the applicability of the PV systems: (1) the cost and the payback period of the system; (2) the availability of a power grid; and (3) the individual inclination to invest in environment-friendly technologies.

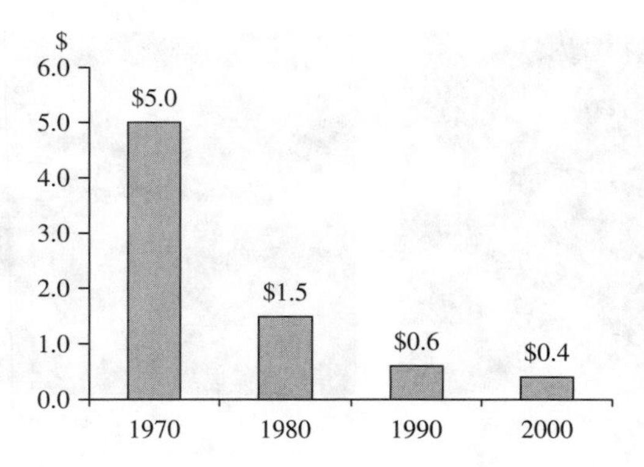

FIGURE 6.10
Prices of kilowatt hour from PV systems.

The price of electricity generated by a PV system varies widely and is based on the size of the PV system, the various design options, the installation cost, the lifetime of the system, the maintenance cost, and the solar power density at the selected site. The estimated average prices of 1 kWh of PV energy using engineering economic models over a period of 30 years are shown in Figure 6.10. As you can see, the price decreases rapidly, although it is still more expensive than most utility rates; the cost of generating 1 kWh of PV electricity is more than three times the price charged by the utilities in the northwest United States region in 2000.

In remote areas without access to power grids, the PV system is often the first choice among the available alternatives. In developing countries and remote areas, thousands of PV systems are already installed and have improved the quality of life tremendously. These solar power systems operate clean water pumps, provide power for lighting, keep food refrigerated, operate emergency radios, and many other critical applications. By the end of the 20th century, the PV systems worldwide had the capacity of more than 900 GWh annually; some of the large systems are shown in Figure 6.11. To put the number into perspective, this energy is enough for about 70,000 homes in the United States, or about 4 million homes in developing countries.

The dream of the solar energy enthusiasts is to replace a large amount of the energy that is currently produced by primary resources with that produced by photovoltaic systems. This is because 1 kW of PV power can reduce the annual discharge of sulfur dioxide (SO_2) from a coal-fired power plant by as much as 30 kg.

Although solar energy is very promising, the PV technology is not free from drawbacks. The following list identifies the main issues associated with solar systems:

1. Although the power generated by the PV system is pollution-free, toxic chemicals are used to manufacture the solar cells such as arsenic and silicon. Arsenic is a generally odorless and flavorless

FIGURE 6.11
(see color insert following Page 208) High-power photovoltiac systems. (Images courtesy of the U.S. Department of Energy.)

semimetallic chemical that is highly toxic and can kill humans quickly if inhaled in large amounts. Small amounts of arsenic with long-term exposure through the skin or inhaled can lead to slow death and a variety of illnesses in the victims. Silicon, by itself, is not toxic. However, when additives are added to make the PV semiconductor material from silicon, the compound can be extremely toxic. Since water is used in the manufacturing process, the runoff could cause the arsenic and silicon compounds to reach local streams. Furthermore, should a PV array catch fire, these chemicals can be released into the environment.

2. Solar power density can be intermittent due to weather conditions (e.g., heavy cloud cover). It is also limited exclusively to use in daytime.
3. For high-power PV systems, the arrays spread over a large area. This makes the system difficult to accommodate in or near cities.
4. The PV systems are considered by some to be visually intrusive (e.g., rooftop arrays are clearly visible to neighbors).
5. The efficiency of the solar panel is still low, making the system more expensive and large in size.
6. Solar systems require continuous cleaning of their surfaces. Dust, shadows, falling leaves, and bird droppings can substantially reduce the efficiency of the system.

6.2 Wind Energy

Wind is one of the oldest forms of energy known to man; it dates back more than 5000 years to Egypt, when sailboats were used for transportation.

Although no one knows exactly who built the first windmill, archaeologists discovered a Chinese vase with a painting on it that resembles a windmill and dates back to the 3rd millennium B.C. By the 2nd millennium B.C., the Babylonians used windmills extensively for irrigation. They are constructed as a revolving door system, similar to the vertical-axis wind machine used today. In the 12th century, windmills were built in Europe to grind grains, pump water, and lift crops. The windmills were also used in Holland to drain lands below the sea level. During this era, working in windmill plants was one of the hazardous jobs in Europe. The workers were frequently injured because the windmills were constructed out of a huge rotating mass with little or no control on its rotation. The grinding or hammering sounds were so loud that many workers became deaf, the grinding dust of certain material such as wood caused respiratory health problems, and the grinding stones often grind against each other causing sparks and fires.

In addition to producing mechanical power, windmills were also used to communicate with neighbors by locking the windmill sails in certain arrangement. During World War II, the Netherlanders used to set windmill sails in certain positions to alert the public of a possible attack by their enemy.

6.2.1 Kinetic Energy of Wind

The wind energy systems harness the kinetic energy of the wind and convert it into electrical energy. The kinetic energy of an object is the energy it possesses while in motion. Starting with Newton's second law, the kinetic energy (KE) of a moving object can be expressed as

$$\mathrm{KE} = \tfrac{1}{2} m v^2 \tag{6.6}$$

where m is the mass of the moving object in kilograms, and v is its velocity in meters per second. The unit of kinetic energy is joule or watts per second. In order to calculate the kinetic energy of the wind, we must calculate the mass of air passing through a given area A. The mass of air depends on the area A, the speed of the wind v, the air density δ, and the time t.

$$m = Av\delta t \tag{6.7}$$

The unit of A is meter square, the unit of v is meter per second, the unit of δ is kilogram per meter cube, and the unit of time is seconds. Substituting the mass in Equation (6.7) into Equation (6.6), yields

$$\mathrm{KE} = \tfrac{1}{2}\, A\delta t v^3 \tag{6.8}$$

Since the energy is power multiplied by time, the wind power in watts is

$$P = \frac{\text{KE}}{t} = \frac{1}{2} A \delta v^3 \tag{6.9}$$

Note that the kinetic energy and the power of the wind are proportional to the cube of the wind's speed. The wind power density ρ in watts per meter square is often used to judge the potential of a given site for wind energy installation.

$$\rho = \frac{P}{A} = \frac{1}{2} \delta v^3 \tag{6.10}$$

The air density is a function of the air pressure, temperature, humidity, elevation, and the gravitational constant. The expression commonly used to compute the air density is

$$\delta = \frac{pr}{\kappa T} \mathrm{e}^{(gh/\kappa T)} \tag{6.11}$$

where pr is the standard atmospheric pressure at sea level (101,325 Pa or N/m^2), T is the air temperature in degree Kelvin (degree Kevin $= 273 +^\circ$ C), κ is the specific gas constant (for air $\kappa = 287$ W s/kg K), g is the gravitational acceleration (9.8 m/s^2), and h is the elevation of the wind above the sea level in meters. Substituting these values into Equation (6.11) yields

$$\delta = \frac{353}{T + 273} \mathrm{e}^{-h/(29.3(T+273))} \tag{6.12}$$

Note that the temperature T in Equation (6.12) is in Celsius. The equation shows that when the temperature decreases, the air is denser. Also, the air is less dense at high altitudes. A wind with higher air density (heavier air) contains more kinetic energy.

EXAMPLE 6.8 The Tehachapi is a desert city in California with an elevation of about 350 m, and is known for its extensive wind farms. Compute the power density of the wind when the air temperature is 30°C and the speed of the wind is 12 m/s.

Solution

To compute the power density of the wind, you need to compute the air density

$$\delta = \frac{353}{30 + 273} \mathrm{e}^{-350/29.3(30+273)} = 1.12\,\text{kg/m}^3$$

The power density, by using Equation (6.10), is

$$\rho = \frac{1}{2}\delta\, v^3 = \frac{1}{2}1.12 \times 12^3 = 967.7\,\text{W/m}^2$$

Compare this power density of wind with the power density of the Sun computed in Example 6.1. Can you draw a conclusion?

6.2.2 Wind Turbine

The wind turbine is a generic name given to the wind energy system that converts the kinetic energy of the wind into electrical energy. The common size of the wind turbine is 300 kW, although as much as 2 MW turbines are available. The wind turbines come in a variety of designs and power ratings, but the most common design is the horizontal axis type shown in Figure 6.12. It consists of the following basic components:

- A tower that keeps the rotating blades at a height that is sufficient to increase the exposure of the blades to the wind.
- Rotating blades that capture the kinetic energy of the wind. They are normally made of fiberglass-reinforced polyester or wood-epoxy material. In most designs, the length of the rotating blades ranges from 5 to 40 m. More advanced blade systems allow the blades to change their pitch angle to maximize their absorption of the wind's kinetic energy. Most wind turbines have three rotor blades.

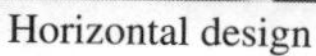

Horizontal design

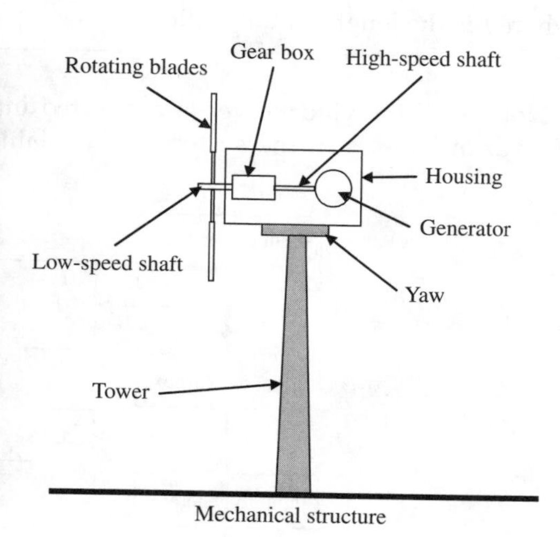

FIGURE 6.12
Basic components of a wind generating system (Image courtesy of the U.S. Department of Energy).

- A yaw mechanism that allows the housing box to rotate and keep the blades perpendicular to the wind speed, thus increasing the exposure of the blades to the wind.
- A gearbox that is used to connect the low-speed rotating blades to the high-speed generator. It also serves as a clutch.
- A generator connected to the high-speed shaft of the gearbox and converts the mechanical energy of the rotating blades into electrical energy.
- A controller that connects the utility system to the wind generator, and locks the blades when the wind speed is below the minimum generation limit or when the wind speed is excessive beyond the design limitations of the system.

EXAMPLE 6.9 A wind turbine at the site discussed in Example 6.8 has three rotating blades; each is 20 m in length. Compute the power captured by the blades.

Solution

The wind power density computed in Example 6.8 is 967.7 W/m^2. The area swept by the blade is a circle of 40 m in diameter. Hence the power captured by the blades is

$$P = A\rho = \pi r^2 \rho = \pi \times 20^2 \times 967.7 = 1.216\,\text{MW}$$

where r is the length of the blade.

Not all of the wind power is converted into electrical power because of the various losses in the system. As explained in Figure 6.13, part of the

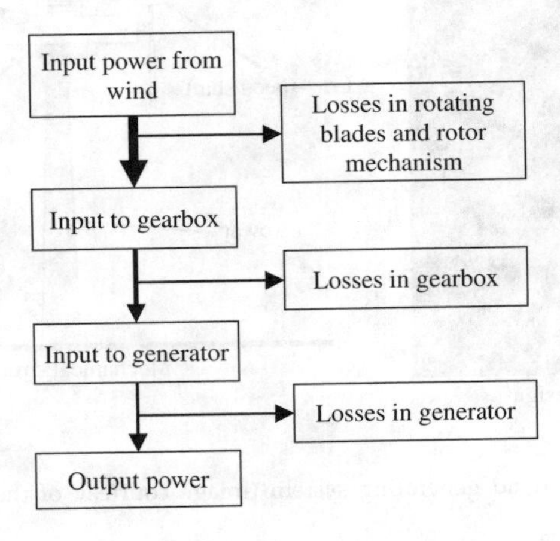

FIGURE 6.13
Power flow of wind turbine.

wind power entering the turbine is wasted as rotational losses (frictional and windage) in the rotating blades and rotor mechanism. The rest of the power enters the gearbox, where part of it is wasted as gearbox rotational losses. The remaining power enters the generator where some of it is wasted in the form of electrical losses in the generator's windings and core as well as the generator's rotational losses. The remaining power is delivered at the output terminals of the generator.

EXAMPLE 6.10 For the wind turbine in Example 6.9 compute the output power of the generator assuming that the efficiency for the rotating blade and rotor mechanism η_b is 40%, the gearbox efficiency η_{gear} is 95% and the generator efficiency η_g is 70%.

Solution

The output power of the wind turbine is the input power multiplied by the total efficiency.

$$P_{out} = \eta_{total}\, P = (\eta_b \eta_{gear} \eta_g)P = (0.4 \times 0.95 \times 0.7) \times 1.126 = 323.4\,\text{kW}$$

The output power is about 27% of the input power. Although the efficiency is low, it is higher than the efficiency of the PV system.

The rotating speed of the blade n (rev/s) is a function of its tip velocity v_{tip} (m/s) and the length of the blade r (m).

$$v_{tip} = \omega r = 2\pi n r \text{ m/s} \tag{6.13}$$

Hence,

$$\begin{aligned} n &= \frac{v_{tip}}{2\pi r} \text{ rev/s} \\ n &= 60\frac{v_{tip}}{2\pi r} \text{ rev/min} \end{aligned} \tag{6.14}$$

The tip velocity of the blades is not equal to the wind speed. The ratio of the tip velocity v_{tip} to the wind speed v is known as the tip speed ratio *TSR*.

$$TSR = \frac{v_{tip}}{v} \tag{6.15}$$

In some advanced types of wind machines, the value of the *TSR* can be adjusted by changing the pitch angle of the blades. At light wind conditions, the pitch angle is set to increase the *TSR*. At excessive wind conditions, the pitch angle is adjusted to reduce the *TSR* and maintain the rotor speed of the generator within its design limits. In some systems, the *TSR* can be reduced to almost zero to lock the blades at excessive wind conditions. The variable *TSR* wind turbines operate for wider ranges of wind speeds, thus they produce more energy as they operate for a longer time compared with the fixed pitch turbines.

EXAMPLE 6.11 A wind turbine is designed to produce power when the speed of the generator n_g is at least 905 r/min, which correlates to a wind speed of 5 m/s. The turbine has a fixed tip speed ratio of 70%, and a sweep diameter of 10 m. Compute the gear ratio.

Solution

Compute the tip speed

$$v_{tip} = TSR\, v = 0.7 \times 5 = 3.5\,\text{m/s}$$

The speed of the shaft at the low-speed side of the gearbox

$$n = 60\frac{v_{tip}}{2\pi\, r} = 60\frac{3.5}{2\pi 5} = 6.7\,\text{rev/min}$$

The gear ratio

$$\frac{n_g}{n} = \frac{905}{6.7} = 135$$

6.2.3 Wind Farm Performance

When a group of wind turbines is installed at one site, the site is called a *wind farm*. Wind farms can have clusters of these machines as shown in Figure 6.14. Major wind farms such as the one located at Tehachapi near Los Angeles, CA, can have hundreds of these machines. The amount of energy generated by

FIGURE 6.14
(see color insert following Page 208) Wind farm located in California. (Images courtesy of the U.S. Department of Energy.)

these farms depends on several factors, among them are:

Wind speed and length of wind season: Wind speed determines how much power can be produced by the wind turbine. As seen in Equation (6.10), the wind power density is proportional to the cube of the wind speed. Assuming all other parameters are fixed, a 10% increase in wind speed will result in a 33% increase in wind power. Most wind turbines start to generate electricity at wind speeds as low as 4 m/s, and reach maximum power output at about 15 m/s.

Diameter of the rotating blades: The power captured by the blades is a function of the area they sweep as shown in Equation (6.9). The area is circular with a radius equal to the length of one blade. The power is then proportional to the square of the radius. Hence, a 10% increase in the blade length will result in a 21% increase in the captured power.

Efficiency of wind turbine components: As seen in Example (6.10), the low efficiency of the various components of the wind turbine reduces the output power of the system. Improvements in the design of the blades, gearbox, and generator should increase the efficiency of the system.

Pitch control: With pitch control, the *TSR* can be adjusted to produce more power at all times. However, controlling the pitch angle adds to the complexity of the system and increase the maintenance cost, so it is more suitable for large turbines.

Arrangement of the turbines in the farm: Layout of individual wind turbines in a wind farm is very crucial. If a wind turbine is placed in the *wind shadow* of other turbines, the amount of generated power can be substantially reduced. Since power is a cubic function of wind speed, good turbine arrangements are essential to maximize the output power of the generators.

Reliability and maintenance: The cost of electricity generated by the wind farm is a function of the capital cost, land use, maintenance, and any other contractual arrangement. The early designs of wind turbines were high-maintenance machines as well as cost-ineffective systems. Newer designs, however, are much better, with a reliability rate around 98%.

6.2.4 Wind Energy and the Environment

Wind power is an environment-friendly form of generating electricity with virtually no contribution to air pollution. For this reason, the wind energy industry is expanding very rapidly worldwide. The country with the most wind power capacity is Germany, followed by the United States, Spain, India, and Austria. In 2003 alone, 8 GW of new wind energy plants were installed, bringing the world's total wind power generating capacity to about 40 GW. This capacity is enough to power over 20 million average European households.

Although it is highly popular, wind energy has some critics who often raise several issues such as (1) noise pollution, (2) aesthetics and visual pollution, and (3) bird collision. In addition, wind turbine has two main technical drawbacks: (1) voltage fluctuations and (2) wind variability.

Noise pollution: There are two potential sources of noise from a wind turbine: mechanical noise from the gearbox and generator, and aerodynamic noise from the rotating blades. The aerodynamic noise is the main component of the total noise, which is similar to the "swish" sound produced by a helicopter. At 400 m downwind, the turbine noise can be as high as 60 decibel (dB), which is equivalent to the noise level from a dishwasher or air conditioner. Although the noise from wind farms is far below the threshold of pain, which is 140 dB, it is a major annoyance to some people. Fortunately, newer and better designs of rotor blades have reduced the aerodynamic noise dramatically.

Aesthetics: Designers of wind turbines believe their creations look pretty, but, unfortunately, not everyone agrees. Critics see wind farms as defacement on the landscape, forcing developers to install them in remote areas, for instance the offshore turbines shown in Figure 6.15.

Bird collisions: Instances of birds begin hit by the moving blades were reported. However, newer turbines are designed for low blade speeds, which should reduce the severity of the problem.

FIGURE 6.15
(see color insert following Page 208) 2 MW off-shore wind turbine farm in Denmark. (Courtesy of LM Glasfiber Group.)

Voltage fluctuations: Most wind turbines use induction machines as generators. These machines are very rugged and require little maintenance and control. They are also self-synchronized with the power grid without the need for additional synchronization equipment. However, because the induction machine has no field circuit, it demands a significant amount of reactive power from the utility system. When used in the generating mode, the induction machine consumes reactive power from the utility while delivering real power. In some cases, the magnitude of the reactive power imported from the utility exceeds the magnitude of the real power generated. Further, the reactive power consumed by the generator is not constant, but is dependent on the speed of its shaft. Figure 6.16 shows the relationship between the reactive power and the speed of the machine. The machine operates as a generator when its speed is higher than the synchronous speed n_S. This is the linear region identified in the figure. Since the speed of wind changes continuously, the reactive power consumed by all induction machines in the farm also changes continuously as shown in Figure 6.17. The voltage at the wind farm is dependent on the reactive power consumed by all wind turbines on the farm; the higher the reactive power, the lower the voltage. Thus the voltage at the site is a mirror image of the reactive power, but at a different scale as shown in Figure 6.17. V_r in the figure is the average voltage. If the voltage variation is slow, it is probably unnoticeable. However, the variations are often fast, creating an annoying flicker.

Variability of wind: Most of the electric loads peak in the morning and early evening. During these times, utilities often fire their fossil fuel power plants to compensate for the extra demands. Renewable energy can play a great role during these periods by supplying the extra energy. However, because the wind schedule is not reliable, energy from wind farms is not always synchronized with the increase in demands.

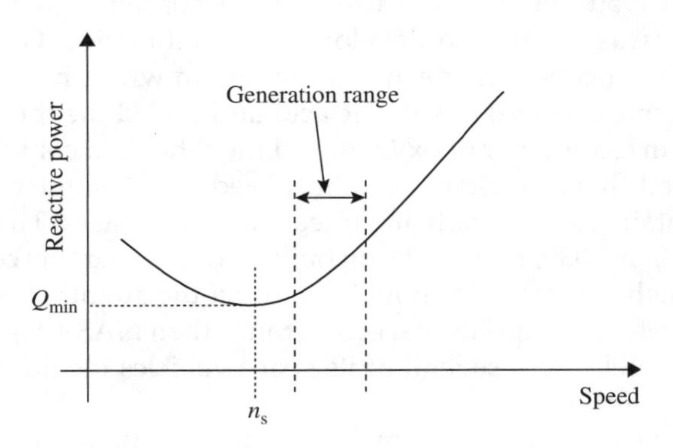

FIGURE 6.16
Reactive power of induction machine.

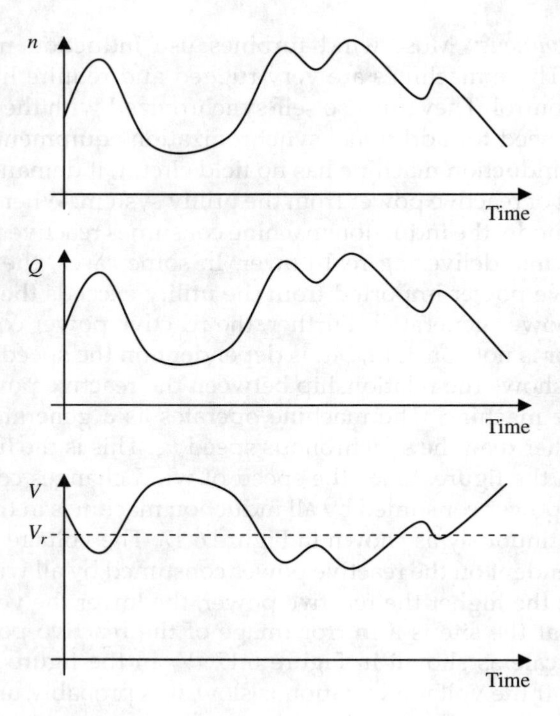

FIGURE 6.17
Wind speed, reactive power, and terminal voltage.

6.3 Fuel Cell

Fuel cell technology was invented over a century ago, but received little attention until 1950s when NASA used it in its space programs. Sir William Grove in 1839 was the first to develop a fuel cell device. Grove, who was an attorney by education, made an experiment in which he immersed parts of two platinum electrodes in sulfuric acid and sealed the remaining part of one electrode in a container of oxygen, and the other in a container of hydrogen. He noticed that a constant current flowed between the electrodes, and the sealed containers eventually included water and gases. This was the first known fuel cell. In 1939, Francis Bacon built a pressurized fuel cell from nickel electrodes which was reliable enough to attract the attention of NASA, who used his fuel cell in its Apollo spacecraft. Since then NASA has used various designs of the fuel cell in several of its space vehicles including the Gemini and the space shuttles.

During the last 30 years, research and developments in fuel cell technology have exploded. Newer material and technologies made fuel cells safer, more reliable, lighter, and more efficient than the earlier models. In

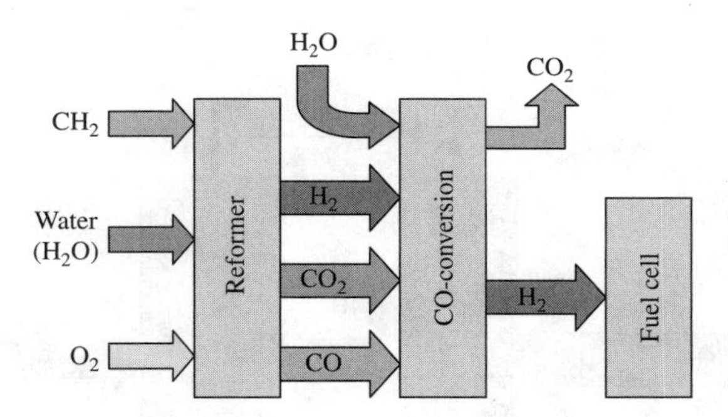

FIGURE 6.18
Generation of hydrogen.

the applications area, fuel cells are used in ground transportation, marine applications, distributed power, cogeneration, and even consumer products.

6.3.1 Hydrogen Fuel

The fuel cell is an electrochemical device that utilizes hydrogen and oxygen to produce electricity. The by-product of the process is just heat and pure water. The hydrogen needed for fuel cells is found in many compounds such as water and fossil fuels. Although it is the most abundant element in the universe, hydrogen is not found alone, but always combined with other elements. To separate hydrogen from the other elements, a reformer process such as the one shown in Figure 6.18 is often used. Inside the reformer, hydrocarbon fuel (CH_2) is chemically treated to produce hydrogen. The hydrocarbon compound, which contains hydrogen and carbon atoms, can be found in fossil fuels such as natural gas and oil. The by-products of the reformer process, however, are carbon dioxide (CO_2) and carbon monoxide (CO). These undesired gases are responsible for the increased hazards to human health and for global warming, as discussed in Chapter 5. Since CO is more hazardous than CO_2, CO is further oxidized by the CO-converter. In this chemical process, water is added to the output of the reformer to convert the CO into CO_2. The carbon dioxide is vented in air, and the hydrogen is used in the fuel cell.

A newer generation of fuel cells uses fuels such as methane directly without the need for reformers. In these systems, which are called *direct fuel cells*, the hydrogen is extracted directly from the methane inside the fuel cell.

6.3.2 Generation of Electricity by Fuel Cells

The basic components of the fuel cell shown in Figure 6.19 are the anode, the cathode, the membrane (electrolyte), and the catalyst. A photo of a system consisting of several fuel cells in series is shown in Figure 6.20. Pressurized hydrogen enters the anode, which is a flat plate with channels built into it to

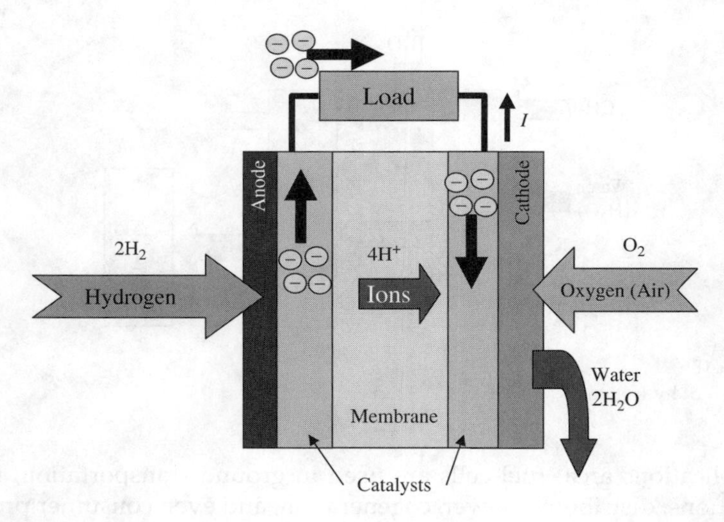

FIGURE 6.19
Main parts of a fuel cell.

FIGURE 6.20
(see color insert following Page 208) PEM fuel cell.

disperse hydrogen equally over the surface of the catalyst. A catalyst (such as platinum) causes 2 hydrogen atoms ($2H_2$) to oxidize into 4 hydrogen ions ($4H^+$) and give up 4 electrons ($4\varepsilon^-$). This is known as the anode reaction.

$$2H_2 \Rightarrow 4H^+ + 4\varepsilon^- \qquad (6.16)$$

The membrane blocks the electrons from passing through, and the freed electrons take the path of least resistance through the external load to the other electrode (cathode). Since the electric current is the flow of electrons, the load is then energized.

The hydrogen ions pass through the membrane from the anode to the cathode. When the electrons enter the catalyst on the side of the cathode, they react with the oxygen from outside air and the hydrogen ions at the cathode to form water. This is known as the cathode reaction.

$$O_2 + 4H^+ + 4\varepsilon^- \Rightarrow 2H_2O \tag{6.17}$$

The overall chemical reaction of the anode and cathode can be represented by Equation (6.18). The equation shows that the fuel cell combines hydrogen and oxygen to produce water and in the process energy is released.

$$2H_2 + O_2 \Rightarrow 2H_2O + \text{energy} \tag{6.18}$$

Keep in mind that the reformer and the CO-converter require water. So it is convenient that the fuel cell produces water, which can be fed back to the reformer and CO-converter. Also, the cathode reaction produces heat that can be used in various applications.

6.3.3 Types of Fuel Cells

Although fuel cells come in various types and operating characteristics, they all work on the same principle discussed earlier. Based on the type of the electrolyte used, fuel cells are classified into five main groups.

1. *Proton exchange membrane (PEM)*: The membrane of the PEM fuel cell is a thin plastic sheet coated with a metal catalyst such as platinum. The PEM fuel cell operates at relatively low temperatures, about 80°C, and its output power can be regulated fairly easy. It is relatively light, has high energy density, and can start very quickly (within a few milliseconds). For these reasons, the PEM fuel cell system, such as the one shown in Figure 6.20, is suitable for a large number of applications including transportations and distributed generation for residential loads.
2. *Alkaline fuel cell (AFC)*: The AFC is the oldest type of fuel cell that was developed by Francis Bacon in the last part of 1930s. Its electrolyte is a liquid alkaline solution of potassium hydroxide. The AFC fuel cell operates at high temperatures, 300 to 400°C; therefore it is slow at starting. Further, it is costly to operate and is very susceptible to contamination.

3. *Phosphoric acid fuel cell (PAFC)*: The phosphoric acid is the electrolyte medium of the PAFC fuel cells. Its operating temperature is high, 160 to 220°C, but is considered suitable for small and midsize generation. The PAFC is among the first generation of modern fuel cells, and is typically used for stationary power generation, but some are used to power large buses.
4. *Solid oxide fuel cell (SOFC)*: The electrolyte of the SOFC is a hard ceramic material such as zirconium oxide. The cell operates at very high temperatures (about 1000°C), which makes the fuel cell start very slowly. It is well suited for large-scale stationary power generation in the MW range.
5. *Molten carbonate fuel cell (MCFC)*: The electrolyte for the MCFC fuel cell is a liquid solution of lithium, sodium, and potassium carbonates. This fuel cell is also a high-temperature type; operates at 600 to 650°C, and is suitable for large systems in the MW range.

6.3.4 Evaluation of Fuel Cells

Experts expect fuel cells to eventually dominate the power generation market, especially for transportation and household use. In automotive applications, using fuel cells to propel an electric motor is an ever-growing trend. The cars powered by fuel cells and using an electric motor as the driving engine do not need the elaborate cooling and lubricating systems that are required by the internal combustion engines, making the car smaller, lighter, more efficient, quieter, and cheaper to maintain. Several generations of fuel cell automobiles and buses are already rooming city streets. In addition to transportation, fuel cells are used as backup systems or independent sources of energy. Several sensitive installations such as hospitals, satellites, and military installations are using fuel cells as backup systems.

The efficiency of the fuel cell alone is relatively high; the range is about 30–70%. If the heat generated by the fuel cell is wasted, the efficiency of the cell is at the low end of the range. When we add the efficiency of the reformer plus CO-converter, the efficiency range of the fuel cell system drops to about 26–40%. If the fuel cell is used to drive an electric motor (electric car), the efficiency of the drive system is about 23–36%. In contrast, modern internal combustion engines have efficiencies around 15–22%.

High-temperature fuel cells produce enough heat to be used in industrial processes, heat buildings, and even in cogeneration applications. These systems are suitable for high-power applications where portability is unnecessary. However, low-temperature fuel cells are suitable for a wider range of applications that require small units with quick startup time such as electric cars.

A single fuel cell produces a dc current at less than 1.5 V. This is barely enough to power small consumer electronic devices. For higher voltage applications, fuel cells are stacked in series as shown in the photo in Figure 6.20. The direct current produced by the fuel cell system cannot be used directly to power ac devices; therefore a converter is needed to convert the dc into ac.

6.3.5 Environmental Effect of Fuel Cells

Hydrogen is a clean gas that is completely nontoxic. So one valid question is why not just burn hydrogen in thermal power plants instead of using it in fuel cells? When hydrogen is burned, it produces nitrous oxide (NO), which is an air pollutant. However, when hydrogen is used in the fuel cell, no pollution is produced besides the carbon dioxide released by the reformer and CO-converter. Furthermore, the burning of hydrogen produces tremendous heat that cannot be efficiently harnessed by the turbines, and the system efficiency is substantially reduced to less than 10%. However, when hydrogen is used in a fuel cell, the overall efficiency is much higher.

Unfortunately, hydrogen cannot be found in nature as a free element, and must be extracted from hydrogen-rich compounds and safely stored. Among the options are the following:

Reformer and co-converter: This process is shown in Figure 6.18. When a reformer alone is used, carbon monoxide and carbon dioxide are produced; both are polluting gases. When a reformer and CO-converter are used, only carbon dioxide is produced, which is one of the greenhouse gases.

Direct fuel cell (DFC): In the DFC, hydrogen can be generated within the fuel cell system by reforming hydrogen-rich fuels such as methanol, ethanol, and hydrocarbon fuels.

Direct methanol fuel cell (DMFC): The DMFC uses pure methanol mixed with water instead of hydrogen. This fuel cell is appealing to engineers since the methanol has a higher energy density than hydrogen, and is easier to transport and supply by using the current gasoline infrastructure. The DMFC is also viewed as the natural successor to the lithium-ion batteries for small load applications such as personal computers and portable electronics devices. For the same size lithium-ion battery, the DMFC can last up to ten times the life span of the battery.

Electrolysis of water: This process sounds very simple since water contains only hydrogen and oxygen. However, the electrolysis of water is an energy intensive process that reduces the overall efficiency of the entire system to less than 10%. Hence, it is not cost effective.

Futuristic methods: Recent research claims that hydrogen can be liberated from a solution of sodium hydride (NaH) in the presence of certain catalysts without using electric energy. Also, researchers believe that old algae can be genetically altered to produce hydrogen instead of oxygen during their photosynthesis process. Another method is based on a phenomenon known in metallurgy where some metal hydrides absorb hydrogen and store it between their molecules when they are cooled. The hydride alloys release the stored hydrogen when heated.

6.3.6 Safety of Fuel Cells

Hydrogen is currently used in several applications such as direct combustion and rocket propulsion. However, because hydrogen is a very explosive gas, it creates tremendous anxiety among the public. Most people associate hydrogen with the well known Hindenburg Zeppelin fire that occurred in 1937. This luxury airship used hydrogen to provide the needed lift and was burned while attempting to dock at Lakehurst, New Jersey. The spectacular fire of the Hindenburg is the main factor affecting the public perception regarding the safety of hydrogen.

The hydrogen used in fuel cell electric cars is stored in pressurized tanks. The tanks are often made from carbon fibers and aluminum cylinders mounted on the top of the car. These tanks can survive a direct force from collisions without exploding. In the worst scenario, if the hydrogen catches fire, it will escape upward leaving the people inside the car below the tank relatively safe.

Another option is to use natural gas, and install a reformer and CO-converter inside the car. Also, the DFC technology uses other gases as a source of hydrogen. However, the storage of pressurized gases such as natural gas has its own safety concerns.

6.4 Small Hydroelectric Systems

Hydroelectric power is one of the oldest forms of generating electricity that dates back to 1882 when the first hydroelectric power plant was constructed across the Fox River in Appleton, Wisconsin. As seen in Chapter 4, the hydroelectric power plants are enormous in size and capacity. These power plants require enormous water resources that are available only in a very limited number of areas worldwide. However, the world has an immense number of small rivers and creeks that can be utilized to generate small amounts of electric energy, enough to power neighborhoods. These types of power plants are called small hydroelectric systems.

Small hydroelectric systems are at the low end of the power ratings — up to a few mega watts. The popularity of these systems is due to their mature and proven technology, reliable operation, suitability for sensitive ecology, and capability to produce electricity even in small streams and rivers. There are two versions of hydroelectric systems: reservoir-type and diversion-type. The reservoir-type may require a small dam to store water at higher elevations.

The diversion-type does not require a dam, and relies on the speed of the current to generate electricity.

6.4.1 Reservoir-Type Small Hydroelectric System

A schematic of a simple reservoir-type hydroelectric plant is shown in Figure 6.21. The system is similar to the impoundment hydroelectric system discussed in Chapter 4, but the head of the reservoir-type small hydroelectric system is much shorter and its reservoir capacity is much smaller. The reservoir-type system consists mainly of a reservoir, a penstock, a turbine, and a generator. The reservoir is either a natural lake or a lake created by a dam at a higher elevation than the downstream river. If the water is allowed to flow to a lower elevation through the penstock, the potential energy of the water is converted into kinetic energy where some of it is captured by the turbine. After passing through the turbine, the water exits to the stream at the lower elevation. The turbine rotates due to its acquired kinetic energy, thus rotating the generator, and electricity is generated.

The power flow inside the reservoir-type small hydroelectric system is shown in Figure 6.22. The input power entering the penstock is P_{p-in}; part of this power is wasted in the penstock due to losses such as water friction. The rest of the power P_{p-out} exits the penstock and enters the turbine. Part of P_{p-out} is converted into mechanical power P_m, which is the input to the electrical generator. Part of P_m is lost due to the various losses inside the generator, and the rest is the output electric power P_g of the small hydroelectric system.

The amount of electric power generated by the reservoir-type small hydroelectric system depends on three key parameters: (1) the water head, which is the vertical distance between the water surface of the reservoir and the

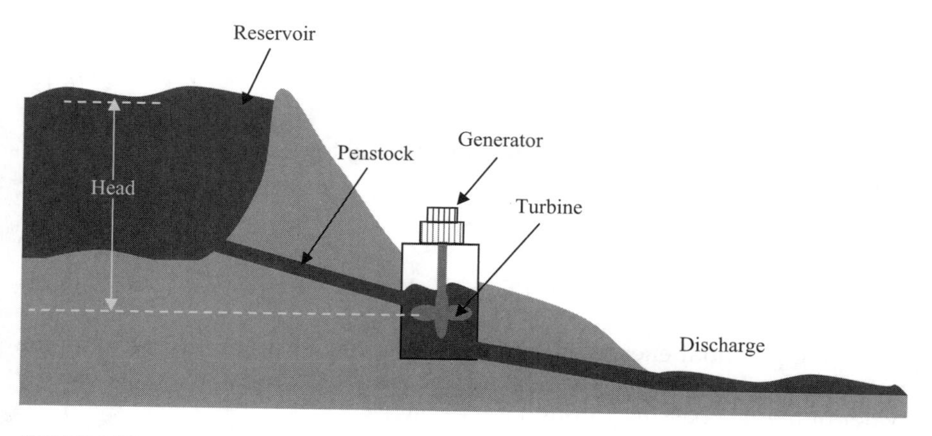

FIGURE 6.21
Small hydroelectric system with reservoir.

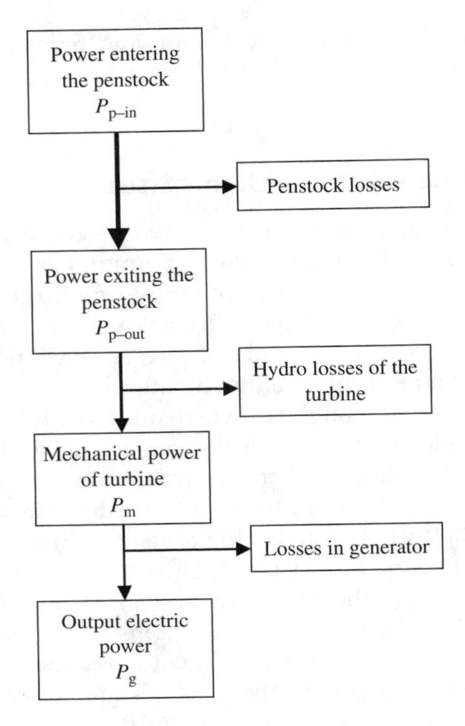

FIGURE 6.22
Power flow in a small hydroelectric system.

turbine; (2) the flow of water, which is the mass of water passing through the penstock per unit time; and (3) the efficiencies of the system components.

The potential energy of the water behind the reservoir PE_r is the weight of the water in the reservoir W multiplied by the average elevation of the water H with respect to the turbine.

$$\mathrm{PE_r} = WH \tag{6.19}$$

The weight of the water behind the dam is equal to its mass M multiplied by the acceleration of gravity g.

$$W = Mg \tag{6.20}$$

$$\mathrm{PE_r} = MgH \tag{6.21}$$

$\mathrm{PE_r}$ is the potential energy of the entire reservoir. If we are to calculate the potential energy of the water entering the penstock $PE_{\mathrm{p-in}}$ we must use the water of mass m passing through the penstock.

$$PE_{\mathrm{p-in}} = mgH \tag{6.22}$$

where $PE_{\text{p-in}}$ is in joules, m is in kilograms, H in meters, and g in meters per second. The water flow f inside the penstock is defined as the mass of water m passing through the penstock during a time interval t.

$$f = \frac{m}{t} \tag{6.23}$$

Hence, Equation (6.22) can be rewritten in terms of water flow as

$$PE_{\text{p-in}} = fgHt \tag{6.24}$$

The potential energy given by Equation (6.22) is converted into kinetic energy inside the penstock. The output kinetic energy of the penstock $\text{KE}_{\text{p-out}}$ is defined as

$$\text{KE}_{\text{p-out}} = \tfrac{1}{2}mv^2 \tag{6.25}$$

where v is the speed of water exiting the penstock. Due to the losses inside the penstock, such as the water frictional, the kinetic energy exiting the penstock $\text{KE}_{\text{p-out}}$ is less than the potential energy at the entrance of the penstock $PE_{\text{p-in}}$.

$$\eta_{\text{p}} = \frac{\text{KE}_{\text{p-out}}}{PE_{\text{p-in}}} \tag{6.26}$$

where η_{p} is the penstock efficiency. The mass m in Equation (6.25) can be substituted by the water volume *vol* times the water density ρ.

$$\text{KE}_{\text{p-out}} = \tfrac{1}{2}\, mv^2 = \tfrac{1}{2}\text{vol}\ \rho v^2 \tag{6.27}$$

The water density is about 1000 kg/m^3. Hence,

$$\text{KE}_{\text{p-out}} = 500\,\text{vol}\ v^2 \tag{6.28}$$

The water volume passing through the penstock during a time interval t can be computed as

$$\text{vol} = Avt \tag{6.29}$$

where A is the area of cross-section of the penstock. Substituting Equation (6.29) into (6.28) yields

$$\text{KE}_{\text{p-out}} = 500Av^3t \tag{6.30}$$

The kinetic energy KE_{p-out} exiting the penstock is not all converted into the kinetic energy KE_m that rotates the turbine. Hence,

$$\eta_h = \frac{KE_m}{KE_{p-out}} \tag{6.31}$$

where η_h is the hydroelectic conversion efficiency defined as the ratio of the energy exiting the penstock to the energy of the rotating shaft. This is the same as the coefficient of performance given in Equation 4.13. The output electric energy E_g of the generator, for example, is equal to the kinetic energy KE_m minus the losses of the generator; hence, the generator efficiency η_g is defined as

$$\eta_g = \frac{E_g}{KE_m} \tag{6.32}$$

Using Equations (6.24), (6.26), (6.31), and (6.32), we can write the equation of the output electrical energy as a function of the water flow, head, and the cross-section of the penstock.

$$E_g = fgHt(\eta_p \eta_h \eta_g) \tag{6.33}$$

EXAMPLE 6.12 A small hydroelectric site with a reservoir has a head of 5 m. The penstock passes 100 kg of water every second. The efficiency of the penstock is 95%, the hydroelectric conversion efficiency is 40%, and the efficiency of the generator is 90%. Assume that the owner of this small hydroelectric system sells the generated energy to the local utility at $ 0.15/kWh and compute his income in one month.

Solution

Using Equation (6.33), we can compute the output electric energy for one month.

$$\begin{aligned} E_g = fgHt(\eta_p \eta_h \eta_g) &= 100 \times 9.8 \times 5 \times (30 \times 24) \\ &\quad \times (0.95 \times 0.4 \times 0.9) \\ &= 1.206\,\text{MWh} \end{aligned}$$

The income in one month is

$$\text{Income} = E_g 0.15 = \$181$$

EXAMPLE 6.13 For the system in Example (6.12), compute the speed of the water exiting the penstock.

Solution

Using Equation (6.22), we can compute the potential energy of the water entering the penstock.

$$PE_{\text{p-in}} = mgH = 100 \times 9.8 \times 5 = 4.75\,\text{kJ}$$

From Equation (6.26), the kinetic energy exiting the penstock is

$$\text{KE}_{\text{p-out}} = \eta_{\text{p}} PE_{\text{p-in}} = 0.95 \times 4.75 = 4.51\,\text{kJ}$$

Now use Equation (6.25) to compute the water speed.

$$v = \sqrt{\frac{2\text{KE}_{\text{p-out}}}{m}} = \sqrt{\frac{2 \times 4.51 \times 10^3}{10}} = 30\,\text{m/s}$$

EXAMPLE 6.14 A man owns a property with a small river. He wants to build a dam to create a reservoir for a small hydroelectric system. The site can accommodate a penstock 3 m in diameter. In order for him to generate 2 MW of electricity, compute the height of the dam. Assume that the penstock efficiency is 97%, the hydroelectric efficiency is 50%, and the generator efficiency is 95%.

Solution

We can solve this problem by using the power chart in Figure 6.22 but in the backward direction. Using Equation (6.32) we can compute the mechanical power of the turbine.

$$\eta_{\text{g}} = \frac{E_{\text{g}}}{\text{KE}_{\text{m}}} = \frac{P_{\text{g}}}{P_{\text{m}}}$$

$$P_{\text{m}} = \frac{P_{\text{g}}}{\eta_{\text{g}}} = \frac{2}{0.95} = 2.105\,\text{MW}$$

Compute the output power of the penstock using Equation (6.31).

$$P_{\text{p-out}} = \frac{P_{\text{m}}}{\eta_{\text{h}}} = \frac{2.105}{0.5} = 4.21\,\text{MW}$$

Use Equation (6.30) to compute the velocity of the water that yields 421.06 kW of power.

$$KE_{p-out} = 500Av^3t$$
$$P_{p-out} = 500Av^3$$

Hence,

$$v = \sqrt[3]{\frac{P_{p-out}}{500A}} = \sqrt[3]{\frac{4.21 \times 10^6}{500(\pi 1.5^2)}} = 10.6\,\text{m/s}$$

Now use Equation (6.26) to compute the head of water.

$$\eta_p = \frac{KE_{p-out}}{PE_{p-in}} = \frac{v^2}{2gH}$$
$$H = \frac{v^2}{2g\eta_p} = \frac{(10.6)^2}{2 \times 9.8 \times 0.97} = 5.91\,\text{m}$$

hence, the height of the dam should be at least 5.91 m.

6.4.2 Diversion-Type Small Hydroelectric System

The diversion-type small hydroelectric system does not require a dam, and therefore is considered more environmentally sensitive. The schematic of this type of small hydroelectric system is shown in Figure 6.23. The river for this small hydroelectric system must have strong enough current for a realistic power generation.

Let us assume that the volume of water entering the turbine vol during a time t is

$$\text{vol} = A_s v t \tag{6.34}$$

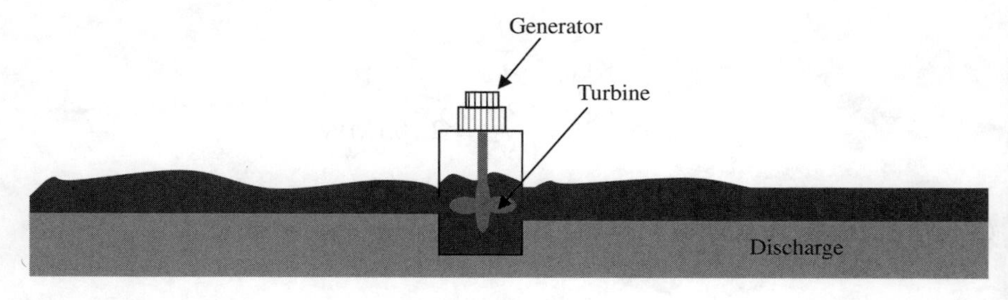

FIGURE 6.23
Diversion-type small hydroelectric system.

where A_s is the sweep area of the turbine's blades, which is the area swept by the turbine blades in one revolution. The kinetic energy captured by the turbine KE_t is

$$KE_t = \frac{1}{2} m v^2 = \frac{1}{2} \text{vol} \rho v^2 = \frac{1}{2} A_s \rho v^3 t \tag{6.35}$$

At a water density of about 1000 kg/m^3, the input power P_t of the turbine is

$$P_t = 500 A_s v^3 \tag{6.36}$$

The mechanical output power of the turbine P_m is equal to P_t minus the turbine losses. Hence,

$$\eta_h = \frac{P_m}{P_t} \tag{6.37}$$

where η_h is the hydroelectric conversion efficiency, which is also known as the coefficient of performance of the turbine. The output electric power of the generator P_g is equal to P_m minus the losses of the generator; hence, the generator efficiency η_g is defined as

$$\eta_g = \frac{P_g}{P_m} \tag{6.38}$$

EXAMPLE 6.15 A diversion-type small hydroelectric system is installed across a small river with current speed of 5 m/s. The diameter of the sweep area of the turbine is 1.2 m. Assume that the hydroelectric conversion efficiency is 40% and the efficiency of the generator is 90%. Compute the output power of the plant and the energy generated in one year. If the price of the energy is $0.05/kWh, compute the income from this small hydroelectric plant in one year.

Solution

Use Equation (6.36) to compute the input power to the turbine.

$$P_t = 500 A_s v^3 = 500(\pi \times 0.6^2) 5^3 = 70.69 \text{ kW}$$

The output electric power is P_t multiplied by the efficiencies.

$$P_g = P_t(\eta_h \eta_g) = 70.69(0.4 \times 0.9) = 25.45 \text{ kW}$$

The energy generated in one year is

$$E_g = 25.45(24 \times 365) = 222.942 \text{ MWh}$$

Income from the plant is

$$\text{Income} = 0.05 \times 222{,}942 = \$11{,}147.00$$

6.5 Geothermal Energy

Geothermal energy, which is the stored heat in the Earth's crust, is an enormous, under-used energy resource. At just a few meters under the Earth's surface, the geothermal temperature in winter is about 10–20°C higher than the ambient temperature, and about 10–20°C lower in the summer. At higher depth, the magma (molten rock) has an extremely high temperature that can produce tremendous amounts of steam suitable for generating large amounts of electricity. As a matter of fact, the thermal energy in the uppermost 10 km of the Earth's crust is much more than the energy from all oil and gas resources known to us so far. Furthermore, because of the tremendous thermal inertia in the molten rocks and the thermal isolation of the magma by the surrounding material, the magma temperature is cooled down at an extremely slow rate—a few degrees every hundreds of years.

The depth at which the magma is located and its surrounding material determine the way we can harness the geothermal energy. In shallow depths, heat pumps can be used to heat houses in the winter and cool them down in the summer. At greater depths, closer to the molten rocks, enough steam can be produced to generate electricity. These two systems are discussed in this section.

6.5.1 Heat Pump

A typical heat pump system is shown in Figure 6.24. It consists of a geo-exchanger, pump, and heat distribution system. The geo-exchanger is a series

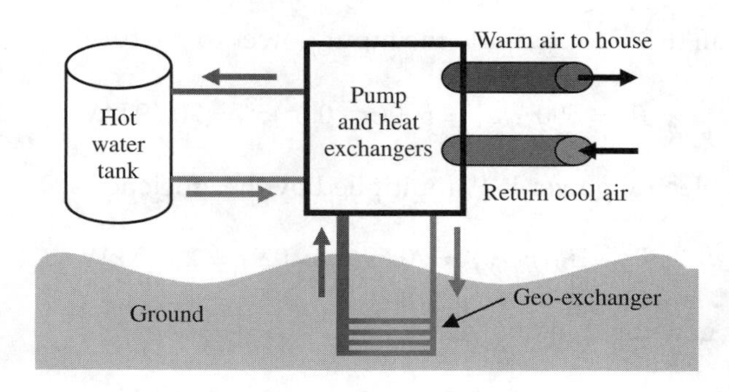

FIGURE 6.24
Heat pump system.

of pipes buried in the ground near the building to be conditioned. The pipes are filled with water, or a mixture of water and antifreeze, to absorb the Earth's heat in the winter, or dissipate heat to the Earth in the summer. Pumps are used to circulate the water mix between the geo-exchanger and the rest of the system. In winter, the building temperature is lower than the geothermal temperature, so the geo-exchanger increases the temperature of its mixture. The pump moves the heated mixture to the building and by using a system of heat exchangers, the air and/or water of the building is warmed up. In summer, when the Earth's temperature is lower than the air temperature, the process is reversed.

6.5.2 Geothermal Power Plant

In nature, steam can be formed in various remarkable ways, for instance, the geyser in Figure 6.25. Besides the intermittent escape of the geyser's steam, other sites around the world have continuous steam venting all year around. Steam is formed when water from rain seeps into the fractured Earth's rocks for miles until it gets near the molten rocks, where the water is turned into steam and pushes its way upward toward any crack on the surface of the Earth. This steam could include other minerals or gases that are mixed with water.

Steam could also be trapped in geothermal reservoirs, which are large pools of steam or hot water in porous rock above the magma. When rain or ground

FIGURE 6.25
(see color insert following Page 208) Steam generated from rain even in cold environment. (Courtesy of the U.S. Department of Energy.)

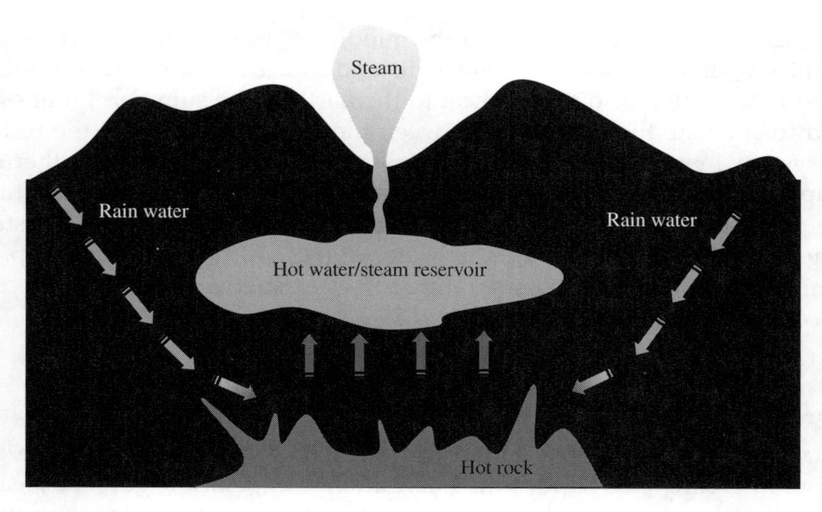

FIGURE 6.26
Geothermal reservoir.

water reaches the molten rocks, the water absorbs the thermal energy and ascends toward the reservoir as shown in Figure 6.26. The steam in some reservoirs is at very high temperatures and can be used directly to generate electricity.

A typical geothermal power plant is shown in Figure 6.27. The steam, which may include hot mist, is drawn from the geothermal reservoir and passed through a mist eliminator, where the hot mist is condensed to water and used to heat buildings. The steam goes to a thermal turbine that converts the steam energy into mechanical energy much like any thermal power plant. This mechanical energy is converted by the generator G to electrical energy. The steam exiting the turbine is cooled down in a cooling tower where external cold water is poured in to extract the heat from the steam to complete the thermal cycle, as explained in Chapter 4. The hot water can be used to heat building, and then injected back into the geothermal reservoir.

The first experimental geothermal power plant was built in Larderello, Italy by Prince Gionori Conti around 1905. The first commercial power plant based on Conti's work was commissioned in 1913 at Larderello. In 1858, New Zealand built its Wairakei's geothermal power plant, followed by Pathe's plant in Mexico in 1959, and the Geysers plant in California, U.S. in 1962. By the end of 2000, 21 countries had operational geothermal power plants with a total capacity of 8.2 GW. The geothermal power plants in the United States have a total generating capacity of 2.7 GW. The Geysers Power Plant in northern California, shown in Figure 6.28, is still the largest geothermal power plant to date with a total generation capacity of 1.7 GW. This is enough power for the San Francisco metropolitan area. Other geothermal power plants in the United States are in Nevada, Utah, and Hawaii.

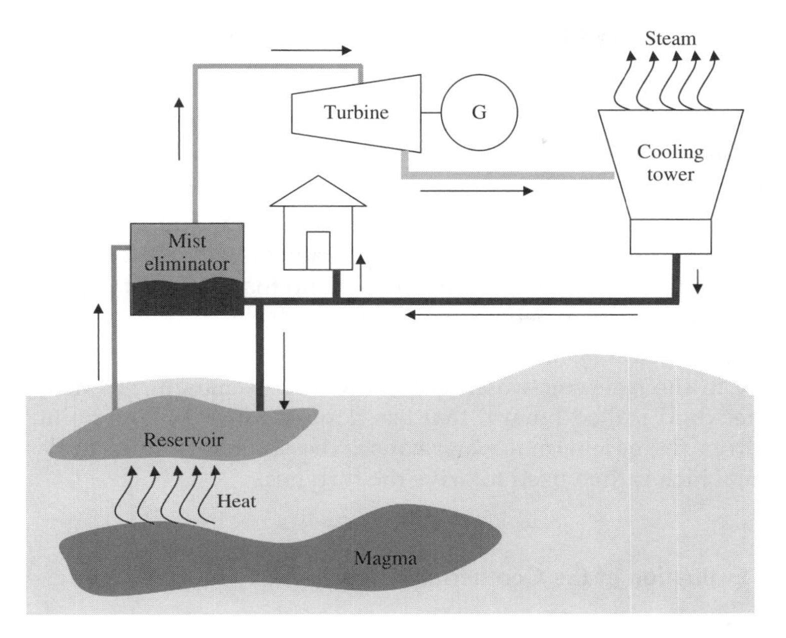

FIGURE 6.27
Geothermal power plant.

FIGURE 6.28
(see color insert following Page 208) The first geothermal power plants in the United States (The Geysers) in northern California. (Courtesy of the U.S. Department of Energy.)

6.5.3 Types of Geothermal Power Plants

The naturally occurring steam in geothermal reservoirs varies in temperatures and pressures. In addition, the depth of the magma and its surrounding material varies from one site to another. These variations make it hard to design one geothermal power plant for all conditions; therefore, there are three basic designs:

Dry steam power plants: This system is used when the steam temperature is very high (300°C) and the steam is readily available. This is the system explained in Figure 6.27.

Flash steam power plants: When the reservoir temperature is above 200°C, the reservoir fluid is drawn into an expansion tank that lowers the pressure of the fluid. This causes some of the fluid to rapidly vaporize (flash) into steam. The steam is then used to generate electricity.

Binary-cycle power plants: At a moderate temperature (below 200°C), the energy in the reservoir water is extracted by exchanging its heat with another fluid (called binary) that has a much lower boiling point. The heat from the geothermal water causes the secondary fluid to flash to steam, which is then used to drive the turbines.

6.5.4 Evaluation of the Geothermal Energy

The Earth's magma can provide clean and reliable energy that can last for thousands of years. However, geothermal electricity is not yet as cost-competitive as the other, more traditional methods. Furthermore, there are several challenges that need to be addressed before the geothermal power is used on a wide scale. Among these challenges are:

- The geothermal power plants can only be constructed in geothermal sites where their magmas are close enough to the surface to heat reservoirs accessible by the current drilling technology. That is why the geothermal power plants are placed in areas with volcanic activity such as the west coast of the United States and Japan. These sites are scarce worldwide.
- The cost of drilling deep wells to reach underground reservoirs surrounded by hard rocks could be very expensive.
- The geothermal fluid of the reservoir may include other minerals and gases besides water that can be pollutant. Geothermal fields could also produce small amount of carbon dioxide.
- The geothermal fluid can produce objectionable odors due to the presence of compounds such as the hydrogen sulfide.

6.6 Tidal Energy

Tides are the daily swell and sag of ocean waters relative to coastlines due to the gravitational pull of the Moon and the Sun. Although the Moon is much smaller in mass than the Sun, it exerts a larger gravitational force because of its relative proximity to Earth. This force of attraction causes the oceans to rise along a perpendicular axis between the Moon and Earth. Because of the Earth's rotation, the rise of water moves opposite to the direction of the Earth's rotation creating the rhythmic rise and sag of coastal water. These tidal waves are slow in frequency (about one cycle every 12 h), but contain tremendous amounts of kinetic energy, which is probably one of the major untapped energy resources of Earth.

The process of harnessing tidal energy was known to the Ancient Egyptians who built grain compressors composed of three stone disks. The middle one was stationary while the other two were attached to a shaft connected to a float and passed through a hole in the middle disk. During the rising tide, the grain was placed between the lower and middle disks as the lower disk pressed against the middle disk. During the falling tide, the grain was placed between the upper and middle disks.

6.6.1 Tidal Energy Systems

There are a very few number of major tidal generating power plants in operation. The major one is the *Le Rance River* in northern France built in 1966 and has a capacity of 240 MW. Other plants are the 20 MW in Nova Scotia, Canada, and the 400 kW near Murmansk in Russia.

The technology required to convert tidal energy into electricity is very similar to the technology used in wind energy or hydroelectric power plants. The common designs are the free flow system and the dam system. In the free flow system, shown in Figure 6.29, the kinetic energy of the tidal current is converted into electrical energy, much the same way as the energy of the wind is converted into electrical energy by wind turbines. The tidal energy turbine has its blades immersed in seawater in the path of the strong tidal currents. The current rotates the blades that are attached to an electrical generator mounted above the water level.

The dam-type tidal energy system is shown in Figure 6.30. It is most suited for inlets where a channel is connecting an enclosed lagoon to the open sea. At the mouth of the channel, a dam is constructed to regulate the flow of tidal water in either direction. A turbine is installed inside a conduit connecting the two sides of the dam. At high tides, the water moves from the sea to the lagoon through the turbine. The turbine plus its generator convert the kinetic energy of the water into electrical energy. When the tide is low, the water stored in the lagoon at high tides goes back to the sea, and in the process electricity is generated.

The difference in hydraulic heads of the tide determines the amount of energy that can be captured. In the United States, large tidal differences of up to 17 m occur in places such as Maine, Alaska, and the Bay of Fundy in Atlantic Canada. These areas can produce large amounts of electrical energy.

FIGURE 6.29
(see color insert following Page 208) Free-flow tidal energy system. (Images courtesy of Marine Current Turbines Limited.)

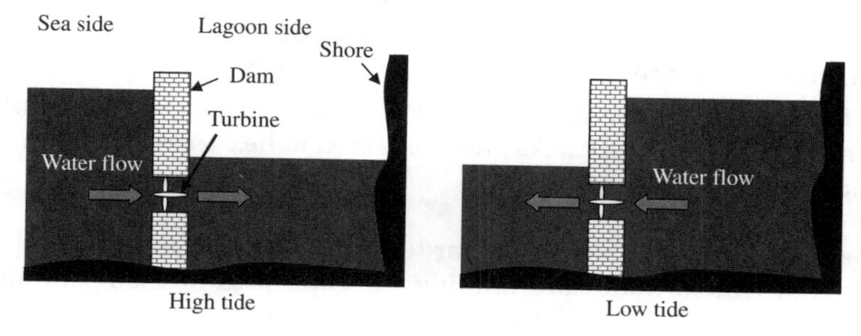

FIGURE 6.30
Dam-type tidal energy system.

6.6.2 Evaluation of Tidal Energy

Tide is a renewable source of energy with no emission of gases. However, tidal energy has some drawbacks and challenges such as the following ones:

- Changing tidal flows by building dams could result in negative effects on the aquatic life, as well as water navigations. It is believed that dams can potentially stimulate the growth of the red tide organism that causes sickness in shellfish.
- Tidal energy is still an expensive method to generate electricity even when compared with the solar and wind systems.
- Although electricity can be generated by water flowing into and out of a lagoon or bay, it can only be generated when the current is strong. Around the peak or the slack of the tide, the current is weak and no electricity can be generated. The system is then active for about 12 h daily.

6.7 Biomass Energy

Garbage is one of the major concerns for modern societies. In the United States, each household produces about 800 kg of garbage every year, followed by Norway (520 kg) and the Netherlands (490 kg). On the low end of the garbage production in the western world are Austria (180 kg) and Portugal (180 kg). The garbage is often collected and dumped in landfills, which are considered by many as a major health hazard because of the following reasons:

- Landfills produce unpleasant odors.
- Leachate, which is the fluid resulting from water mixed with garbage, contaminates underground water.

- Gases such as ethanol and methanol can be generated in landfills, increasing the fire hazards.
- Housing developments are often expanded to areas closer to landfills; the landfills and the trucks are considered to be excessive pollutions and safety hazards.

For these reasons, it is becoming harder to build new landfills, and cities all over the United States are limiting the capacities of their landfills, or eliminating them altogether. Therefore, the burning of garbage to produce electricity (biomass energy) sounds like a reasonable idea.

The biomass consists of garbage, agriculture products, and tree crops. When burned in incinerators, the biomass volume is reduced by 90%, and in the process steam can be produced to generate electricity. One of these systems is shown in Figure 6.31. The biomass material is fed to a furnace where it is burned and the heat produced is used to generate steam by heating water pipes. The steam is used to generate electricity through the turbine–generator system much like the regular thermal power plant. The steam exiting the turbine is condensed to water to complete the thermal cycle. The ash produced in the furnace is collected and sent to a landfill. The volume of the ash is about 10% of the original volume of the biomass material. The gases generated by combustion enter the filtering stage where more ash is extracted, collected

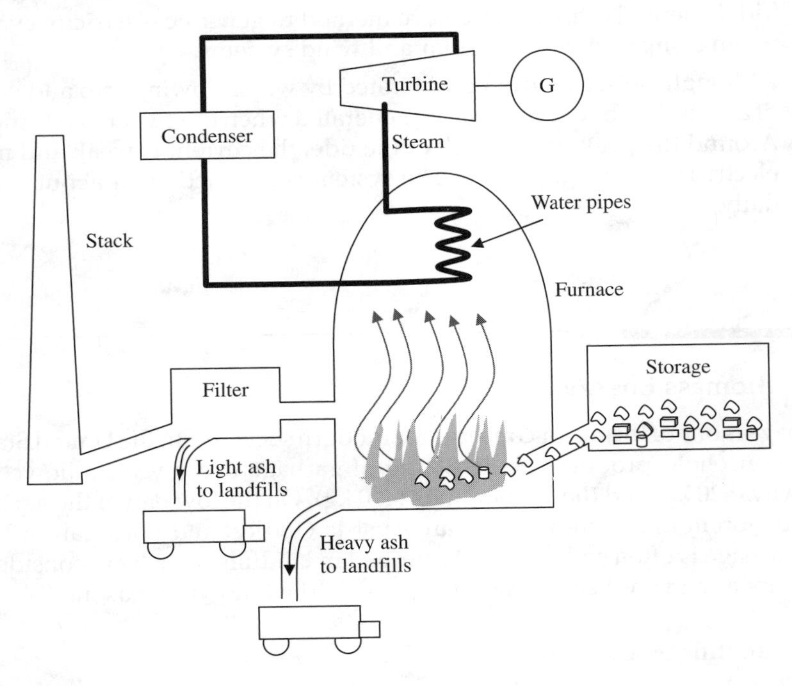

FIGURE 6.31
Biomass incineration power plant.

by trucks, and sent to landfills. The remaining gases enter the stack and are released into the air.

The biomass power plants generate electricity and, at the same time, reduce the amount of material sent to landfills. However, biomass incineration is not free from polluting material. Heavy metal and dioxins are formed during the various stages of the incinerations. Dioxin is highly carcinogenic and can cause cancer and genetic defects.

Exercise

6.1. A solar panel consists of four parallel columns of PV cells. Each column has ten PV cells in series. Each cell produces 2 W at 0.5 V. Compute the voltage and current of the panel.

6.2. State three factors that determine the amount of power generated by wind machines.

6.3. What is the function of the reformer and CO-conversion used in fuel cell technology?

6.4. Compute the amount of sulfur dioxide produced when a coal-fired power plant generates 50 MW of electricity.

6.5. A solar power density for a given area has a standard deviation of 3 h, and a maximum power of 200 W at noon. Compute the solar energy in one day.

6.6. An area located near the equator has the following parameters:

$$\alpha_{\mathrm{dt}} = 0.82; \qquad \alpha_p = 0.92; \qquad \beta_{\mathrm{wa}} = 0.06$$

If, the solar power density measured at 11:00 A.M. is 890 $\mathrm{W/m^2}$, compute the solar power density at 4:00 P.M.

6.7. An investor wishes to install a wind farm in the Snoqualmie pass area located in Washington State, U.S. The pass is about 920 m above the sea level. The average low temperature of the air is −4°C, and the average high is 18°C. Compute the power density of the wind in winter and summer assuming that the average wind speed is 15 m/s.

6.8. For the site in the above problem, compute the length of the blades to capture 200 kW of wind energy during the summer.

6.9. A wind turbine with a gearbox ratio of 200 produces electric energy when the generator speed is at least 910 rpm. The length of each blade is 5 m. The turbine has a variable tip speed ratio (TSR). At a wind speed of 10 m/s, compute the minimum TSR of the wind turbine.

6.10. Generate an idea to reduce the voltage flickers associated with wind energy.

6.11. What are the main advantages of direct fuel cells?

6.12. What are the main components used in fuel cell electric vehicles? Explain the operation of the vehicle.

6.13. What are the different types of fuel cells?

6.14. Which fuel cell type is suitable for high power?

6.15. Which fuel cell type is suitable for mobile energy?

6.16. What are the advantages of using a fuel cell?

6.17. Is using hydrogen in fuel cells safe?

6.18. Why not just burn hydrogen in a thermal power plant instead of using it in fuel cells?

6.19. A reservoir-type small hydroelectric system has a penstock that is 2 m in diameter. The speed of water at the exit end of the penstock is 10 m/s. Assume that the hydroelectric efficiency is 0.5 and the generator efficiency is 92%. Compute the output electric power of the generator.

6.20. Assume that the penstock efficiency in the previous problem is 95%. Compute the water head.

6.21. A man wants to build a reservoir-type small hydroelectric system on his property. The site can accommodate a penstock of 2 m in diameter. In order for him to generate 1 MW of electricity, compute the height of the dam. Assume that the penstock efficiency is 95%, the hydroelectric efficiency is 45%, and the generator efficiency is 96%.

6.22. A person wishes to build a diversion-type small hydroelectric system on his property. To measure the speed of the water, he dropped a ping-pong ball in the river and found the ball to travel 10 m in 4 s. He also selected a turbine with a sweep diameter of 0.8 m. Assume that the hydroelectric conversion efficiency is 45% and the efficiency of the generator is 96%. Compute the output power of the plant and the energy generated in one year. If the price of the energy sold to the neighboring utility is $0.06/kWh, compute the annual income from this small hydroelectric plant.

6.23. What are the benefits of using geothermal energy?

6.24. Why is geothermal energy a renewable resource?

6.25. What are the environmental effects of geothermal power plants?

6.26. What makes a geothermal site good for power generation?

6.27. What is biomass?

6.28. How electricity is generated from biomass?

6.29. Is biomass a renewable source of energy? Why?

7

Alternating Current Circuits

Nikola Tesla proposed the *alternating current* (ac) system as a better alternative to the *direct current* (dc) system. His idea was initially met with fierce resistance from Thomas Edison as explained in Chapter 1. Eventually, alternating current was selected for power systems worldwide because of three main reasons:

1. The voltage of the ac system can be adjusted by transformers, which are simple devices.
2. The voltage of the long transmission lines can be increased to high levels, thus reducing the current of the line, reducing the transmission line losses, and reducing the cross-sections of the conductors of the transmission lines conductors.
3. The ac systems produce rotating magnetic fields that spin electric motors. Keep in mind that electric motors represent the majority of the electrical loads.

Nomenclature

In this chapter, the following nomenclatures are used:

Instantaneous current	i	Instantaneous voltage	v
Average current	I_{ave}	Average voltage	V_{ave}
Maximum (peak) current	I_{max}	Maximum (peak) voltage	V_{max}
Current magnitude in root mean square	I	Voltage magnitude in root mean square	V
Phasor current (complex form)	$\overline{I}$	Phasor voltage (complex form)	$\overline{V}$
Complex impedance	$\overline{Z}$	Complex power	$\overline{S}$
Magnitude of impedance	Z	Magnitude of complex power	S

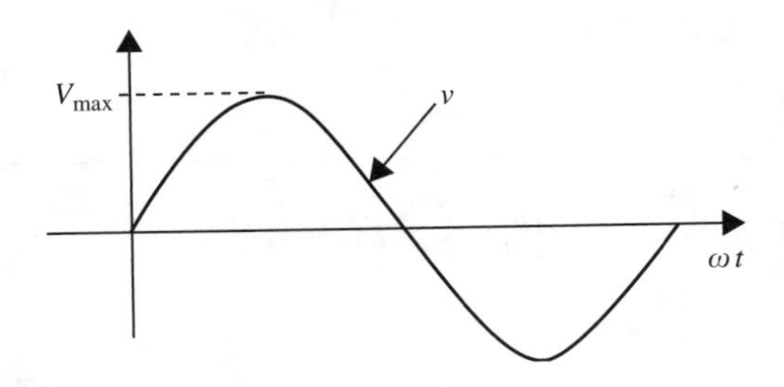

FIGURE 7.1
Sinusoidal waveform of ac voltage.

7.1 Alternating Current Waveform

The ideal ac circuit has its voltage and current in sinusoidal forms. The voltage waveform is shown in Figure 7.1 and expressed mathematically by

$$v = V_{max} \sin \omega t \tag{7.1}$$

where v is the instantaneous voltage, V_{max} is the peak (maximum) value of the sine wave, ω is the angular frequency, and t is the time. The unit of ω is radians/second (rad/s) and is expressed by

$$\omega = 2\pi f \tag{7.2}$$

where f is the frequency of the ac waveform. In North, Central, and South America, South Korea, Taiwan, Philippines, Saudi Arabia, and Western Japan, the frequency of the power supply is 60 cycles/second or 60 Hz. For the rest of the world, the supply frequency is 50 Hz, including Eastern Japan.

During the 19th century, 60 Hz was chosen by Westinghouse in the United States. Meanwhile, in Germany, the giant power equipment manufacturers AEG and Siemens chose 50 Hz. These companies had a virtual monopoly in Europe, and the 50 Hz standard spread to the rest of Europe and most of the world.

7.2 Root Mean Square

The waveform in Figure 7.1 has its magnitude changing with time. For 60 Hz systems, the waveform is repeated 60 times every second. Quantifying the value of this sinusoidal voltage is tricky. If we quantify the sinusoidal waveform by its maximum value it would be unreliable, since the peak is often skewed by transients and harmonics. Using the average value is not a good

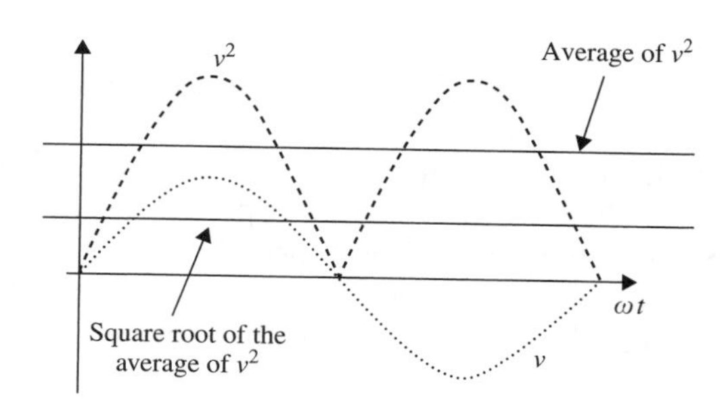

FIGURE 7.2
Concept of rms.

idea either, as the average value is always zero because of the symmetry of the waveform around the x-axis.

To address this problem, engineers developed an effective method to quantify ac waveforms; it employs *root mean square* (rms) values. The concept is shown in Figure 7.2 where the rms value is computed in three steps:

1. Compute the square of the waveform.
2. Compute the average of the squared waveform.
3. Compute the root of the average value of the squared waveform.

The idea behind step 1 (squaring the waveform) is to create another waveform that is always positive so that its average value in step 2 is nonzero. The square root in step 3 is imposed to somehow compensate for the initial squaring in step 1. Today, all ac waveforms in power circuits are quantified by their rms values.

The rms value can be computed by following the three steps mentioned above. Let us consider the waveform in Equation (7.1). The first step is to square this waveform.

$$v^2 = V_{max}^2 \sin^2 \omega t = \frac{V_{max}^2}{2}(1 - \cos 2\omega t) \tag{7.3}$$

The second step is to find the average value (Ave) of Equation (7.3).

$$\text{Ave} = \frac{1}{2\pi}\int_0^{2\pi} \frac{V_{max}^2}{2}(1 - \cos 2\omega t)\, d\omega t = \frac{V_{max}^2}{2} \tag{7.4}$$

The final step is to find the rms voltage (V) by computing the square root of Equation (7.4).

$$V = \sqrt{\frac{V_{max}^2}{2}} = \frac{V_{max}}{\sqrt{2}} \tag{7.5}$$

The household rms voltage in North America is 120 V. For Europe and the Middle East, the rms voltage is 220 to 240 V. There are a few exceptions, such as Japan, where the voltage is 100 V.

As mentioned in Chapter 2, it appears that 120 V was chosen somewhat arbitrarily. Actually, Thomas Edison came up with a high-resistance lamp filament that operated well at 120 V. Since then, 120 V was selected in the United States. Other nations chose a higher voltage, 220 to 240 V, to reduce the current in the electric wires, and therefore use wires with smaller cross-sections.

EXAMPLE 7.1 The outlet voltage measured by an rms voltmeter is 120 V. Compute the maximum value of the voltage waveform.

Solution

$$V_{\text{max}} = \sqrt{2}\,V = \sqrt{2} \times 120 = 169.7\text{ V}$$

Note that the voltage of all household equipment is given in rms, not the maximum value.

EXAMPLE 7.2 Compute the rms voltage of the waveform in Figure 7.3.

Solution

The general equation of the rms value of v is

$$V = \sqrt{\frac{1}{\tau}\int_0^{\tau} v^2\,dt}$$

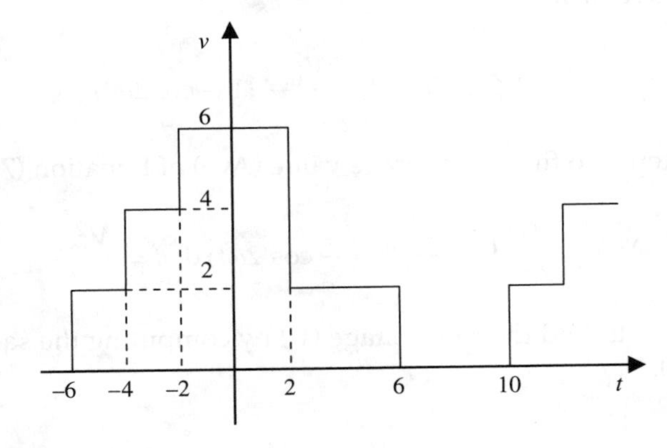

FIGURE 7.3
Voltage waveform.

where τ is the period. Since the waveform in Figure 7.3 is discrete, the rms voltage can be expressed by

$$V = \sqrt{\frac{1}{\tau}\sum_{k}\left(v_k^2 t_k\right)}$$

$$V = \sqrt{\frac{1}{16}\sum 2^2 \times 2 + 4^2 \times 2 + 6^2 \times 4 + 2^2 \times 4} = 3.535\ V$$

7.3 Phase Shift

The waveforms for currents and voltages in ac circuits are sinusoidal. For a purely resistive load R, the current waveform is in phase with the voltage waveform as shown in Equation (7.6) and Figure 7.4.

$$\begin{aligned} v &= V_{\max}\sin(\omega t) = i_R R \\ i_R &= \frac{V_{\max}}{R}\sin(\omega t) \end{aligned} \tag{7.6}$$

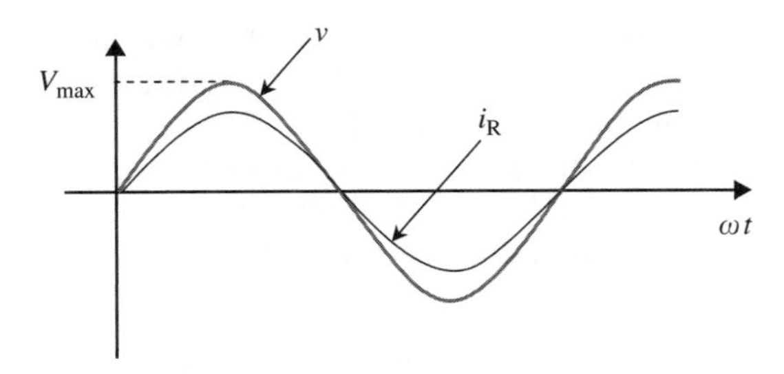

FIGURE 7.4
Voltage and current waveforms of purely resistive load.

If the circuit has a purely inductive load (no resistance or capacitance), the voltage–current relationship is

$$v = V_{\max}\sin(\omega t) = L\frac{di_L}{dt} \tag{7.7}$$

where L is the inductance of the load. The current in the purely inductive load i_L can be obtained from Equation (7.7).

$$i_L = \frac{1}{L}\int v\,dt = \frac{V_{\max}}{L}\int \sin(\omega t)\,dt = -\frac{V_{\max}}{\omega L}\cos(\omega t) = -\frac{V_{\max}}{X_L}\cos(\omega t) \tag{7.8}$$

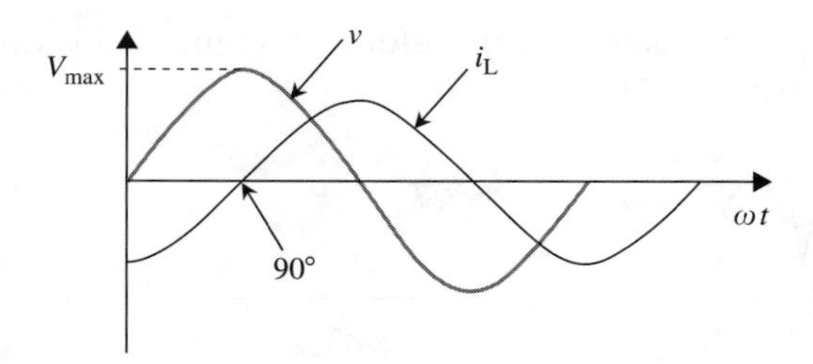

FIGURE 7.5
Voltage and current waveforms of purely inductive load.

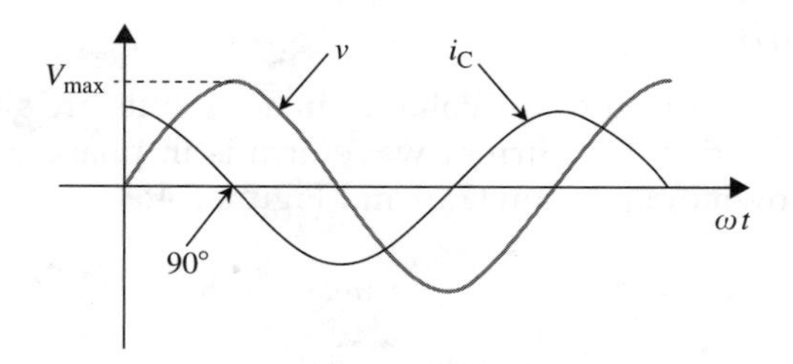

FIGURE 7.6
Voltage and current waveforms of purely capacitive load.

where $X_L = \omega L$ is the magnitude of the inductive reactance of the load. As seen in Equation (7.8), the current i_L of a purely inductive load is a negative cosine waveform, which is shown in Figure 7.5. The positive peak of the current occurred 90° after the voltage reaches its own positive peak. In this case, the current is said to be *lagging* the voltage by 90°.

For a purely capacitive load, the voltage–current relationship is given by

$$v = V_{max} \sin(\omega t) = \frac{1}{C} \int i_C \, dt$$
$$i_C = C \frac{dv}{dt} \tag{7.9}$$

where C is the load capacitance. The current i_C can be computed as

$$i_C = C \frac{d}{dt} [V_{max} \sin(\omega t)] = \omega C V_{max} \cos(\omega t) = \frac{V_{max}}{X_C} \cos(\omega t) \tag{7.10}$$

where $X_C = 1/\omega C$ is the magnitude of the capacitive reactance of the load. As seen in Equation (7.10), the current of a purely capacitive load i_C is a cosine waveform. Figure 7.6 shows the voltage across the capacitor and the current i_C. Note that the positive peak of the current waveform occurs 90° before the voltage peak. Hence, the current is said to be *leading* the voltage by 90°.

If the load is composed of elements such as resistances, capacitances, and inductances, the phase shift angle θ of the current can be any value between

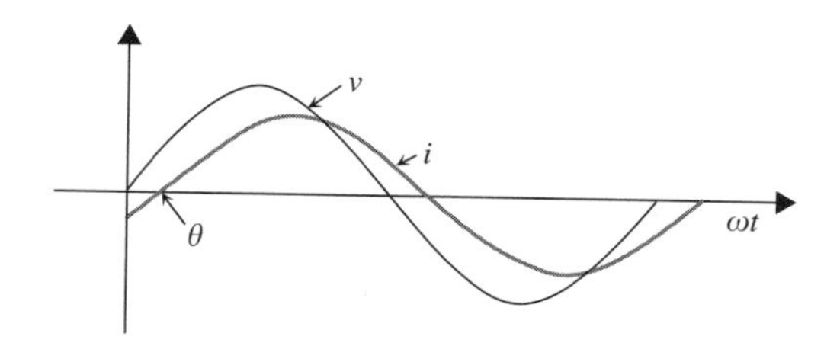

FIGURE 7.7
Voltage and current waveforms of lagging current.

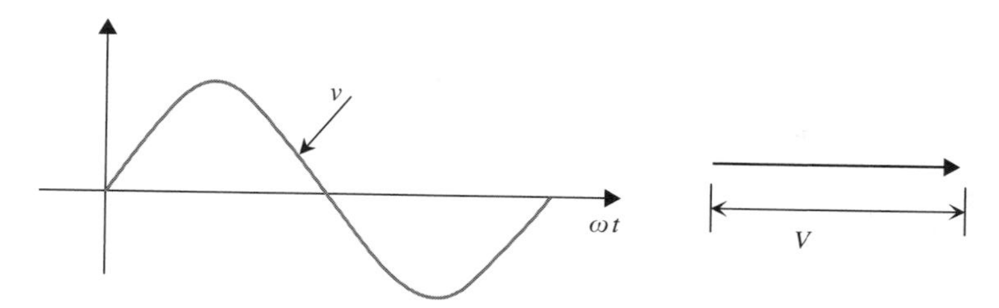

FIGURE 7.8
Phasor representation of a sinusoidal waveform.

−90° and 90°. Figure 7.7 shows the current and voltage waveforms of an inductive load that is composed of a resistance and inductance.

The current waveform in Figure 7.7 can be expressed mathematically by

$$i = I_{\max} \sin(\omega t - \theta) \tag{7.11}$$

7.4 The Concept of Phasors

Phasors are handy mathematical representations in ac circuits. The phasors give quick information on the magnitude and phase shift of any waveform. Consider Figure 7.8. The left side of the figure shows a sinusoidal waveform without a phase shift. The waveform is represented by the phasor on the right side of the figure. The length of the phasor is proportional to the rms value of the waveform, and its phase angle represents the phase shift of the sinusoidal waveform. Since the waveform has zero phase shift, the phasor of the voltage is aligned with the x-axis.

Consider the circuit in Figure 7.9 for a purely resistive load. Since the resistance does not cause any phase shift in the current, as shown in Figure 7.4, the current phasor is *in phase* with the voltage as shown on the right side of Figure 7.9 .

For a purely inductive load, the current lags the voltage by 90°, as given in Equation (7.8) and shown in Figure 7.5. The phasor diagram in this case

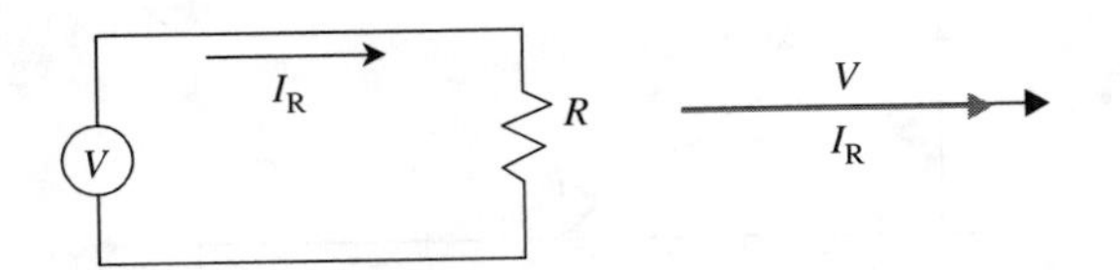

FIGURE 7.9
Phasor representation of current and voltage of a purely resistive load.

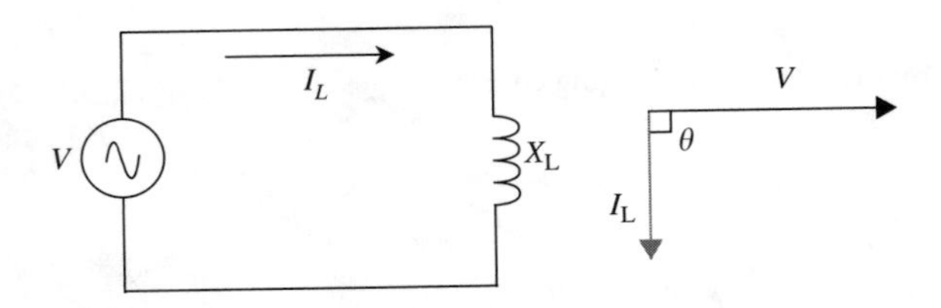

FIGURE 7.10
Phasor representation of current and voltage of a purely inductive load.

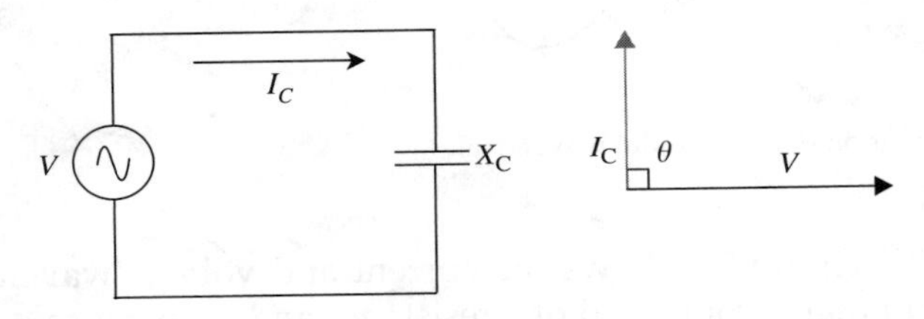

FIGURE 7.11
Phasor representation of current and voltage of a purely capacitive load.

is depicted on the right side of Figure 7.10. The lagging angles are in the clockwise direction.

For a purely capacitive load, the current leads the voltage by 90°, as given in Equation (7.10) and shown in Figure 7.6. The phasor diagram in this case is given on the right side of Figure 7.11. The leading angles are in the counterclockwise direction.

Now let us consider the waveforms in Figure 7.7. As seen in the figure, the current is lagging the voltage by an angle θ. These waveforms can be represented by the phasor diagram in Figure 7.12. The voltage is considered the reference with no shift, so it is aligned with the x-axis. Since the current lags the voltage by θ, the phasor for the current lags the reference by θ in the clockwise direction.

7.5 Complex Number Analysis

It is inconvenient to analyze electric circuits graphically as done so far. An alternative method is to use the complex number analysis, where any phasor

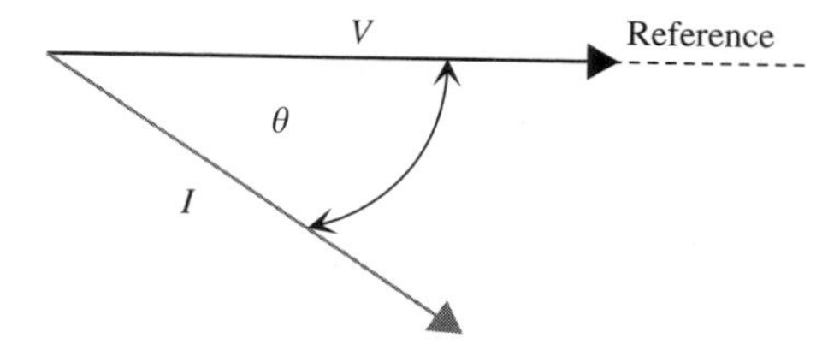

FIGURE 7.12
Phasor representation of the current and voltage in Figure 7.7.

is represented by a magnitude and an angle. For example, the voltage and current in Figure 7.7 or Figure 7.12 can be represented by the following complex variable equations:

$$\overline{V} = V \angle 0^\circ \tag{7.12}$$

$$\overline{I} = I \angle {-\theta}^\circ \tag{7.13}$$

where $\overline{V}$ is the phasor representation of the voltage. It has a bar on top, indicating that the variable is a complex number. V is the rms value (magnitude) of the voltage. $\angle$ is followed by the angle of the phasor, which is zero for voltage and $-\theta$ for the lagging current.

The complex forms in Equations (7.12) and (7.13) are called *polar* forms (also known as *trigonometric* forms). The math of the polar forms is very simple for multiplication and division. Let us assume that we have two complex numbers:

$$\overline{A} = A \angle \theta_1 \tag{7.14}$$

$$\overline{B} = B \angle \theta_2 \tag{7.15}$$

The multiplication and division of these two complex numbers are

$$\overline{A}\,\overline{B} = A \angle \theta_1 \; B \angle \theta_2 = AB \angle(\theta_1 + \theta_2) \tag{7.16}$$

$$\frac{\overline{A}}{\overline{B}} = \frac{A \angle \theta_1}{B \angle \theta_2} = \frac{A}{B} \angle(\theta_1 - \theta_2) \tag{7.17}$$

The addition and subtraction of complex numbers in polar forms are harder to implement. However, if we switch to the *rectangular* form (also known as *cartesian* form), we can perform these operations very quickly. To show how to convert from polar to rectangular form, consider the phasor $\overline{A}$ in Figure 7.13. The reference in the figure represents the x-axis. The component of $\overline{A}$ projected on the reference is $X = A \cos \theta$, and the vertical projection is $Y = A \sin \theta$. These two components can be represented by

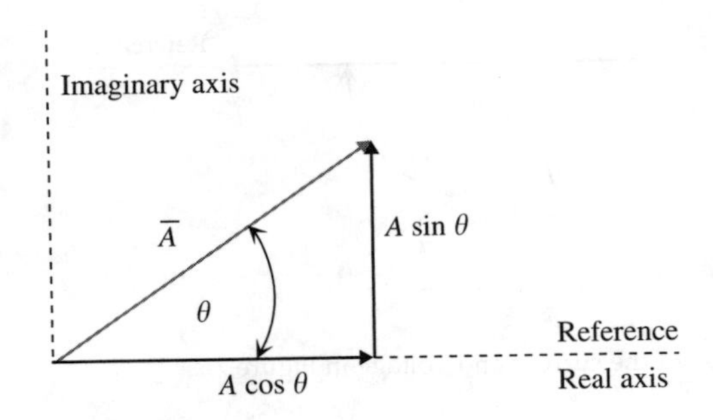

FIGURE 7.13
X and Y components of Phasor $\overline{A}$.

Equation (7.18).

$$\overline{A} = A\angle\theta = A[\cos\theta + j\sin\theta] = A\cos\theta + jA\sin\theta = X + jY \qquad (7.18)$$

The first component along the x-axis is known as the *real* component. The second component, preceded by an operator j, is known as the *imaginary* component. Keep in mind that there is nothing imaginary about the quantity $A\sin\theta$; we simply use the label to distinguish between the directions of the real and imaginary vectors. The operator j is used to represent the counterclockwise rotation of a vector by 90°; any quantity preceded by j is shifted in the counterclockwise direction by 90°. The basic relations of the operator j are summarized in Equation (7.19).

$$\begin{aligned} j &= 1\angle 90^\circ \\ j^2 &= 1\angle 180^\circ = -1 \\ j^3 &= 1\angle 270^\circ = 1\angle -90^\circ = -j \\ j^4 &= 1\angle 0^\circ = 1 \\ j^5 &= 1\angle 90^\circ = j \end{aligned} \qquad (7.19)$$

and so on.

The mathematical expressions in Equation (7.19) can also be achieved by defining the complex value $j = \sqrt{-1}$.

By using Figure (7.13), the conversion of $\overline{A}$ from the rectangular form to the polar form is

$$\overline{A} = X + jY = \sqrt{X^2 + Y^2}\angle\tan^{-1}\frac{Y}{X} \qquad (7.20)$$

The addition (subtraction) of complex numbers in rectangular form is done by adding (subtracting) the real and imaginary components independently.

$$\begin{aligned}\overline{A}+\overline{B} &= A(\cos\theta_1 + j\sin\theta_1) + B(\cos\theta_2 + j\sin\theta_2)\\ &= (A\cos\theta_1 + B\cos\theta_2) + j(A\sin\theta_1 + B\sin\theta_2) \qquad (7.21)\\ \overline{A}-\overline{B} &= A(\cos\theta_1 + j\sin\theta_1) - B(\cos\theta_2 + j\sin\theta_2)\\ &= (A\cos\theta_1 - B\cos\theta_2) + j(A\sin\theta_1 - B\sin\theta_2) \qquad (7.22)\end{aligned}$$

One of the convenient operations used in complex number analysis is the conjugate of a vector. Two complex numbers are conjugates if the sign of their imaginary components are different. If $\overline{A}$ is a complex vector and $\overline{A}^*$ is its conjugate, then

$$\begin{aligned}\overline{A} &= X + jY = \sqrt{X^2+Y^2}\,\angle\theta\\ \overline{A}^* &= X - jY = \sqrt{X^2+Y^2}\,\angle{-\theta}\end{aligned} \qquad (7.23)$$

The conjugate is useful in complex operations, such as when a phasor is inverted. Let us assume that the phasor $\overline{A}$ is in rectangular form. To compute $1/\overline{A}$, follow the process below.

$$\begin{aligned}\frac{1}{\overline{A}} &= \frac{1}{X+jY}\frac{\overline{A}^*}{\overline{A}^*}\\ \frac{1}{\overline{A}} &= \frac{1}{X+jY}\frac{X-jY}{X-jY} = \frac{X-jY}{X^2+Y^2}\\ \frac{1}{\overline{A}} &= \frac{X}{X^2+Y^2} - j\frac{Y}{X^2+Y^2}\end{aligned} \qquad (7.24)$$

EXAMPLE 7.3 Assume $\overline{A} = 10\angle 60^\circ$ and $\overline{B} = 5\angle 40^\circ$. Compute the following:

1. $\overline{C} = \overline{A}\,\overline{B}$
2. $\overline{C} = \overline{A}/\overline{B}$
3. $\overline{C} = \overline{A} + \overline{B}$
4. $\overline{C} = \overline{A} - \overline{B}$

Solution

1. $\overline{C} = \overline{A}\,\overline{B} = 10\angle 60^\circ\, 5\angle 40^\circ = (10 \times 5)\angle(60^\circ + 40^\circ) = 50\angle 100^\circ$
2. $\overline{C} = \dfrac{\overline{A}}{\overline{B}} = \dfrac{10\angle 60^\circ}{5\angle 40^\circ} = \dfrac{10}{5}\angle(60^\circ - 40^\circ) = 2\angle 20^\circ$

3. $\overline{C} = \overline{A} + \overline{B} = 10\angle 60° + 5\angle 40° = 10(\cos 60° + j \sin 60°) + 5(\cos 40° + j \sin 40°)$

$$\overline{C} = (5 + j8.66) + (3.83 + j3.21) = 8.83 + j11.87$$

$$\overline{C} = \sqrt{(8.83)^2 + (11.87)^2}\,\angle \tan^{-1}\left(\frac{11.87}{8.83}\right) = 14.79\,\angle 53.35°$$

4. $\overline{C} = \overline{A} - \overline{B} = 10\angle 60° - 5\angle 40° = 10(\cos 60° + j \sin 60°) - 5(\cos 40° + j \sin 40°)$

$$\overline{C} = (5 + j8.66) - (3.83 + j3.21) = 1.17 + j5.45$$

$$\overline{C} = \sqrt{(1.17)^2 + (5.45)^2}\,\angle \tan^{-1}\left(\frac{5.45}{1.17}\right) = 5.57\,\angle 77.88°$$

7.6 Complex Impedance

The impedance in ac circuits is composed of any combination of resistances, inductances, and capacitances. These elements are also complex quantities and can be computed as given in the following three equations:

$$\overline{R} = \frac{\overline{V}}{\overline{I}_R} = \frac{V\,\angle 0}{I_R\,\angle 0} = R\,\angle 0 \tag{7.25}$$

$$\overline{X}_L = \frac{\overline{V}}{\overline{I}_L} = \frac{V\,\angle 0}{I_L\,\angle{-90°}} = \frac{V}{I_L}\,\angle 90° = X_L\,\angle 90° \tag{7.26}$$

$$\overline{X}_C = \frac{\overline{V}}{\overline{I}_C} = \frac{V\,\angle 0}{I_C\,\angle 90°} = \frac{V}{I_C}\,\angle{-90°} = X_C\,\angle{-90°} \tag{7.27}$$

where the magnitudes are $X_L = \omega L$ and $X_C = 1/\omega C$ as given in Equations (7.8) and (7.10).

7.6.1 Series Impedance

When the load is composed of different elements, the total ohmic value of the load is called *impedance*. In Figure 7.14 a resistance is connected in series with an inductive reactance. Each of these components is a complex parameter with a magnitude and phase angle as given in Equations (7.25) and (7.26).

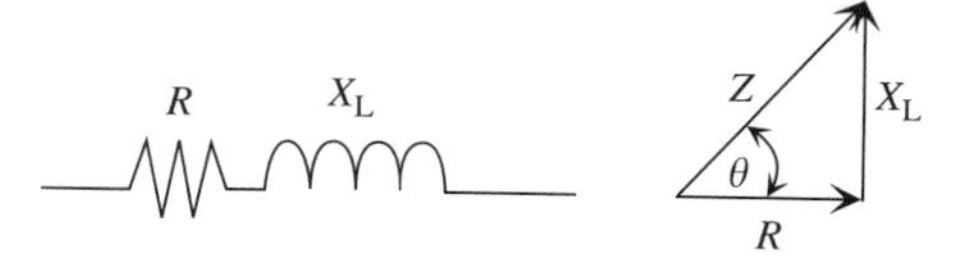

FIGURE 7.14
Series impedance of inductive load.

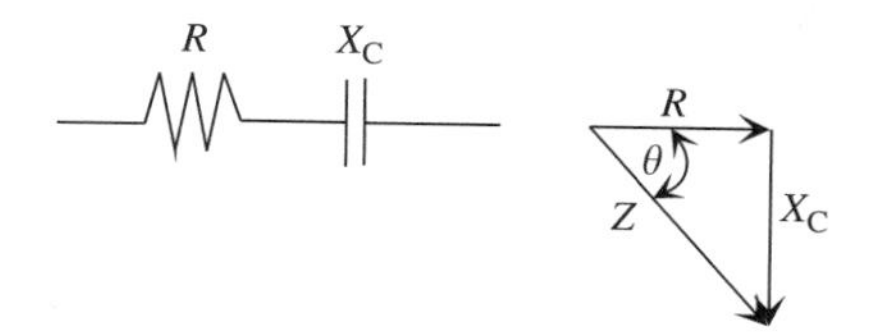

FIGURE 7.15
Series impedance of capacitive load.

The impedance $\overline{Z}$ of this load is

$$\overline{Z} = \overline{R} + \overline{X}_L = R\angle 0 + X_L\angle 90^o = R + jX_L \tag{7.28}$$

The phasor diagram of the impedance is shown on the right side of Figure 7.14. Note that the impedance angle θ is positive and is equal in magnitude to the phase angle of the current in Figure 7.12.

Similarly, the capacitive load in Figure 7.15 is composed of a resistor and a capacitor connected in series. The impedance of this load can be expressed by Equation (7.29).

$$\overline{Z} = \overline{R} + \overline{X}_C = R\angle 0 + X_C\angle{-90^\circ} = R - jX_C \tag{7.29}$$

The phasor diagram of the capacitive load impedance is shown on the right side of Figure 7.15. Note that the capacitive reactance X_C lags the resistance by 90°.

Now let us assume a general case where a resistor is in series with a capacitor and an inductor. The circuit is shown in Figure 7.16. The impedance of the load can be expressed by Equation (7.30).

$$\overline{Z} = \overline{R} + \overline{X}_L + \overline{X}_C = R\angle 0 + X_L\angle 90^\circ + X_C\angle{-90^\circ} = R + j(X_L - X_C) \tag{7.30}$$

Depending on the magnitude of the inductive reactance X_L with respect to the capacitive reactance X_C, the impedance diagram of the circuit can be any one of the three phasor diagrams in Figure 7.16. In the first one on the left side, $X_L > X_C$ and the impedance angle is positive. Hence the circuit acts as if it contains the resistance and an equivalent inductive reactance of a magnitude equal to $(X_L - X_C)$. When $X_L = X_C$, the impedance angle of the circuit is zero, and the circuit impedance is equivalent to the resistance only. In the third case

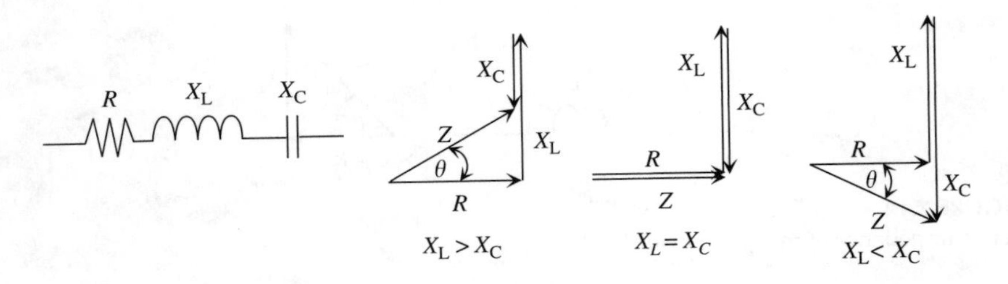

FIGURE 7.16
Series impedance of complex load.

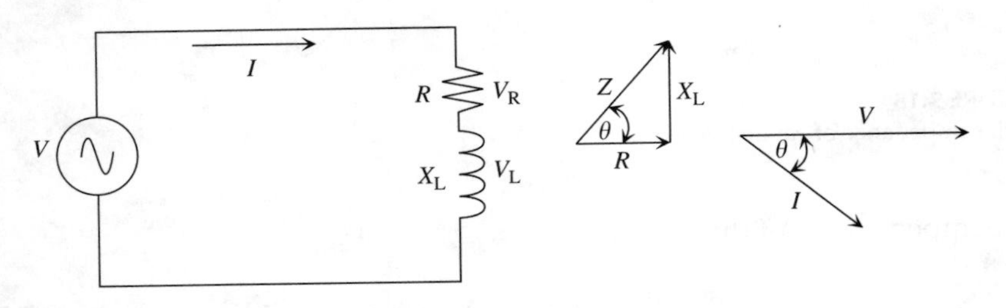

FIGURE 7.17
Circuit connection and phasor diagrams.

when $X_L < X_C$, the impedance angle is negative, and the circuit acts as if it contains the resistance and an equivalent capacitive reactance of a magnitude equal to $(X_C - X_L)$.

EXAMPLE 7.4 A load is composed of a 4 Ω resistance connected in series with a 3 Ω inductive reactance. The load is connected across a 120 V source as shown in Figure 7.17.

1. Compute the current of the circuit.
2. Compute the voltage across the resistance.
3. Compute the voltage across the inductive reactance.

Solution

1. The current of the circuit is

$$\bar{I} = \frac{\bar{V}}{\bar{Z}}$$

where $\bar{Z} = R\angle 0° + X_L\angle 90° = R + jX_L = 4 + j3 = 5\angle 36.87°\ \Omega$.

Hence, $\bar{I} = \dfrac{120\angle 0}{5\angle 36.87°} = 24\angle -36.87°\,\text{A}$.

2. The voltage across the resistance (V_R) is

$$\overline{V}_R = \overline{I}\overline{R} = 24\angle{-36.87°}\ 4\angle 0 = 96\angle{-36.87°}\ \text{V}$$

3. The voltage across the inductive reactance (V_L) is

$$\overline{V}_L = \overline{I}\overline{X}_L = 24\angle{-36.87°}\ 3\angle 90° = 72\angle 53.13°\ \text{V}$$

Note that the sum of the voltage across the resistance and across the inductive reactance yields the value of the source voltage. This, however, must be done using complex mathematics. Try it!

Also note that the impedance angle and the angle of the current are the same in magnitude, but opposite in sign, as shown in Figure 7.17.

EXAMPLE 7.5 A 120 V adjustable frequency ac source is connected in series with a resistor, an inductor and a capacitor. At 60 Hz, the resistance is 5 Ω, the inductive reactance is 3 Ω, and the capacitive reactance is 4 Ω.

1. Compute the load impedance at 60 Hz.
2. Compute the frequency at which the total impedance is equal to the load resistance.

Solution

1. As given in Equation (7.30), the load impedance is

$$\overline{Z} = R + j(X_L - X_C) = 5 + j(3 - 4) = 5 - j = 5.1\angle{-11.31°}\ \Omega$$

2. First, let us compute the inductance and capacitance of the load.

$$X_L = 2\pi f L$$

$$L = \frac{3}{2\pi 60} = 7.96\ \text{mH}$$

Similarly,

$$X_C = \frac{1}{2\pi f C}$$

$$C = \frac{1}{2\pi 60 \times 4} = 663.15\ \mu\text{F}$$

If $\overline{Z} = R$, then $X_L = X_C$. This can occur by changing the frequency of the supply voltage until $X_L = X_C$. The frequency in this case is called the *resonance frequency* or f_0.

$$X_L = X_C$$

$$2\pi f_0 L = \frac{1}{2\pi f_0 C}$$

$$f_0 = \frac{1}{2\pi\sqrt{LC}} = \frac{10^3}{2\pi\sqrt{7.96 \times 0.66315}} = 69.27 \text{ Hz}$$

7.6.2 Parallel Impedance

In parallel circuits, the current of the source is divided among the parallel elements of the circuit according to their impedance values. As shown in Figure 7.18, the phasor sum of all branch currents is equal to the current of the sources. This can be represented mathematically as shown in Equation (7.31).

$$\bar{I} = \bar{I}_R + \bar{I}_L + \bar{I}_C = \frac{\overline{V}}{\overline{R}} + \frac{\overline{V}}{\overline{X}_L} + \frac{\overline{V}}{\overline{X}_C} = \overline{V}\left[\frac{1}{\overline{R}} + \frac{1}{\overline{X}_L} + \frac{1}{\overline{X}_C}\right] = \frac{\overline{V}}{\overline{Z}} \tag{7.31}$$

Hence, the total impedance $\overline{Z}$ is defined as

$$\frac{1}{\overline{Z}} = \frac{1}{\overline{R}} + \frac{1}{\overline{X}_L} + \frac{1}{\overline{X}_C} \quad \text{or} \quad \overline{Y} = \overline{G} + \overline{B}_L + \overline{B}_C \tag{7.32}$$

where

$$\begin{aligned}
\overline{Y} &= \frac{1}{\overline{Z}} \\
\overline{G} &= \frac{1}{\overline{R}} = \frac{1}{R\angle 0^\circ} = \frac{1}{R}\angle 0^\circ \\
\overline{B}_L &= \frac{1}{\overline{X}_L} = \frac{1}{X_L\angle 90^\circ} = \frac{1}{X_L}\angle{-90^\circ} \\
\overline{B}_C &= \frac{1}{\overline{X}_C} = \frac{1}{X_C\angle{-90^\circ}} = \frac{1}{X_C}\angle 90^\circ
\end{aligned} \tag{7.33}$$

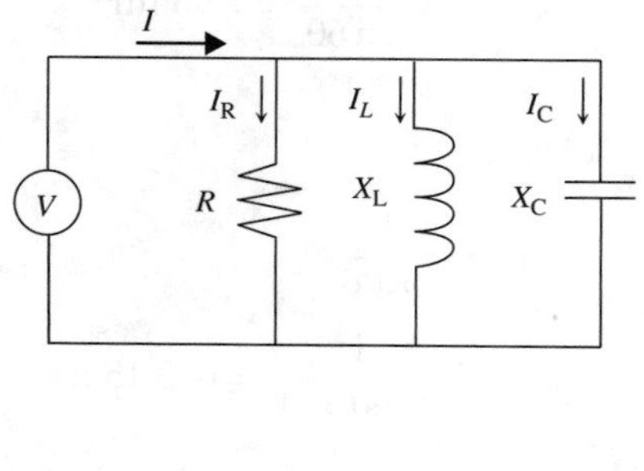

FIGURE 7.18
Parallel impedance.

G is called the *conductance*, and its angle is zero, $\overline{B}_L$ is called the *inductive susceptance*, and $\overline{B}_C$ is called the *capacitive susceptance*. Note that the angle of B_L is $-90°$ as opposed to the angle of the inductive reactance X_L, which is $+90°$. Similarly the angle of B_C is $+90°$ as opposed to the angle of the capacitive reactance X_C which is $-90°$. The sum of all conductances and susceptances is called the *admittance* $\overline{Y}$.

$$\overline{Y} = \overline{G} + \overline{B}_L + \overline{B}_C = G + j(B_C - B_L) \tag{7.34}$$

The unit of G, B_L, B_C, and Y is the mho, which is the reverse spelling of the word "ohm." The impedance diagram of the parallel circuit is shown in Figure 7.19. Compare these phasor diagrams with the ones shown in Figure 7.16 for the series impedance.

EXAMPLE 7.6 A 120 V adjustable frequency ac source is connected in parallel with a resistor, an inductor, and a capacitor as shown in Figure (7.18). At 60 Hz, the resistance is 5 Ω, the inductive reactance is 10 Ω, and the capacitive reactance is 2 Ω.

1. Compute the load impedance at 60 Hz.
2. Compute the frequency at which the total impedance is equivalent to the load resistance only.

Solution

1. As given in Equation (7.34), the load admittance Y is

$$\overline{Y} = G + j(B_C + B_L) = \tfrac{1}{5} + j(\tfrac{1}{2} - \tfrac{1}{10}) = 0.2 + j0.4 \text{ mho}$$

The load impedance Z is

$$\overline{Z} = \frac{1}{\overline{Y}} = \frac{1}{0.2 + j0.4} = 1.0 - j2.0 \text{ ohm}$$

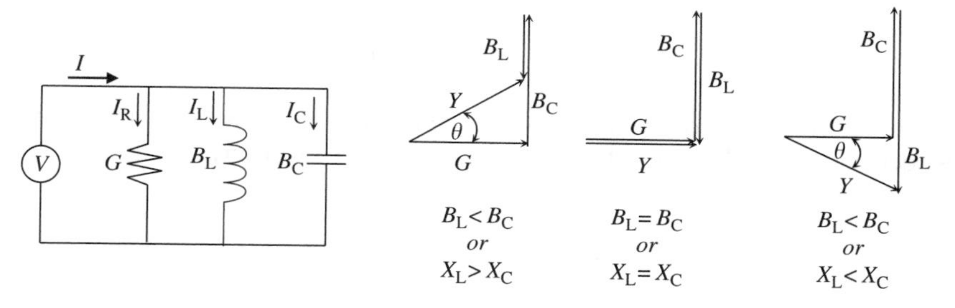

FIGURE 7.19
Parallel circuit.

2. First, let us compute the inductance and capacitance of the load.

$$X_L = 2\pi f L$$

$$L = \frac{10}{2\pi 60} = 26.5 \text{ mH}$$

Similarly,

$$X_C = \frac{1}{2\pi f C}$$

$$C = \frac{1}{2\pi 60 \times 2} = 1.326 \text{ mF}$$

If $\overline{Z} = R$, then $B_L = B_C$. This can occur by changing the frequency of the supply voltage.

$$f_0 = \frac{1}{2\pi\sqrt{LC}} = \frac{10^3}{2\pi\sqrt{26.5 \times 1.326}} = 26.85 \text{ Hz}$$

7.7 Electric Power

The instantaneous power ρ is defined as the multiplication of the instantaneous voltage v by the instantaneous current i.

$$\rho = vi \tag{7.35}$$

Let us assume that the waveforms of the voltage and current are sinusoidal as described by Equations (7.1) and (7.11). Hence, the instantaneous power is

$$\begin{aligned}
\rho &= vi = [V_{\max}\sin(\omega t)]\,[I_{\max}\sin(\omega t - \theta)] \\
\rho &= \frac{V_{\max} I_{\max}}{2}\,[\cos(\theta) - \cos(2\omega t - \theta)] \\
\rho &= \frac{V_{\max}}{\sqrt{2}}\frac{I_{\max}}{\sqrt{2}}\,[\cos(\theta) - \cos(2\omega t - \theta)] \\
\rho &= VI\,[\cos(\theta) - \cos(2\omega t - \theta)]
\end{aligned} \tag{7.36}$$

where V and I are the rms values of the sinusoidal waveforms v and i, respectively. θ is the impedance angle. Now, let us rewrite Equation (7.36) as follows:

$$\begin{aligned}
\rho &= VI\cos(\theta) - VI\cos(2\omega t - \theta) \\
\rho &= P + h
\end{aligned} \tag{7.37}$$

The first term P in Equation (7.37) is time-invariant since V, I, and θ are all time-independent. The second term h is time-varying sinusoidal with a frequency equal to twice the frequency of the supply voltage and shifted by θ. P and h are plotted in Figure 7.20.

For a purely resistive load, θ is zero, hence

$$\begin{aligned} \rho &= vi = VI\cos(\theta) - VI\cos(2\omega t - \theta) \\ \rho &= VI[1 - \cos(2\omega t)] \end{aligned} \tag{7.38}$$

The power waveform for this case is shown in Figure 7.21. Note that the instantaneous power is always positive.

For a purely inductive load, the impedance angle $\theta = 90°$, and its instantaneous power is given by Equation (7.39) and shown in Figure 7.22. Note that the average value of the instantaneous power is zero, which indicates that the inductor consumes power during the first quarter of the voltage cycle, and then returns the power back during the second quarter of the cycle.

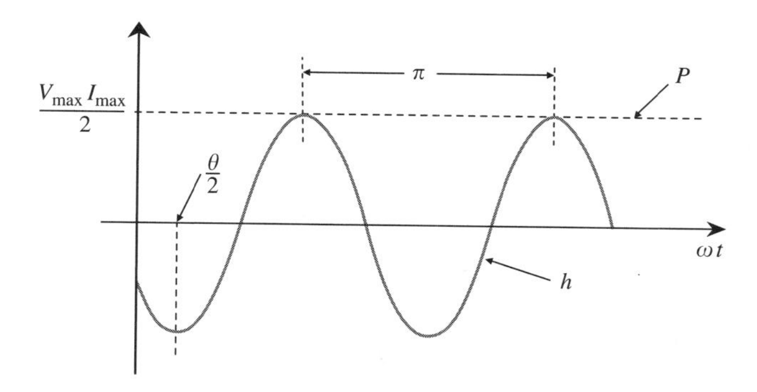

FIGURE 7.20
The two terms of the instantaneous power in Equation (7.37).

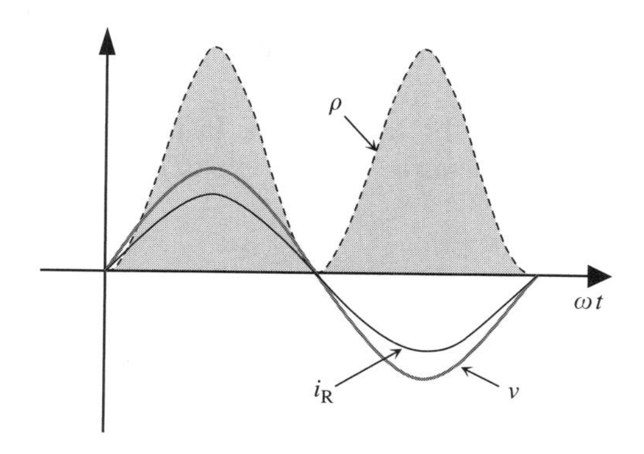

FIGURE 7.21
Instantaneous power of a purely resistive load.

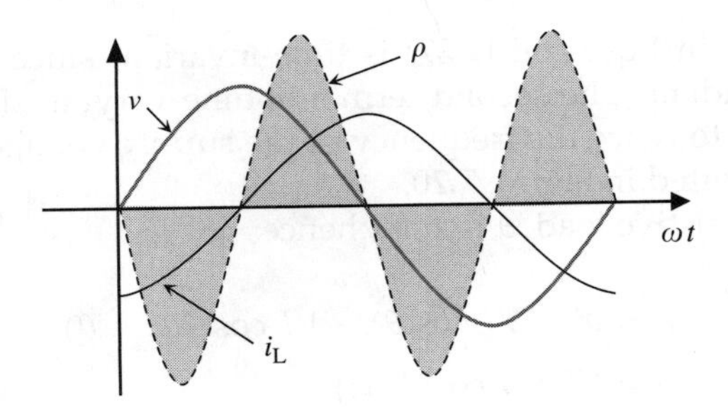

FIGURE 7.22
Instantaneous power of a purely inductive load.

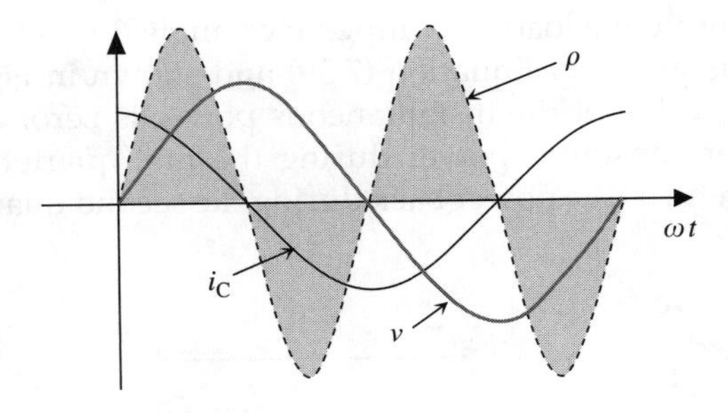

FIGURE 7.23
Instantaneous power of a purely capacitive load.

The inductor, in this case, does not store any energy.

$$\rho = vi = VI\,[\cos(\theta) - \cos(2\omega t - \theta)]$$
$$\rho = -VI\,\sin(2\omega t) \tag{7.39}$$

For a purely capacitive load, the impedance angle $\theta = -90°$, and its instantaneous power is given by Equation (7.40) and shown in Figure 7.23. Note that the average value of the instantaneous power for a capacitor is zero, which is similar to that for the inductance. In this case, the capacitor also exchanges energy with the source. In one cycle, the total energy consumed by the capacitor is zero.

$$\rho = vi = VI\,[\cos(\theta) - \cos(2\omega t - \theta)]$$
$$\rho = VI\,\sin(2\omega t) \tag{7.40}$$

For the general case of loads with any phase shift, Equation (7.37) can be used. For the case of lagging current, the waveforms are shown in Figure 7.24. Note that the net sum of the instantaneous power is nonzero.

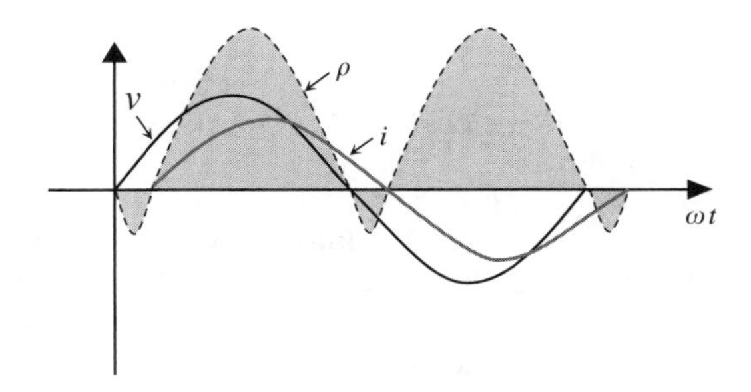

FIGURE 7.24
Instantaneous power of a complex load.

7.7.1 Real Power

The power that produces energy is the average value of ρ given in Equation (7.37). Since the first term, P, in the equation is time-invariant, its average value is the term itself. For the second term, h, its average value is zero since it is sinusoidal (i.e., symmetric across the time axis). P is called *active power* or *real power*, and its units are the watt (W), kilowatt (kW), megawatt (MW), etc.

$$P = \frac{1}{2\pi}\int_0^{2\pi} \rho \, d\omega t = \frac{1}{2\pi}\int_0^{2\pi} (P + h) \, d\omega t = \frac{VI}{2\pi}\int_0^{2\pi} [\cos(\theta) - \cos(2\omega t - \theta)] \, d\omega t$$
$$P = VI\cos(\theta) \tag{7.41}$$

7.7.2 Reactive Power

The reactive power is the power exchanged between the source and the inductive or capacitive elements of the circuit. Recall that the inductive and capacitive elements have their instantaneous powers in sinusoidal form, as shown in Figure 7.22 and Figure 7.23. This means the power is drawn from the source then delivered back to the source.

To identify the reactive power mathematically, let us rewrite Equation (7.37) by expanding the term h.

$$\begin{aligned} \rho &= VI\cos(\theta) - VI[\cos(\theta)\cos(2\omega t) + \sin(\theta)\sin(2\omega t)] \\ \rho &= VI\cos(\theta)[1 - \cos(2\omega t)] - VI\sin(\theta)\sin(2\omega t) \end{aligned} \tag{7.42}$$

Define

$$\begin{aligned} P &\equiv VI\cos(\theta) \\ Q &\equiv VI\sin(\theta) \end{aligned} \tag{7.43}$$

Hence,

$$\rho = P[1 - \cos(2\omega t)] - Q\sin(2\omega t) \tag{7.44}$$

The term Q is called *imaginary power* or *reactive power*. It is time-invariant and its units are VoltAmpere reactive (VAr), kiloVoltAmpere reactive (KVAr), etc.

Another method to compute the reactive power is by using the current and reactance. Assume that an inductive load is composed of a resistance in series with an inductive reactance. Since the magnitude of the voltage across the load is

$$V = IZ \tag{7.45}$$

and

$$\sin(\theta) = \frac{X_L}{Z} \tag{7.46}$$

then

$$\begin{aligned} Q_L &= VI\,\sin(\theta) = (IZ)I\frac{X_L}{Z} \\ Q_L &= I^2 X_L \end{aligned} \tag{7.47}$$

Similarly for a capacitive load, it can be shown that the reactive power is

$$Q_C = I^2 X_C \tag{7.48}$$

7.7.3 Complex Power

The complex power is the phasor sum of the real and reactive powers. Let us first define the complex power S as

$$\overline{S} \equiv \overline{V}\overline{I}^* \tag{7.49}$$

where $\bar{I}^*$ is the conjugate of the current $\bar{I}$. If $\bar{I} = I\angle{-\theta}, \bar{I}^* = I\angle\theta$. Hence, for lagging current the complex power is

$$\overline{S} = \overline{V}\overline{I}^* = V\angle 0\, I\angle\theta = VI\angle\theta = VI\cos(\theta) + j\,VI\sin(\theta) \tag{7.50}$$

substituting the values of P & Q in Equation (7.43) into Equation (7.50), yields

$$\overline{S} = P + jQ \tag{7.51}$$

The complex power S is also known as the *apparent power*. Its units are VoltAmpere (VA), kiloVoltAmpere (KVA), etc. The phasor quantity $\overline{S}$ in Equation (7.51) can be represented by the phasor diagram in Figure 7.25.

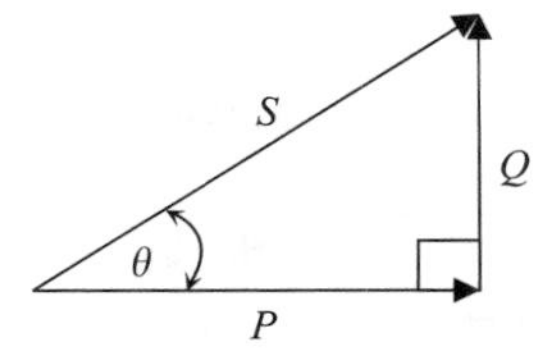

FIGURE 7.25
Phasor diagram of the complex power.

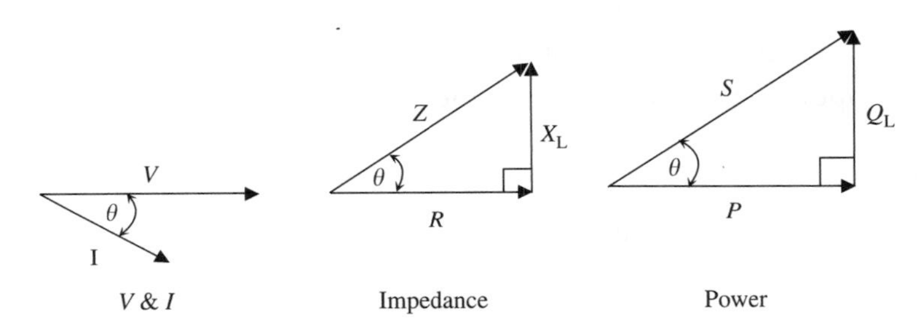

FIGURE 7.26
Phasor diagrams of an inductive load.

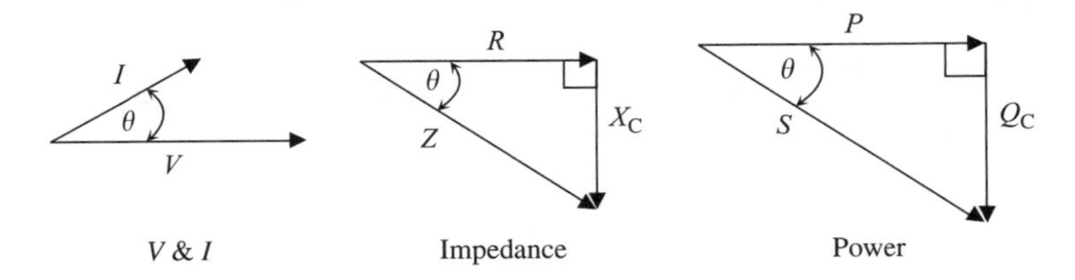

FIGURE 7.27
Phasor diagrams of a capacitive load.

7.7.4 Summary of ac Phasors

For the inductive load in Figure 7.17, the phasor diagrams of volt–current, impedance, and power are shown in Figure 7.26. In the volt–current diagram, the voltage is taken as the reference and the current lags the voltage by θ. In the impedance diagram, the resistance is always the reference and the inductive reactance leads the resistance by 90°. For the power diagram, the real power is always the reference, and the inductive reactive power leads the real power by 90°.

For a capacitive load, the phasor diagrams are shown in Figure 7.27. Note that the reactive power of the capacitive load is lagging the real power by 90°, while the reactive power of the inductive load is leading.

EXAMPLE 7.7 The voltage and current waveforms of an electric load are

$$v = 150\sin(377t + 0.2)$$
$$i = 25\sin(377t - 0.5)$$

1. Compute the frequency of the supply voltage.
2. Compute the phasor voltage.
3. Compute the phasor current.
4. Compute the real power of the load.
5. Compute the reactive power of the load.

Solution

1. $\omega = 2\pi f = 377$ rad/sec

$$f = \frac{\omega}{2\pi} = \frac{377}{2\pi} = 60 \text{ Hz}$$

2. The phasor voltage is represented by the magnitude in rms and the phase angle in degrees. Note that the angles in the waveform equations are in radians.

$$\overline{V} = \frac{V_{\max}}{\sqrt{2}} \angle\theta_V = \frac{150}{\sqrt{2}} \angle\left(0.2\frac{180}{\pi}\right) = 106.06 \angle 11.46^\circ \text{ V}$$

3. Similarly, the phasor current is

$$I = \frac{I_{\max}}{\sqrt{2}} \angle\theta_I = \frac{25}{\sqrt{2}} \angle\left(-0.5\frac{180}{\pi}\right) = 17.67 \angle{-28.65^\circ} \text{ A}$$

4. The angle between the voltage and current is

$$\theta = \theta_V - \theta_I = 11.46 + 28.65 = 40.11^\circ$$

The real power of the load is

$$P = VI\cos(\theta) = 106.06 \times 17.67 \times \cos(40.11) = 1.433 \text{ kW}$$

5. The reactive power of the load is

$$Q = VI\sin(\theta) = 106.06 \times 17.67 \times \sin(40.11) = 1.207 \text{ kVAr}$$

7.7.5 Power Factor

The apparent power S and the real power P are linearly dependent, related by the term $\cos(\theta)$.

$$P = VI\,\cos(\theta) = S\,\cos(\theta) \tag{7.52}$$

The ratio $P/S = \cos(\theta)$ is known as the *power factor* (pf), and θ is known as the *power factor angle*. The power factor can be computed by various methods. Consider Figure 7.26 and Figure 7.27. The pf in these figures is

$$\text{pf} = \cos(\theta) = \frac{P}{S} = \frac{R}{Z} \tag{7.53}$$

The power factor can be either *lagging* or *leading* depending on the angle of the current with respect to the voltage. When the current lags the voltage, the power factor is lagging. When the current leads the voltage, the power factor is leading.

7.7.6 Problems Related to Reactive Power

The reactive power does not produce energy, and its presence in the system can create several problems. Consider the simple system in Figure 7.28 where the source is assumed to be at a distance from a purely reactive load whose inductive reactance is X_L. To transmit the power from the source to the load, a cable (wire or transmission line) is used. The wire has a resistance R_{wire}, an inductive reactance X_{wire}, and a capacitive reactance. However, for simplicity, let us ignore the capacitive reactance, as its impact is relatively minor. The voltage of the source is V_S, and the voltage across the load is V_{load}.

The current in the wire is

$$\bar{I} = \frac{\overline{V}_S}{R_{wire} + j(X_{wire} + X_L)} \tag{7.54}$$

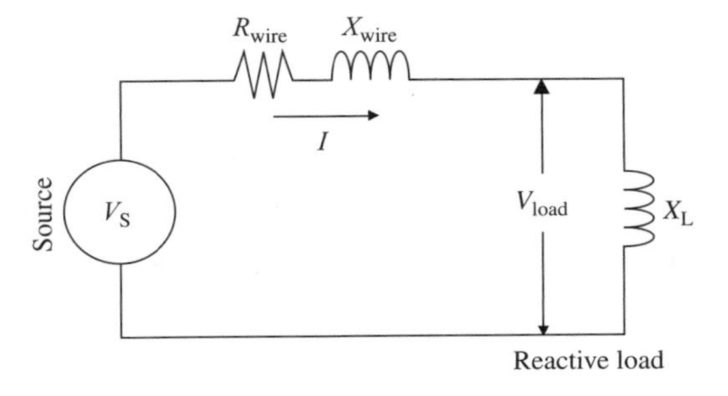

FIGURE 7.28
Currents due to inductive loads.

and the reactive power of the load is

$$Q_L = I^2 X_L \tag{7.55}$$

Keep in mind that only real power produces energy and almost all utilities charge residential customers for the real power only. It is highly unlikely that you pay for any reactive power that you consume in your house. Hence, the inductive load in Figure 7.28 is not generating any revenue to the utility; besides, it creates several problems such as:

1. It increases the losses of the transmission line.
2. It reduces the spare capacity of the line.
3. It reduces the voltage across the load.

The first problem can be understood by examining Equation (7.54). Because of the inductive load, the line current that feeds the inductive load results in real power losses in the wire P_{loss}, where

$$P_{loss} = I^2 R_{wire} \tag{7.56}$$

For the second problem, each transmission line has a capacity defined by the maximum current of the line, which is determined by the wire's cross-section and its material. The spare capacity is defined as the line capacity minus the actual current in the line. Because the reactive current is transmitted through the line, the spare capacity is reduced. This reduction in the spare capacity limits the ability of the utility to use the line to deliver real power to additional customers.

The third problem can be explained by considering the system in Figure 7.28, where the voltage on the load side $\overline{V}_{load}$ can be expressed by

$$\overline{V}_{load} = \overline{I} jX_L = \overline{V}_S \frac{jX_L}{R_{wire} + j(X_{wire} + X_L)} \tag{7.57}$$

Hence, the magnitude of the load voltage V_{load} is

$$V_{load} = \frac{V_S}{\sqrt{(R_{wire}/X_L)^2 + (1 + (X_{wire}/X_L))^2}} \tag{7.58}$$

Note that the load voltage decreases when the inductive reactance X_L decreases (i.e., more reactive load is added). When $X_L = \infty$ (i.e., no reactive load), the load voltage is equal to the source voltage.

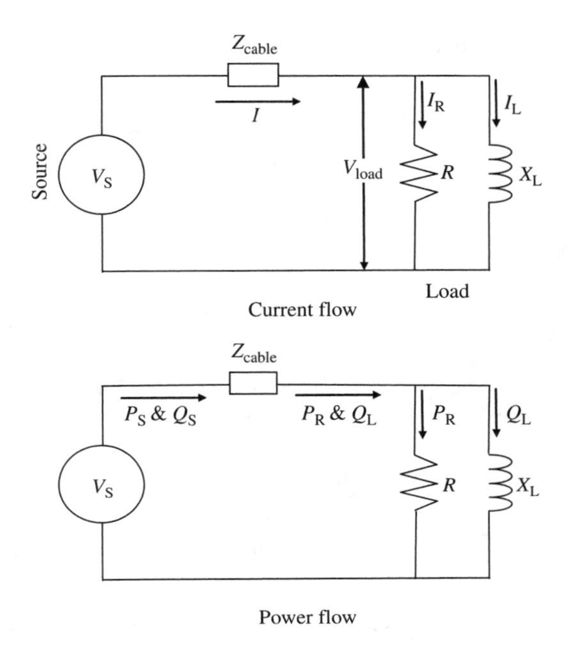

FIGURE 7.29
Current and power flow.

EXAMPLE 7.8 The circuit in Figure 7.29 has a source voltage of 110 V at 60 Hz. The load, which is heavily inductive, consists of a 20 Ω resistance in parallel with a 10 Ω inductive reactance. The cable resistance is 1 Ω and its inductive reactance is 5 Ω. Compute the following:

1. The load impedance.
2. The line current.
3. The load voltage.
4. The real and imaginary components of the load current.
5. The real and reactive powers of the load.
6. The real and reactive losses of the cable.
7. The real and reactive powers delivered by the source.

Solution

1. The load admittance Y is

$$\overline{Y} = G - jB_L = \frac{1}{20} - j\frac{1}{10} = 0.05 - j0.1 \text{ mho}$$

The load impedance Z is

$$\overline{Z} = \frac{1}{\overline{Y}} = \frac{1}{0.05 - j0.1} = 4 + j8 \text{ ohm}$$

2. To compute the line current, we need to compute the total impedance of the circuit.

$$\overline{Z}_{\text{total}} = \overline{Z} + \overline{Z}_{\text{cable}} = (4 + j8) + (1 + j5) = 5 + j13 \text{ ohm}$$

$$\overline{I} = \frac{\overline{V}_S}{\overline{Z}_{\text{total}}} = \frac{110 \angle 0^\circ}{5 + j13} = 2.835 - j7.37 = 7.9 \angle{-68.96^\circ} \text{ A}$$

3. The load voltage can be computed by

$$\overline{V}_{\text{load}} = \overline{I}\,\overline{Z} = (7.9 \angle{-68.96^\circ})(4 + j8) = 70.63 \angle{-5.52^\circ} \text{ V}$$

Note that the load voltage in this case is about 64% of the source voltage. This low voltage is mainly due to the voltage drop across the cable.

4. The real and imaginary components of the load current are

$$\overline{I}_R = \frac{\overline{V}_{\text{load}}}{R} = \frac{70.63 \angle{-5.52^\circ}}{20} = 3.531 \angle{-5.52^\circ} \text{ A}$$

$$\overline{I}_L = \frac{\overline{V}_{\text{load}}}{jX_L} = \frac{70.63 \angle{-5.52^\circ}}{10 \angle 90^\circ} = 7.06 \angle{-95.52^\circ} \text{ A}$$

Note that the reactive current is higher than the real current. This is because X_L is smaller in magnitude than R.

5. The real and reactive power of the load are

$$P_R = I_R^2 R = (3.531)^2 \times 20 = 249.36 \text{ W}$$

$$Q_L = I_L^2 X_L = (7.06)^2 \times 10 = 498.437 \text{ VAr}$$

6. The real power loss of the cable is

$$P_{\text{loss}} = I^2 R_{\text{cable}} = 7.9^2 \times 1 = 62.41 \text{ W}$$

The reactive power loss of the cable is

$$Q_{\text{loss}} = I^2 X_{\text{cable}} = 7.9^2 \times 5 = 312.05 \text{ VAr}$$

7. The real power delivered by the source is the load's real power plus the real power loss of the line.

$$P_s = P_R + P_{loss} = 249.36 + 62.41 = 311.77 \text{ W}$$

Similarly, the reactive power delivered by the source is

$$Q_s = Q_L + Q_{loss} = 498.437 + 312.05 = 810.487 \text{ VAr}$$

EXAMPLE 7.9 Consider the system in Example 7.8, but assume that the load is purely resistive with no inductive elements. Compute the following:

1. The load impedance.
2. The line current.
3. The load voltage.
4. The real and imaginary components of the load current.
5. The real and reactive powers of the load.
6. The real and reactive losses of the cable.
7. The real and reactive powers delivered by the source.

Solution

1. The load impedance Z is nothing but the resistance

$$\overline{Z} = R = 20 \text{ ohm}$$

2. To compute the line current, we need to compute the total impedance of the circuit.

$$\overline{Z}_{total} = \overline{Z} + \overline{Z}_{cable} = 20 + (1.0 + j5) = 21 + j5 \text{ ohm}$$

$$\overline{I} = \frac{\overline{V}_S}{\overline{Z}_{total}} = \frac{110\angle 0^\circ}{21 + j5} = 5.09\angle -13.39^\circ \text{ A}$$

3. The load voltage is

$$\overline{V}_{load} = \overline{I}\,Z = 5.09\angle -13.39^\circ \times 20 = 101.8\angle -13.39^\circ \text{ V}$$

Note that the load voltage in this case is about 92% of the source voltage. This is a much higher magnitude compared with the case in Example 7.8.

4. The real component of the load current is the same as the line current, since the imaginary current of the load is zero.

$$\overline{I}_R = \overline{I} = 5.09\angle -13.39^\circ \text{ A}$$

5. The real power of the load is

$$P_R = I_R^2 R = 5.09^2 \times 20 = 518.16 \text{ W}$$
$$Q_L = 0 \text{ VAr}$$

6. The real power loss of the cable is

$$P_{loss} = I^2 R_{cable} = 5.09^2 \times 1 = 25.91 \text{ W}$$

The reactive power loss of the cable is

$$Q_{loss} = I^2 X_{cable} = 5.09^2 \times 5 = 129.54 \text{ VAr}$$

7. The real power delivered by the source is the real power of the load plus the line losses.

$$P_s = P_R + P_{loss} = 518.16 + 25.91 = 544.07 \text{ W}$$

The reactive power delivered by the source is

$$Q_s = Q_L + Q_{loss} = 0 + 129.54 = 129.54 \text{ VAr}$$

7.7.7 Power Factor Correction

As seen in the previous section, the inductive load increases the cable losses, reduces the voltage across the load, and reduces the spare capacity of the transmission lines. For these reasons, most utilities install capacitors near the service areas to compensate for the reactive power consumed by the load. Figure 7.30 shows two compensation methods: the picture on the left side shows three rectangular capacitors mounted on a distribution pole

On a distribution pole

To a distribution substation

FIGURE 7.30
(see color insert following Page 208) Capacitors for power factor correction.

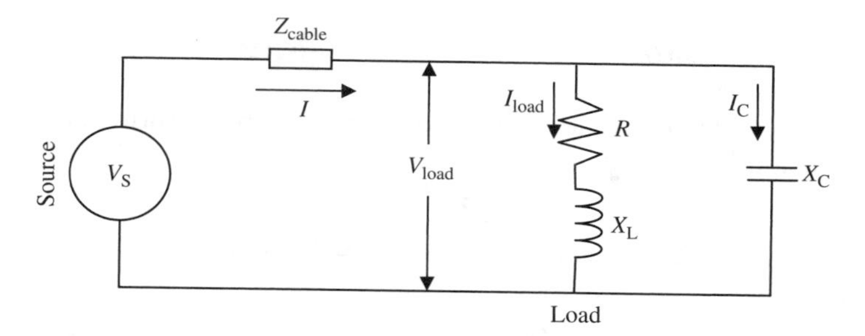

FIGURE 7.31
Phasor diagrams of resistive and inductive loads.

close to a residential area; and the picture on the right side shows a much larger compensation at a substation where the capacitors are mounted on racks.

The circuit in Figure 7.31 shows a source feeding a load through a cable. Across the load, a compensation capacitor is installed. Since the reactive power of the purely inductive load is positive and that of the purely capacitive load is negative, the reactive power of the capacitor annuls the reactive power of the inductor. This process is called *power factor correction*, or *reactive power compensation*. If the two reactive powers sum up to zero, the source delivers only real power to the load, and the negative effects of the reactive current on the transmission line are eliminated.

EXAMPLE 7.10 The circuit in Figure 7.31 has a voltage source of 110 V, 60 Hz. The source is connected to an inductive load through a cable. The load impedance is $\overline{Z} = 3 + j4\ \Omega$, the cable resistance is 1 Ω, and the cable inductive reactance is 2 Ω.

Assume that the capacitor is not connected to the circuit and compute the following:

1. The power factor of the load.
2. The line current.
3. The load voltage.
4. The real and reactive powers of the load.
5. The losses of the cable.

If a 6 Ω capacitive reactance is connected in parallel with the load, compute the following:

1. The power factor of the load plus the capacitor.
2. The line current.
3. The load voltage.
4. The real and reactive powers of the load plus the capacitor.

Solution

Without the capacitor

1. The power factor angle of the load is equal in magnitude to the angle of the load impedance.

$$\bar{Z} = R + jX_L = 3.0 + j4.0 = 5\angle 53.13^\circ \text{ ohm}$$

The power factor angle is

$$\text{pf} = \cos(53.13) = 0.6 \text{ lagging}$$

Note that the power factor must be identified with either lagging or leading. Lagging is when the current lags the voltage, which is the case of the inductive load.

2. To compute the line current, we need to compute the total impedance of the circuit.

$$\bar{Z}_{\text{total}} = \bar{Z} + \bar{Z}_{\text{cable}} = (3.0 + j4.0) + (1.0 + j2.0) = 4.0 + j6.0 \text{ ohm}$$

$$\bar{I} = \frac{\bar{V}_S}{\bar{Z}_{\text{total}}} = \frac{110\angle 0^\circ}{4.0 + j6.0} = 8.461 - j12.69 = 15.254\angle -56.31^\circ \text{ A}$$

3. The load voltage can be computed as

$$\bar{V}_{\text{load}} = \bar{I}Z = (15.254\angle -56.31^\circ)(3 + j4) = 76.27\angle -3.18^\circ \text{ V}$$

Note that the load voltage in this case is just about 69.3% of the source voltage.

4. The real and reactive powers of the load can be computed as

$$P_R = I^2 R = 15.254^2 \times 3 = 698.05 \text{ W}$$

$$Q_L = I^2 X_L = 15.254^2 \times 4 = 930.74 \text{ VAr}$$

5. The losses of the cable are

$$P_{\text{loss}} = I^2 R_{\text{cable}} = 15.254^2 \times 1 = 232.68 \text{ W}$$

$$Q_{\text{loss}} = I^2 X_{\text{cable}} = 15.254^2 \times 2 = 465.36 \text{ VAr}$$

With the capacitor

1. The load impedance Z_l is the load impedance in parallel with the capacitive reactance.

$$\overline{Y}_l = \frac{1}{\overline{Z}} + \frac{1}{\overline{X}_C} = \frac{1}{3.0 + j4.0} + \frac{1}{-j6.0} = 0.12 + j0.007 \text{ mho}$$

$$\overline{Z}_l = \frac{1}{\overline{Y}_{\text{total}}} = 8.32\,\angle{-3.18^\circ} \text{ ohm}$$

The power factor angle is

$$\text{pf} = \cos(3.18) = 0.9985 \text{ leading}$$

Note that the power factor is near unity.

2. Compute the total impedance of the circuit.

$$\overline{Z}_{\text{total}} = \overline{Z}_l + \overline{Z}_{\text{cable}} = 8.32\,\angle{-3.18^\circ} + (1.0 + j2.0) = 9.307 + j1.538 \text{ ohm}$$

$$\bar{I} = \frac{\overline{V}_S}{\overline{Z}_{\text{total}}} = \frac{110\,\angle 0^\circ}{9.307 + j1.538} = 11.505 - j1.901 = 11.66\,\angle - 9.38^\circ \text{ A}$$

Note that the line current with the capacitor installed is 11.66 A, which is smaller than the line current computed in part 2 (15.254 A). This is about 23.5% reduction in the line current, which increases the spare capacity of the cable.

3. The load voltage is

$$\overline{V}_{\text{load}} = \bar{I}\,\overline{Z}_l = 11.66\,\angle{-9.38^\circ}(8.32\,\angle{-3.18^\circ}) = 97.01\,\angle{-12.56^\circ} \text{ V}$$

Note that the load voltage in this case is higher than the case without the capacitor (in part 3).

4. Before we compute the powers, we need to compute the load current I_{load} and the capacitor current I_C.

$$\bar{I}_{\text{load}} = \frac{\overline{V}_{\text{load}}}{\overline{Z}} = \frac{97.01\,\angle{-12.56^\circ}}{5.0\,\angle 53.13^\circ} = 19.4\,\angle{-65.69^\circ} \text{ A}$$

$$\bar{I}_C = \frac{\overline{V}_{\text{load}}}{\overline{X}_C} = \frac{97.01\,\angle{-12.56^\circ}}{6.0\,\angle{-90^\circ}} = 16.17\,\angle 77.44^\circ \text{ A}$$

The real and reactive powers of the load are

$$P_R = I_{load}^2 R = 19.4^2 \times 3 = 1.129 \text{ kW}$$

$$Q_L = I_{load}^2 X_L = 19.4^2 \times 4 = 1.505 \text{ kVAr}$$

$$Q_C = I_C^2 X_C = 16.17^2 \times 6 = 1.569 \text{ kVAr}$$

The total real power P_{total} of the load with the capacitor installed is the real power of the load alone. This is because the capacitor and the inductor consume no real power.

$$P_{total} = P_R = 1.129 \text{ kW}$$

The total reactive power Q_{total} is the phasor sum of the inductive and capacitive powers.

$$\overline{Q}_{total} = \overline{Q}_L + \overline{Q}_C = j1.505 - j1.569 = -j64 \text{ VAr}$$

Note that the total reactive power on the load side without the capacitor is 930.74 VAr.

EXAMPLE 7.11 The voltage source of the circuit in Figure 7.32 is 240 V, 60 Hz. The load consists of 4 Ω resistance in series with a 3 Ω inductive reactance. A capacitor is installed to improve the power factor on the source side to unity. Compute the value of X_C.

Solution

We can solve this problem by several methods. One of them is to compute the reactive power of the load, then compute the value of X_C that produces capacitive reactive power equal in magnitude to the inductive power. Another method is to compute the value of X_C that would set the imaginary component of the total impedance to zero.

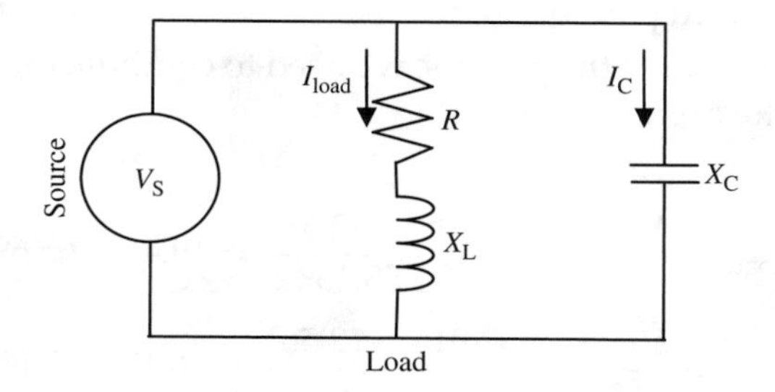

FIGURE 7.32
Phasor diagrams of resistive and inductive loads.

First method:
The current of the load is

$$\bar{I}_{load} = \frac{\overline{V}_S}{\overline{Z}} = \frac{240 \angle 0^\circ}{4.0 + j3.0} = 38.4 - j28.8 = 48 \angle -36.87^\circ \text{ A}$$

The reactive power of the load is

$$Q_L = I_{load}^2 X_L = (48)^2 \times 3 = 6.912 \text{ kVAr}$$

The reactive power of the capacitor must be equal (in magnitude) to the reactive power of the load.

$$Q_C = I_C^2 X_C = \frac{V_S^2}{X_C} = \frac{240^2}{X_C} = 6.912 \text{ kVAr}$$

Hence,

$$X_C = \frac{240^2}{6912} = 8.33\ \Omega$$

Second method:
The other method is to set the imaginary component of the total impedance equal to zero.

$$\text{Im}\left(\frac{1}{\overline{Z}} + \frac{1}{\overline{X}_C}\right) = 0$$

$$\text{Im}\left(\frac{1}{4.0 + j3.0} + \frac{1}{-jX_C}\right) = 0$$

$$0.12 - \frac{1}{X_C} = 0$$

$$X_C = 8.33\ \Omega$$

7.8 Electric Energy

The energy E consumed by a load is the power delivered to the load P over a period of time. The units of energy are watt hours (Wh), kWh, MWh, etc. For constant power during a period τ, the energy is the shaded area under the power line in Figure 7.33.

$$E = P\tau \tag{7.59}$$

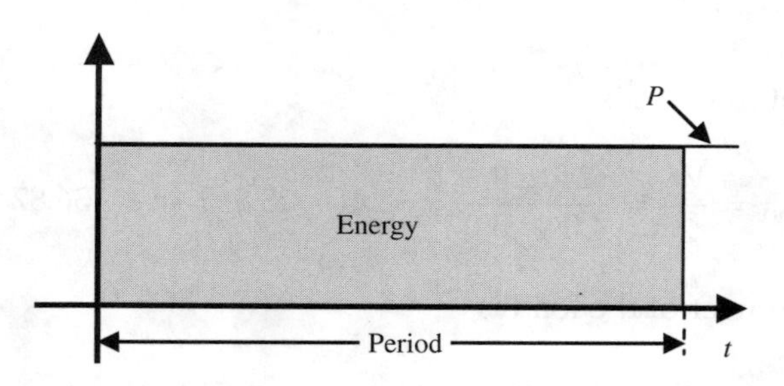

FIGURE 7.33
Energy of constant power.

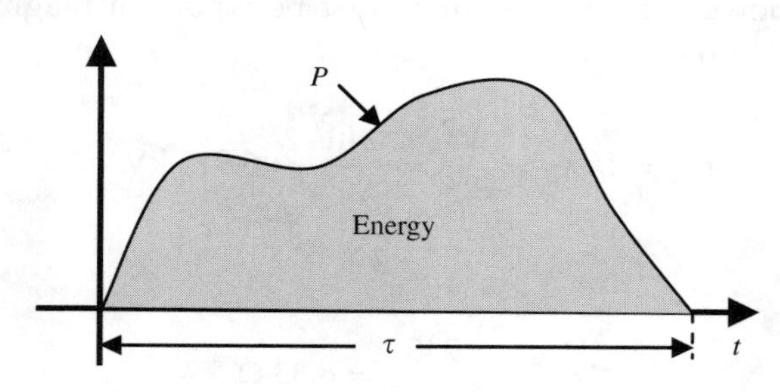

FIGURE 7.34
Energy of variable power.

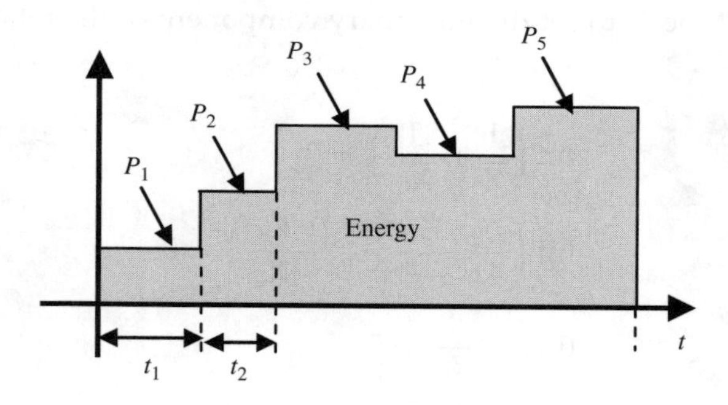

FIGURE 7.35
Energy of discrete power.

For time varying power, such as the one shown in Figure 7.34, the energy during a time period τ is the integral of the power over the period τ.

$$E = \int_0^{\tau} P \, dt \tag{7.60}$$

For discrete power, as in Figure 7.35, the energy is the summation of the power for each time segment.

$$E = \sum_i P_i t_i \tag{7.61}$$

EXAMPLE 7.12 An electric load is connected across a 120 V source. The load impedance is changing over a period of 24 h as follows:

Time Period	Load Impedance (Ω)	Power Factor Angle (degrees)
8:00–10:30 A.M.	10	30
11:00 A.M.–1:00 P.M.	20	0
3:00–5:00 P.M.	15	60
5:00–8:00 P.M.	5	45

Compute the energy consumed by the load during this 24-h period.

Solution

Time Period	Load Impedance Z (Ω)	Pf Angle θ (degrees)	Load Current V_S/Z (A)	Load Power $V_S I \cos\theta$ (W)	Time Period (h)	Energy (Wh)
8:00–10:30 A.M.	10	30	12	1247.08	2.5	3117.7
11:00 A.M.–1:00 P.M.	20	0	6	720	2	1440
3:00–5:00 P.M.	15	60	8	480	2	960
5:00–8:00 P.M.	5	45	24	2036.47	3	6109.41
Total Energy						11627.11

EXAMPLE 7.13 A load is connected across a 120 V source. The load power is represented by

$$P = 25 + 100\sin(20t) \text{ kW}$$

where t is the time in hours. Compute the energy consumed by the load after 1 h and after 2 h.

Solution

$$E = \int_0^{\tau} P\,\mathrm{d}t = \int_0^{\tau} (25 + 100\sin(20t))\,\mathrm{d}t = \left[25t - 100\frac{\cos(20t)}{20}\right]_0^{\tau}$$

$$E = 25\tau - 5\cos(20\tau) + 5$$

After 1 h, $\tau = 1$.

$$E = 25(1) - 5\cos(20) + 5 = 25.3 \text{ kWh}$$

After 2 h, $\tau = 2$.

$$E = 25(2) - 5\cos(40) + 5 = 51.17 \text{ kWh}$$

Exercise

7.1. Given the following two vectors:

$$\overline{A} = 12 + j12$$
$$\overline{B} = -6 + j10$$

compute $\overline{A} + \overline{B}$, $\overline{A} - \overline{B}$, $\overline{A}\,\overline{B}$, and $\overline{A}/\overline{B}$.

7.2. The voltage and current equations of an electric load are

$$v = V_{\max} \sin \omega t$$
$$i = I_{\max} \cos(\omega t - 30^0)$$

Compute the phase shift of the current.

7.3. The voltage and current of an electric load can be expressed by the following equations:

$$v = 340 \sin(628.318t + 0.5236)$$
$$i = 100 \sin(628.318t + 0.87266)$$

Calculate the following:

a. The rms voltage.
b. The frequency of the current.
c. The phase shift between current and voltage in degrees.
d. The average voltage.
e. The load impedance.

7.4. The current and voltage waveforms of an electric circuit are

$$i(t) = 25 \sin\left(377t + \frac{\pi}{3}\right) A; \qquad v(t) = 169.7 \sin\left(377t - \frac{\pi}{6}\right) \text{V}$$

Compute the following:

a. The rms voltage.
b. The rms current.
c. The frequency of the supply voltage.

d. The phase angle of the current with respect to voltage (indicate leading or lagging).
e. The real power consumed by the circuit.
f. The reactive power consumed by the circuit.
g. The impedance of the circuit.

7.5. An electric load consists of a 4 Ω resistance, a 6 Ω inductive reactance, and an 8 Ω capacitive reactance connected in series. The total impedance of the load is connected across a voltage source of 120 V.

a. Compute the power factor of the load.
b. Compute the source current.
c. Compute the real power of the circuit.
d. Compute the reactive power of the circuit.

7.6. A load impedance consists of a 25 Ω resistance in series with a 38 Ω inductive reactance. The load is connected across a 240 V source. Compute the real and reactive powers consumed by the load.

7.7. A voltage source $\overline{V} = 120\angle 30°$ is connected across a circuit consisting of $R = 3.5\ \Omega$ in series with $L = 0.0833$ H. If the circuit is energized in the United States, compute the following:

a. The load current.
b. The load power factor.
c. The real power consumed by the load.
d. The reactive power consumed by the load.

7.8. A voltage source $\overline{V} = 120\angle 30°$ is connected to a circuit consisting of $R = 3.5\ \Omega$ in series with $L = 0.0833$ H. If the circuit is energized in Europe, compute the following:

a. The load current.
b. The load power factor.
c. The real power consumed by the load.
d. The reactive power consumed by the load.

7.9. An inductive load consists of $R_1 = 3\ \Omega$ in series with $X_{L1} = 4\ \Omega$. The load is connected in parallel with another load of unknown impedance. The voltage source of the system is 120 V. The total real and reactive powers delivered by the source are 3 kW and 2 kVAr, respectively. Compute the impedance of the unknown load.

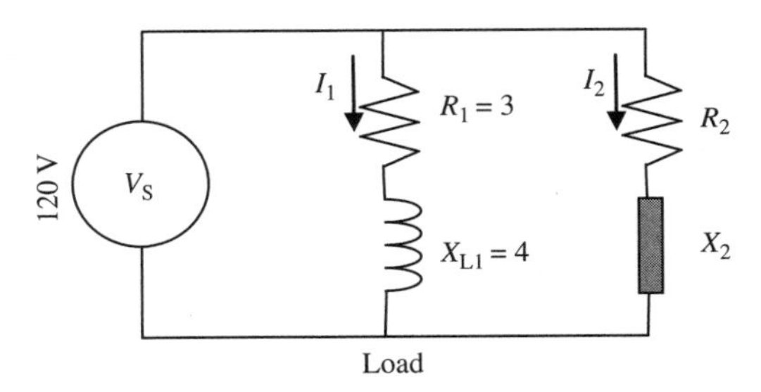

7.10. The rms voltage and current of an inductive load are 110 V and 10 A, respectively. The frequency of the voltage waveform is 60 Hz. The instantaneous power consumed by the load has zero average value. Calculate the following:

a. The real (active) power consumed by the load.
b. The reactive power consumed by the load.
c. The power factor.
d. The frequency of the reactive power.

7.11. An electric load consists of a 5 Ω resistance, a 20 Ω capacitive reactance, and a 10 Ω inductive reactance, all connected in parallel. A voltage source of 100 V is applied across the load. Compute the following:

a. The real power of the load.
b. The reactive power of the load.
c. The current of the load.

7.12. An inductive load consists of a 1 Ω resistance and a 1.0 mH inductance connected in series. The load is connected across a 120 V, 60 Hz source.

a. Compute the power factor of the load.
b. A capacitor is connected in parallel with the entire load so that no reactive power is delivered by the source. Compute the value of the capacitance.

7.13. The power consumption of an electric load is represented by $P(t) = 100(1 - e^{-t/10})$ kW, where t is the time of the day in hours. Compute the energy consumed by the load in a period of 24 h.

8

Electric Safety

Electricity is probably the best form of energy known to man; it is clean, distributed, readily available, quiet, and highly reliable. The popularity of electricity makes it available all around and hence, it can be acquired by just flipping a switch. Furthermore, the equipment powered by electricity is pollution-free and is more compact than those powered by other energy forms such as gas or oil.

Electricity is completely safe if used properly. However, when the electrical equipment is partially damaged or improperly used, hazardous conditions can develop. For example, loose-fitting plugs can overheat and lead to fire, water intrusion inside a plugged-in appliance can lead to electrical shock, and cracks or damage in the electrical insulations can lead to fires and/or electrical shock.

Since the early days of the *electrical revolution*, electricity was recognized as hazardous to humans and animals. Today, even with the safe products in the market, more than 1000 people are killed each year in the United States due to electric shocks and several thousands more are injured. The popular myth that only high voltage is dangerous to humans makes some of us unfortunately careless when we use regular household equipment. Indeed, most electric shocks occur at household low-voltage levels.

The potential hazards of electricity make the manufacturers of electrical equipment very sensitive to the safety of their products. Several regulations and standards are developed to address every electric safety issue known to man so far. In the United States, the *Occupational Safety and Health Administration* (OSHA) imposes and enforces the basic safety standards. On a more global level, the *Institute of Electrical and Electronics Engineers* (IEEE) sets several standards for various electric safety issues which are normally adopted by OSHA and other similar agencies all over the world.

8.1 Electric Shock

Electric shocks occur when people become part of electrical circuits and accordingly electric currents flow through their bodies. The electric current can cause a wide range of harmful effects on humans and animals ranging from minor sensations to death. The current flowing inside a body can overheat the cells leading to internal and external burns. The most sensitive organs to this effect are the lungs, brain, and heart. The degree of injury depends on several factors such as the magnitude of the current, the duration of the shock, and the pathway of the current and its frequency. These factors are discussed in detail later in this chapter.

Generally, electric shocks are divided into two categories: *secondary shocks* and *primary shocks*. The secondary shock is due to low currents that may cause pain without direct physical harm. The primary shock, however, is due to higher currents that produce direct physical harm and death.

Most of us probably experience secondary shock when we use some types of appliance or equipment. Sometimes, when we lightly touch a charged object, we may experience tingling effects due to the small current passing through our fingers. If we grip that object, the current is spread out over a wide contact area and we may not feel anything.

The incredible question is how to find out the safe limit of the electrical current; it is hard to find volunteers for primary shock experiments! Early researchers, however, did their secondary shock studies on human volunteers and the primary shock experiments on animals with similar weight and general biological characteristics as humans. It was also reported that some bizarre research on primary shocks was done on inmates condemned to death by the electric chair.

Among the highly respected researchers in this area was Charles Dalziel from the University of California, Berkeley. Dalziel did extensive lab tests during the middle of the 20th century. In one of his papers, published in 1972, Dalziel summarizes the results of his various lab tests, which until today are used as the de facto limits. In his study, he had several male and fewer female volunteers. Figure 8.1 shows one of Dalziel's tests published in his 1972 paper. The caption of the figure states that "Muscular reaction at the let-go current value is increasingly severe and painful, as shown by the subject during laboratory tests." Today, many people find this type of test incredible.

8.1.1 Current Limits of Electric Shocks

The research results of Dalziel and other researchers are summarized in Tables 8.1 and 8.2. Table 8.1 shows the levels of dc and ac currents that produce various types of secondary shocks on men and women. Although the secondary shock can be painful, it is not life threatening. Table 8.2 shows the effects of primary shocks and their ac current limits.

As seen in Table 8.1, the following conclusions regarding the secondary shocks can be made:

- Women are more sensitive to electric shocks than men.
- It takes less of ac current than dc to produce the same effect on humans.

FIGURE 8.1
A volunteer during secondary electric shock test. (Charles F. Dalziel, "Electric Shock Hazard," IEEE spectrum, 1972—courtesy of IEEE.)

TABLE 8.1
Effects of ac and dc Secondary Shock Currents (IEEE Standard 524a–1993)

	Current (mA)			
	dc		ac	
Reaction	**Men**	**Women**	**Men**	**Women**
No sensation on hand	1.0	0.6	0.4	0.3
Tingling (threshold of perception)	5.2	3.5	1.1	0.7
Shock: uncomfortable, muscular control not lost	9.0	6.0	1.8	1.2
Painful shock, muscular control is not lost	62.0	41.0	9.0	6.0

TABLE 8.2
Threshold Limit of ac Primary Shock Current (IEEE Standard 1048–1990)

	Current (mA)			
	0.5% of Population		50% of Population	
Threshold	**Men**	**Women**	**Men**	**Women**
Let-go: worker cannot release wire	9	6	16	10.5
Respiratory tetanus: breathing is arrested	–	–	23	15
Ventricular fibrillation: heart stops	100	67	–	–

- It takes as little as 1.2 mA for women and 1.8 mA for men to produce uncomfortable effects.
- For a current less than 6 mA in women and 9 mA in men, the person, while in pain, can still control his/her muscles.

The more critical statistics are the ones related to the primary shock shown in Table 8.2. In this table, the population is divided into sensitive and average sets. The sensitive population, about 0.5% of the total population, is harmed at lower currents than the rest of the population. The average population forms 50% of the total population. The table shows three key effects for the primary shocks: *let-go*, *respiratory tetanus*, and *ventricular fibrillation*.

When a person grips an energized conductor, and the current passing through his muscles is about 16 mA (for 50% of the male population), the person may not be able to control his muscles and hence may not release the gripped conductor. In this case, the current passes through the person's body for long time and may increase to a lethal limit unless the circuit is interrupted by a protective device or by another person.

A current above the let-go threshold passing through the chest can cause an involuntary contraction of the muscles, which will arrest breathing as long as the current continues to flow. This is known as *respiratory tetanus*. If the current passing through the chest disturbs the heart's own electrical stimulation, the heart experiences an uncontrolled vibration, and may even cease to beat. This is known as *ventricular fibrillation*.

8.1.2 Factors Determining the Severity of Electric Shocks

The harmful effect of electric shock depends mainly on six factors:

1. The voltage level of the gripped or touched equipment.
2. The amount of current passing through the person's body.
3. The resistance of the person's body.
4. The pathway of the current inside the body.
5. The duration of the shock.
6. The frequency of the source.

8.1.2.1 Effect of Voltage

Most electric shocks occur at 100 to 400 V because it is readily available to everyone, it is high enough to produce a significant current in the body, and can cause muscles to contract tightly to the energized object.

The higher voltage, in the kilovolts range, is even more lethal because it causes high currents to pass through the person, but the access to this voltage is normally limited to professionals such as linemen. The general public rarely comes in contact with high-voltage wires, unless a wire is downed, or a person climbs a high object and touches the wire or something like that. Therefore, the number of deaths from high voltage is much less than that from low voltage.

When a person touches a high-voltage wire, fierce involuntary muscle contractions may throw the person away from the hazard. However, the high current may cause fatal deep burns.

8.1.2.2 Effect of Current

Contrary to popular belief, it is current, not voltage, that causes death. However, the current is a function of the voltage and the impedance of the body; the current through the body is equal to the voltage applied to the body divided by the impedance of the body.

Electric currents passing through tissues produce thermal heating proportional to the square of the current ($P = I^2R$). Excessive thermal heating can permanently damage tissues and organs in the person's body. The electric currents can also interfere with the normal operation of the heart and lungs, causing the heart to beat out of step and the lungs to function irregularly.

8.1.2.3 Effect of Body Resistance

Body resistance is highly nonlinear and is a function of several factors such as the hydration condition of the body, skin condition, and fat concentration. The palm resistance, for example, can range from 100 Ω to 1 MΩ depending on the skin condition of the person. Dry skins tend to have higher resistance and sweat tends to lower the resistance. Nerves, arteries, and muscles are low in resistance, while bone, fat, and tendons are relatively high in resistance.

The IEEE established the ranges for various body resistances shown in Table 8.3. These numbers can be used to roughly estimate the current through the body. However, the variability in human body resistance could make the results inaccurate for people with body resistances outside the specified range. Therefore, when designing electric safety equipment, lower values of body resistance should always be used.

TABLE 8.3
Body Resistance in Ohms (IEEE Standard 1048–1990)

	Hand-to-Hand		Hand-to-Feet
Resistance	**Dry Condition**	**Wet Condition**	**Wet Condition**
Maximum	13,500	1,260	1,950
Minimum	1,500	610	820
Average	4,838	865	1,221

8.1.2.4 Effect of Current Pathway

Current passing through the skin is not as harmful as current passing through vital organs. The fatal currents often pass through the heart, lungs, and brain. A mere 10 μA passing directly through the heart can cause cardiac arrest. The heart muscles could beat out of step resulting in insufficient blood being pumped through the body. A current in the spinal cord may also alter the respiratory control mechanism.

8.1.2.5 Effect of Shock Duration

The longer the duration of the shock the higher is the likelihood of death. This is because the thermal heat inside the tissues is a form of energy, and therefore proportional with time. Also, if the current interferes with the operation of the heart or lungs, the longer the duration, the longer is the chance for death from respiratory or cardiac arrest. When the current is above the let-go threshold, the person is incapable of releasing his/her grip on the wire and the shock duration is often long.

Charles Dalziel carried out research on the time–current relationship for primary shocks. He and other researchers obtained their results by experimenting on animals of weights and organ sizes similar to humans. Although inconclusive, the data of these experiments are the best available information to date. Based on these studies, Dalziel developed the following current duration formula for ventricular fibrillation.

$$I = \frac{K}{\sqrt{t}} \tag{8.1}$$

I is the ventricular fibrillation current in milliamperes, t is the time duration of the current in seconds, and K is a constant value that depends on the weight of the test subject: for people weighing less than 70 kg (154 lb), $K = 116$; and for people weighing more than 70 kg, $K = 157$.

Figure 8.2 is a graph of Dalziel's formula. As seen in the figure, it takes a very short time to induce ventricular fibrillation. For instance, if the current is about 80 mA, it takes about 2 s to kill a person weighing less than 70 kg, while a person weighting over 70 kg may survive for about 4 s. In either case, the survival duration is very short for such a little current.

8.1.2.6 Effect of Frequency

As seen in Table 8.1, ac current is more dangerous than dc current. The current at power frequencies (50 to 60 Hz) has a much greater ability to cause ventricular fibrillation than dc current. Also, at power frequencies, involuntary muscle contractions could be so severe that the individual cannot let go of the energized object. At frequencies above 500 kHz, little energy passes through the internal organs. Higher frequencies at the gigahertz range (microwaves, x-rays, and gamma rays) are extremely dangerous as they

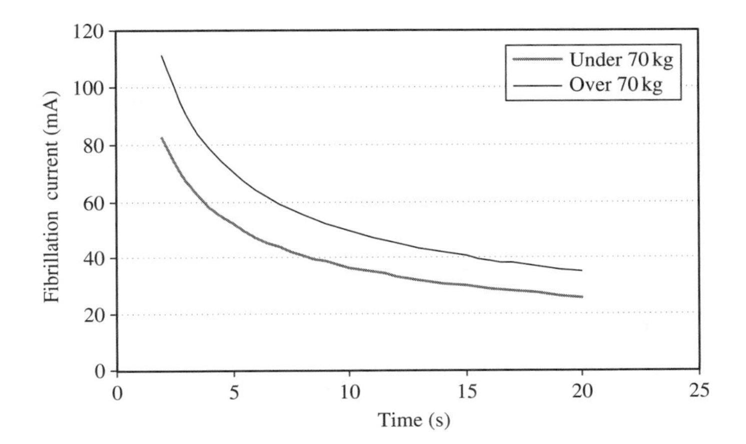

FIGURE 8.2
Ventricular fibrillation current as a function of shock duration.

are ionizing frequencies, which means they have enough energy to break molecules apart.

EXAMPLE 8.1 A child climbs a tree to retrieve his kite which is tangled on a utility line. The voltage of the line is 240 V. The resistance of the wire from the source to the location of the kite is 0.2 Ω. The tree has a resistance of 500 Ω. The soil resistance is 300 Ω. Assume the child resistance is 2000 Ω. If the child touches the power line with his hand, estimate the current through his body. Also, estimate the time it takes to induce ventricular fibrillation.

Solution

If you trace the path of the current starting from the source, you can obtain the electrical circuit shown in Figure 8.3. The current passing through the child starts at the source, passes through the line to the child, then to the tree, and finally to the ground. The current in this case is

$$I = \frac{V}{R_l + R_b + R_t + R_g} = \frac{240}{0.2 + 2000 + 500 + 300} = 85.7 \text{ mA}$$

According to Dalziel's formula, the child will only survive for

$$t = \left(\frac{K}{I}\right)^2 = \left(\frac{116}{85.7}\right)^2 = 1.83 \text{ s}$$

Note that we used $K = 116$, as the child is assumed to weigh less than 70 kg. Also notice that the wire resistance has very little effect on the current since it is much smaller than the other resistances. One important resistance is the

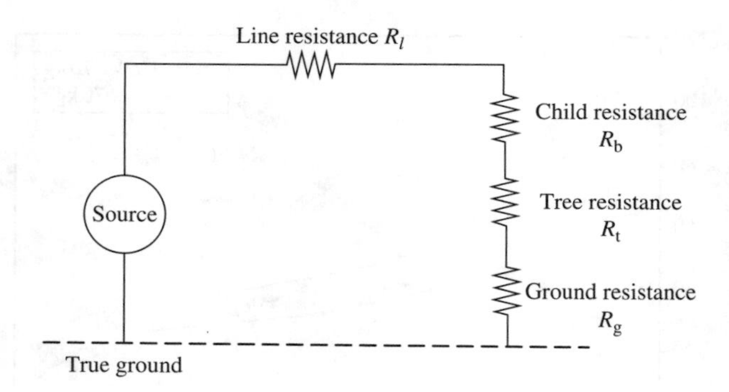

FIGURE 8.3
Model representation of the accident.

ground resistance. It varies widely and can have a major effect on the value of the current.

8.2 Ground Resistance

The center of the Earth is the only absolute zero potential spot; any other location inside the Earth has nonzero potential. Hence, any object on the surface of the Earth such as water pipes, building foundations, and steel structures has nonzero potential. Figure 8.4 shows a simple schematic of the Earth with an object at the earth's surface. The object, which is in contact with the soil, has a potential higher than zero. Hence, the potential difference V between the object and the center of the Earth can be represented by Ohm's law $V = IR$, where I is the current flowing from the object to the center of the Earth, and R is the resistance between the object and the center of the Earth. This resistance R is called the *ground resistance* of the object.

The ground resistance of an object is a function of several factors, such as the shape of the object, the type of ground soil, and the dampness of the soil. Furthermore, the ground resistance is nonlinear and is frequency dependent. An exact computation of the ground resistance of an object is a very tedious task and requires the knowledge of highly varying parameters that are hard to measure under all possible conditions. Nevertheless, a simplified computation of the ground resistance of an object can be made with reasonable degree of accuracy as shown below.

Assume the simple case in Figure 8.5 of a hemisphere buried in the soil. Assume that the soil is homogeneous with a resistivity ρ. If the hemisphere is connected to a conductor that carries a current I, the current enters the hemisphere and is then distributed uniformly through the Earth. The current

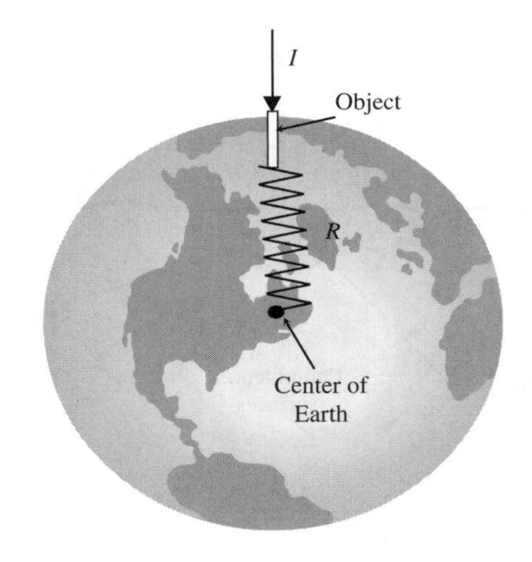

FIGURE 8.4
Ground resistance.

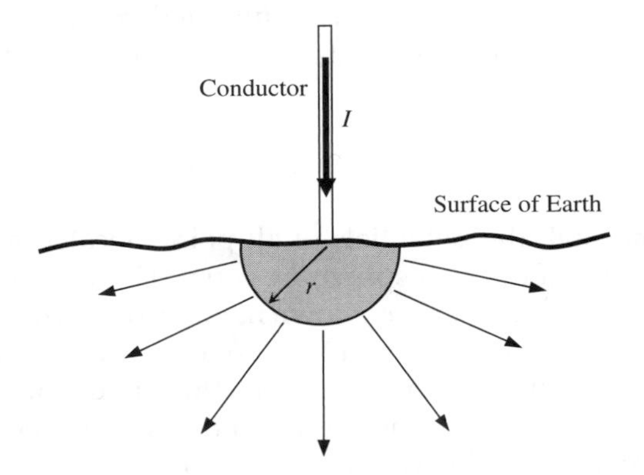

FIGURE 8.5
Current distribution of a hemisphere.

density at the surface of the hemisphere J is the current leaving the hemisphere divided by the surface area of the hemisphere.

$$J = \frac{I}{2\pi r^2} \tag{8.2}$$

Assuming the soil to be homogeneous, the current density at any point x from the center of the hemisphere, and outside the hemisphere, can be computed as

$$J(x) = \frac{I}{2\pi x^2}; \quad x \geq r \tag{8.3}$$

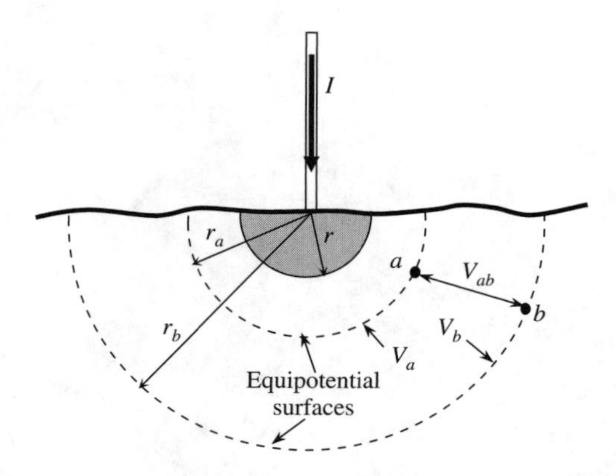

FIGURE 8.6
Equipotential surface.

Ohm's law states that the current in a medium creates an electric field intensity E that is equal to the current density multiplied by the resistivity of the medium. Hence, the electric field intensity $E(x)$ at any distance x outside the hemisphere is

$$E(x) = \rho J(x); \quad x \geq r \tag{8.4}$$

For homogeneous soil, the potentials of all points located at the same distance from the center of the hemisphere are equal. The surface of these points is known as the *equipotential surface*, shown in Figure 8.6. The potential difference between any two points (a and b) inside the Earth can be computed by integrating the electric field intensity between these two points. Hence, V_{ab} in the figure is the potential difference between the two equipotential surfaces where a and b are located. This voltage can be computed as

$$V_{ab} = \int_{x=a}^{x=b} E(x)\,\mathrm{d}x = \int_{x=a}^{x=b} \rho J(x)\,\mathrm{d}x = \frac{\rho I}{2\pi}\left[\frac{1}{r_a} - \frac{1}{r_b}\right] \tag{8.5}$$

where r_a and r_b are the distances of points a and b from the center of the hemisphere, respectively. Hence, the resistance between a and b is

$$R_{ab} = \frac{V_{ab}}{I} = \frac{\rho}{2\pi}\left[\frac{1}{r_a} - \frac{1}{r_b}\right] \tag{8.6}$$

The ground resistance of the hemisphere R_g is the resistance between the surface of the hemisphere and the center of the Earth. This can be computed by Equation (8.6) by setting r_a equal to the radius of the hemisphere, and r_b to ∞.

$$R_g = \frac{\rho}{2\pi r} \tag{8.7}$$

TABLE 8.4

Ground Resistance of Common Objects

Object	Ground Resistance	Parameters
Rod	$\frac{\rho}{2\pi l}\ln\left(\frac{2l+r}{r}\right)$	l is the length of the rod. r is the radius of the rod.
Circular plate at the surface	$\frac{\rho}{4r}$	r is the radius of the disk.
Buried wire	$\frac{\rho}{2\pi l}\left(\ln\left(\frac{l}{r}\right)+\ln\left(\frac{l}{2d}\right)\right)$	l is the length of the wire. r is the radius of the wire. d is the depth at which the wire is buried.

TABLE 8.5

Soil Resistivity

	Soil composition			
	Wet organic	Moist	Dry	Bedrock
Resistivity ρ (Ωm)	10	100	1000	10,000

The ground resistance can also be computed for other shapes; however, the process is more involved. Table 8.4 shows the ground resistance of some common objects.

Table 8.5 shows the resistivity of various soil compositions. Note that the value of the resistivity is substantially reduced if the soil is wet and/or organic.

EXAMPLE 8.2 A hemisphere 2 m in diameter is buried in wet organic soil. Compute the ground resistance of the hemisphere. Also compute the ground resistance at 2, 10, and 100 m away from the center of the hemisphere.

Solution

$$R_g = \frac{\rho}{2\pi r} = \frac{10}{2\pi 1} = 1.6\ \Omega$$

The resistance between two points can be computed by using Equation (8.6).

$$R_{ab} = \frac{\rho}{2\pi}\left[\frac{1}{r_a} - \frac{1}{r_b}\right]$$

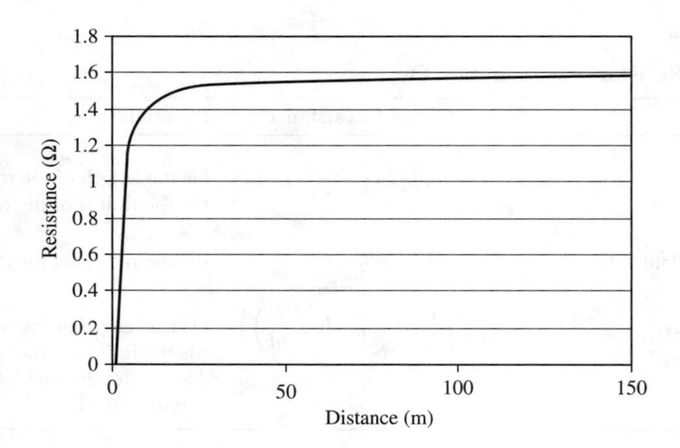

FIGURE 8.7
Ground resistance between the center of the hemisphere and points at various distances.

At 2 m,

$$R_{ab1} = \frac{10}{2\pi}\left[\frac{1}{1} - \frac{1}{2}\right] = 0.8\ \Omega$$

At 10 m,

$$R_{ab10} = \frac{10}{2\pi}\left[\frac{1}{1} - \frac{1}{10}\right] = 1.43\ \Omega$$

At 100 m,

$$R_{ab100} = \frac{10}{2\pi}\left[\frac{1}{1} - \frac{1}{100}\right] = 1.57\ \Omega$$

Figure 8.7 shows the ground resistance between the hemisphere and points at various distances. Note that the change in ground resistance is insignificant when the distance from the center of the hemisphere increases beyond 10 m. So practically, the ground resistance of an object is a function of the immediate distance, rather than the distance to the center of the Earth.

8.2.1 Measuring Ground Resistance

The ground resistance can be measured by using a technique known as the *fall-of-potential* method. The method, which is also known as the *three-point test,* is explained in Figure 8.8. The setup consists of the object whose ground resistance is to be determined, a current electrode, and a potential probe. The electrode is a small copper rod that is driven into the soil. A voltage source is connected between the object and the current electrode. A voltmeter is

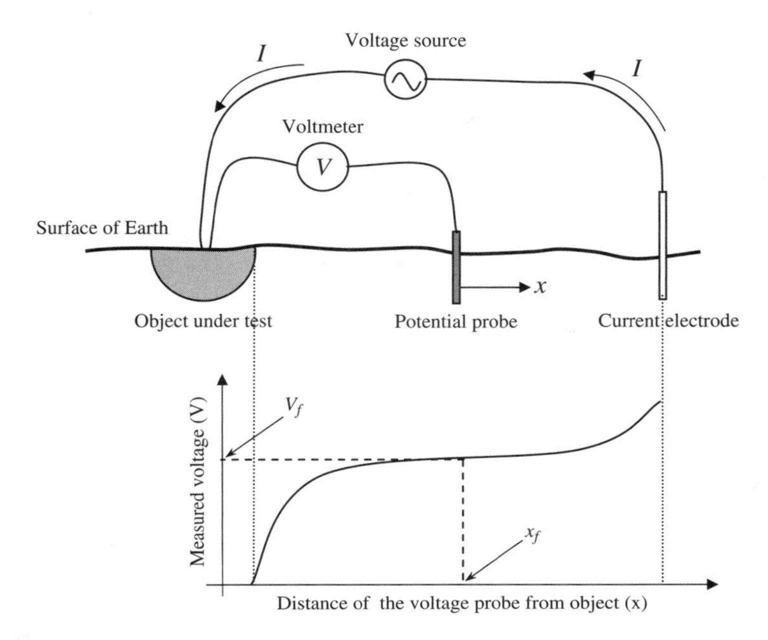

FIGURE 8.8
Fall-of-potential method to measure ground resistance.

connected between the object and the potential probe. The potential probe moves along the line between the object and the current electrode. At each position, the current (of the current electrode) and the voltage (of the potential probe) are recorded and plotted as shown at the bottom of Figure 8.8. When the potential probe touches the object under test, the measured voltage is zero, and when it touches the current electrode, the voltage is equal to the source voltage. The magnitude of the measured voltage is a nonlinear function with respect to the distance x. However, the voltage is often unchanged for a wide range of x as shown in the flat region in the figure. At the middle of this region, V_f is recorded at a distance x_f. The ratio V_f/I is the ground resistance of the object.

8.2.2 Ground Resistance of People

A person standing on the ground has a ground resistance under his/her feet as shown in Figure 8.9. Each foot has a ground resistance R_f between the bottom of the foot and the center of the Earth. The trick is how to model the bottom of a shoe. An approximation is often made by using a circular plate that has the same area as the footprint of an average person. Using the ground resistance equation of the circular plate given in Table 8.4, the ground resistance of a single foot R_f is

$$R_f = \frac{\rho}{4r} \tag{8.8}$$

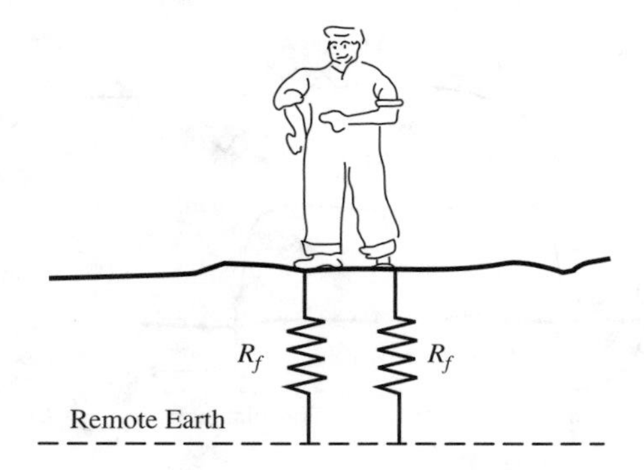

FIGURE 8.9
Feet resistance.

The area of the plate is

$$A = \pi r^2 \tag{8.9}$$

Assuming an approximate area for a footprint of about 0.02 m^2, we can compute the ground resistance in Equation (8.8) as

$$R_f = \frac{\rho}{4\sqrt{A/\pi}} = \frac{\rho}{4\sqrt{0.02/\pi}} \approx 3\rho \tag{8.10}$$

The total ground resistance of a walking person and standing person are different. If you assume that the standing person has his feet close enough to each other, then the total ground resistance R_g is the parallel combination of two R_f.

$$R_g = \frac{R_f \times R_f}{R_f + R_f} = 0.5R_f \approx 1.5\rho \tag{8.11}$$

8.3 Touch and Step Potentials

The direct hazard of electricity is due to an energized object touching a person. However, there are indirect hazards that are less obvious but equally dangerous, such as that due to excessive touch and step potentials. The touch potential could be present if a person touches a metallic structure discharging current to the ground. If the person is standing on the ground, he may have a potential difference between his hand and feet. This voltage could be hazardous.

When a person is walking adjacent to an object that discharges current into the ground, the person could have a potential difference between his two feet. This voltage could also be hazardous.

8.3.1 Touch Potential

To explain the hazards of touch potentials, let us study a simple but well known structure — the tower of the power line shown in Figure 8.10. The tower is built out of steel trusses with several cross arms. On the cross arms, insulators are mounted. On the other side of each insulator, a high-voltage wire is attached. In areas known for their lightning activities, ground wires (also known as static wires) are installed on top of the tower to protect the energized lines from being hit by lightning bolts, thus save the system from severe damages and blackouts. The ground wires are often bonded to the tower structure, which is grounded through a local ground under the tower footage.

Although made of steel and grounded locally, the tower structure is not always at zero potential, because the structure could carry current owing to a number of reasons, as explained by the following three scenarios:

Scenario 1: Because of the presence of energized lines in the vicinity of a ground wire, a voltage is induced on the ground wire. The ground wire discharges its acquired energy through all adjacent grounds including the tower ground. This may result in currents passing through the structure hardware to the local ground underneath the tower.

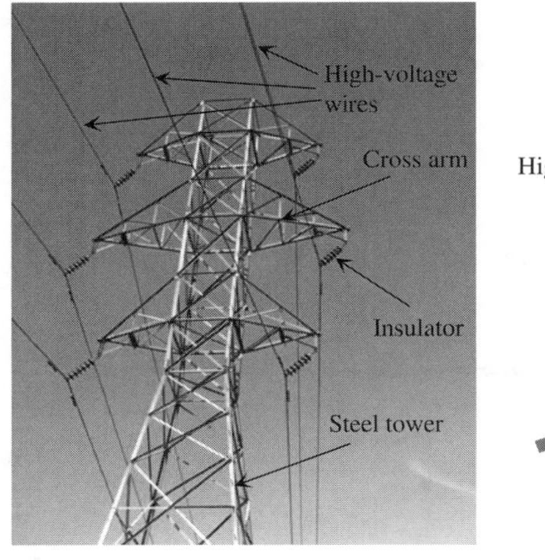

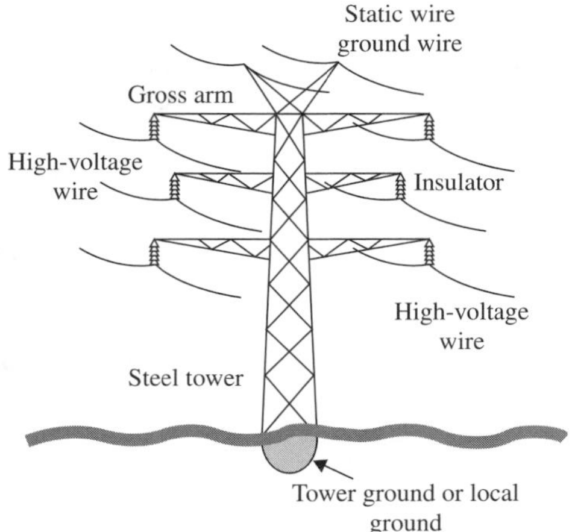

FIGURE 8.10
(see color insert following Page 208) Main components of a steel power line.

Scenario 2: In humid environments, especially close to seas, the surface of an insulator could become contaminated with salty moisture. This mix can provide a path for the current from the energized line to the tower structure, as shown in Figure 8.11. This is known as insulator flashover.

Scenario 3: While linemen are working on power lines, accidents could happen when an energized conductor comes in contact with the structure of the tower.

Now assume a person is touching the steel tower while standing on the ground, as shown in Figure 8.12. If the structure carries a current I, the person and part of the tower are in parallel. Part of the current passes through the structure and goes to the ground of the tower (I_{tg}); the other part goes through the man touching the tower (I_{man}). Keep in mind that the ground resistance of the man is 0.5 R_f as given in Equation (8.11).

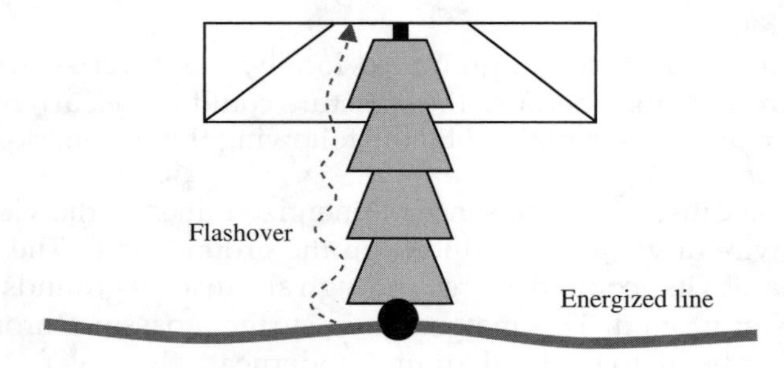

FIGURE 8.11
Flashover due to conductive depositions on the insulators.

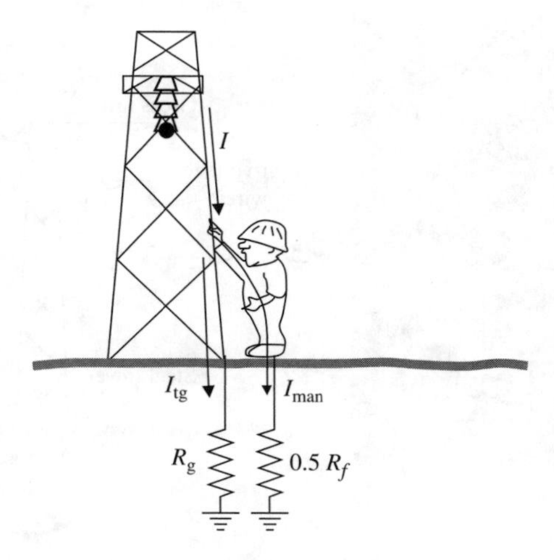

FIGURE 8.12
Touch potential.

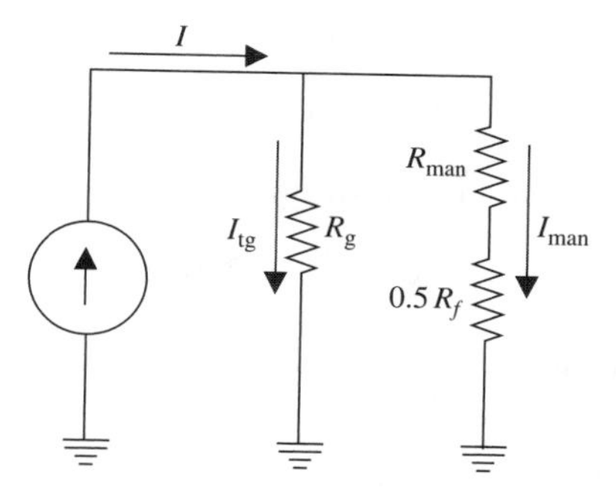

FIGURE 8.13
Circuit diagram of Figure 8.12.

Figure 8.13 shows a circuit diagram representing the case in Figure 8.12. Let us assume that we need to compute the current going through the man. This can be done by using the current divider equation

$$I_{\text{man}} = I \frac{R_g}{R_g + R_{\text{man}} + 0.5R_f} \tag{8.12}$$

The potential of the structure under the above condition is known as the *ground potential rise* (GPR). It is the voltage that the grounded object attains with respect to the remote earth.

$$\text{GPR} = I_{\text{tg}} R_g \tag{8.13}$$

EXAMPLE 8.3 A power line insulator partly fails and 10 A pass through the tower structure to the ground. Assume that the tower ground is a hemisphere with a radius of 0.5 m, and the soil surrounding the hemisphere is moist.

1. Compute the voltage of the tower.
2. Assume that a man with a body resistance of 3 kΩ touches the tower while standing on the ground. Compute the current passing through the man.
3. Use Dalziel's formula and compute the man's survival time.

Solution

1. Compute the ground resistance of the hemisphere.

$$R_g = \frac{\rho}{2\pi r} = \frac{100}{2\pi \times 0.5} = 32\ \Omega$$

The voltage of the tower is the ground potential rise of the tower.

$$V = IR_g = 10 \times 32 = 320 \text{ V}$$

2. To compute the current through the man, first compute R_f.

$$R_f = 3\rho = 300\ \Omega$$

The current through the man is given in Equation (8.12).

$$I_{man} = I\frac{R_g}{R_g + R_{man} + 0.5R_f} = 10\frac{32}{32 + 3000 + 150} = 100 \text{ mA}$$

3. According to Dalziel's formula, the man can survive for

$$t = \left(\frac{K}{I_{man}}\right)^2 = \left(\frac{157}{100}\right)^2 = 2.5 \text{ s}$$

EXAMPLE 8.4 Repeat Example 8.3 assuming that a grounding rod is used instead of the hemisphere. Assume that the rod is 4 cm in diameter and is driven 1 m into the ground.

Solution

1. Compute the ground resistance of the rod.

$$R_g = \frac{\rho}{2\pi l}\ln\left(\frac{2l + r}{r}\right) = \frac{100}{2\pi \times 1}\ln\left(\frac{2 + 0.02}{0.02}\right) = 73.45\ \Omega$$

The voltage of the tower is the ground potential rise of the tower.

$$V = IR_g = 10 \times 73.45 = 734.5 \text{ V}$$

Note that the GPR of a ground rod is much higher than the GPR of a hemisphere.

2. $I_{man} = I\frac{R_g}{R_g + R_{man} + 0.5R_f} = 10\frac{73.45}{73.45 + 3000 + 150} = 227.9 \text{ mA}$

3. According to Dalziel's formula, the man can survive for

$$t = \left(\frac{K}{I_{man}}\right)^2 = \left(\frac{157}{227.9}\right)^2 = 475 \text{ ms}$$

If the ground rod is driven 1 m into the soil, it provides less protection than the hemisphere. Repeat the problem by assuming that the ground rod is driven 2 m into the soil. Can you develop a conclusion?

8.3.2 Step Potential

A person walking adjacent to a structure that passes large currents into the ground could be vulnerable to electric shocks. This is particularly hazardous in substations, near power line towers that carry high currents, or during lightning storms.

Figure 8.14 shows a person walking near a structure that discharges current into the ground. The ground current may cause a high enough potential difference between the person's feet resulting in current through his legs and abdomen. This current can be computed using Thevenin's theorem, where Thevenin's voltage V_{th} is the open circuit voltage between point a and b, which is the same as V_{ab} in Equation (8.5) for a hemisphere ground. r_a in this case is the distance from the center of the hemisphere to the person's rear foot, and r_b is the distance from the center of the hemisphere to his front foot. Thevenin's resistance R_{th} is the open circuit resistance between points a and b (without the person), as shown in Figure 8.15.

$$R_{\text{th}} = 2R_f \tag{8.14}$$

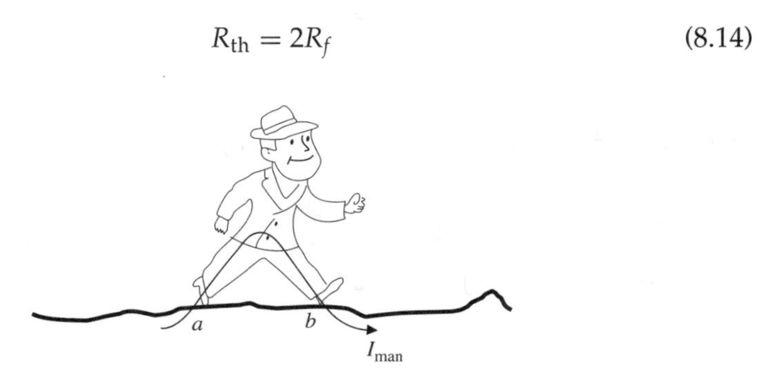

FIGURE 8.14
Current through a person due to a step potential.

FIGURE 8.15
Thevenins's impedance.

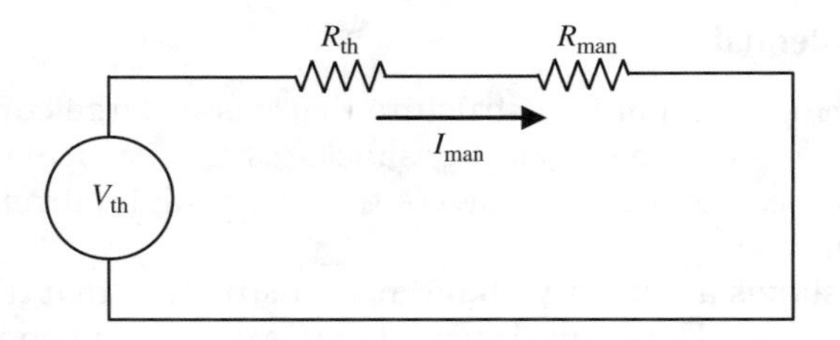

FIGURE 8.16
Equivalent circuit for the step potential.

Figure 8.16 shows the equivalent circuit for the step potential calculations. The body resistance of the person is the resistance of his legs and abdomen. The current through the man can be computed by using Thevenin's theorem as shown in Equation (8.15).

$$I_{man} = \frac{V_{th}}{R_{th} + R_{man}} = \frac{V_{th}}{2R_f + R_{man}} \tag{8.15}$$

EXAMPLE 8.5 A short circuit current of 1000 A passes through a hemisphere ground. A person is walking 5 m away from the center of the hemisphere. Assume that the leg-to-leg body resistance of the person is 2 kΩ. Assume that the soil surrounding the hemisphere is moist. Compute the current through the person and his step potential.

Solution

For moist soil, $\rho = 100\ \Omega$m. Assume that the step of the person is about 0.6 m. Thevenin's voltage can be computed using Equation (8.5).

$$V_{th} = \frac{I\rho}{2\pi}\left[\frac{1}{r_a} - \frac{1}{r_b}\right] = \frac{1000 \times 100}{2\pi}\left[\frac{1}{5} - \frac{1}{5.6}\right] = 341\ \text{V}$$

$$R_f = 3\rho = 300\ \Omega$$

The current through the man can be computed by Equation (8.15).

$$I_{man} = \frac{V_{th}}{2R_f + R_{man}} = \frac{341}{600 + 2000} = 131\ \text{mA}$$

The step voltage is the voltage between the person's feet.

$$V_{step} = I_{man} R_{man} = 131 \times 2000 = 262\ \text{V}$$

EXAMPLE 8.6 Repeat Example 8.5 assuming that a grounding rod is used instead of the hemisphere. Assume that the rod is inserted 1 m into the ground.

Solution

Using the rod equation in Table 8.4 we can obtain Thevenin's voltage.

$$V_{\text{th}} = \frac{I\rho}{2\pi l}\left[\ln\left(\frac{2l + r_a}{r_a}\right) - \ln\left(\frac{2l + r_b}{r_b}\right)\right]$$
$$= \frac{1000 \times 100}{2\pi}\left[\ln\left(\frac{2+5}{5}\right) - \ln\left(\frac{2+5.6}{5.6}\right)\right] = 495 \text{ V}$$

The current through the man is given in Equation (8.15).

$$I_{\text{man}} = \frac{V_{\text{th}}}{2R_f + R_{\text{man}}} = \frac{495}{600 + 2000} = 190 \text{ mA}$$

Note that the ground rod provides less protection than the ground hemisphere.

The step potential of the man is

$$V_{\text{step}} = I_{\text{man}} R_{\text{man}} = 190 \times 2000 = 380 \text{ V}$$

As seen in Examples 8.5 and 8.6, the current through the person is very high. But it may not be fatal since it does not pass through the heart, lungs, or brain. However, it could be painful enough to cause the person to fall; then a lethal current can flow through his vital organs.

EXAMPLE 8.7 During a weather storm, an atmospheric discharge hits a lightning pole. The pole is grounded through a hemisphere, and the maximum lightning current through the pole is 20,000 A.

1. A person is playing golf 30 m away from the center of the hemisphere. The distance between his feet is 0.3 m, and his leg-to-leg resistance is 2 kΩ. Assume the soil surrounding the hemisphere is moist. Compute the current through the person and his step potential.
2. Another person is 3 m away from the center of the hemisphere. The distance between his feet is also 0.3 m, and his leg-to-leg resistance is also 2 kΩ. Compute the current through the person and his step potential.

Solution

For moist soil, $\rho = 100\ \Omega\text{m}$. Thevenin's voltage can be computed using Equation (8.5).

1. $$V_{\text{th}} = \frac{I\rho}{2\pi}\left[\frac{1}{r_a} - \frac{1}{r_b}\right] = \frac{20{,}000 \times 100}{2\pi}\left[\frac{1}{30} - \frac{1}{30.3}\right] = 105 \text{ V}$$
   $R_f = 3 \times \rho = 300\ \Omega$

The current through the man can be computed by using Equation (8.15).

$$I_{\text{man}} = \frac{V_{\text{th}}}{2R_f + R_{\text{man}}} = \frac{105}{600 + 2000} = 40.4 \text{ mA}$$

The step voltage is the voltage between the person's feet.

$$V_{\text{step}} = I_{\text{man}} R_{\text{man}} = 40.4 \times 2000 = 80.8 \text{ V}$$

2. $$V_{\text{th}} = \frac{I\rho}{2\pi}\left[\frac{1}{r_a} - \frac{1}{r_b}\right] = \frac{20{,}000 \times 100}{2\pi}\left[\frac{1}{3} - \frac{1}{3.3}\right] = 9.646 \text{ kV}$$

The current through the man can be computed by using Equation (8.15).

$$I_{\text{man}} = \frac{V_{\text{th}}}{2R_f + R_{\text{man}}} = \frac{9646}{600 + 2000} = 3.71 \text{ A}$$

$$V_{\text{step}} = I_{\text{man}} R_{\text{man}} = 3.71 \times 2000 = 7.42 \text{ kV}$$

The step voltage for the second person is extremely high, and it is unlikely the person can maintain his balance. If he falls to the ground, a lethal current could pass through his vital organs.

8.4 Electric Safety at Home

Before we discuss electric safety at home, let us examine the most common power outlet in the United States, shown in Figure 8.17. The outlet (also

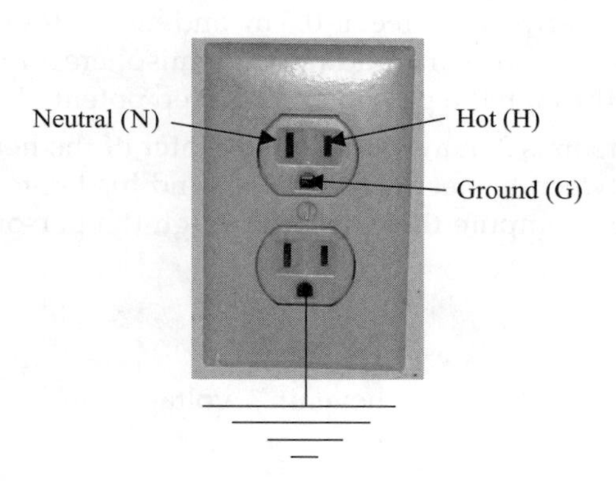

FIGURE 8.17
Household power outlet.

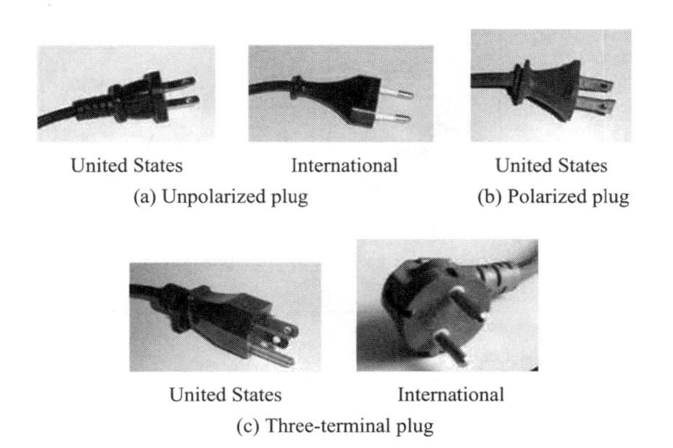

FIGURE 8.18
Common household plugs.

known as a receptacle) has three terminals: hot, neutral, and ground. Their wires are color coded — hot is black, neutral is white, and ground is either green or uninsulated. The hot terminal is connected to the 120 V side of the service transformer. The neutral terminal is grounded at the service transformer. The ground terminal is grounded locally at the house through ground rods or water pipes. The hot and neutral terminals accept flat prongs, and the ground terminal accepts a rounded prong. The neutral terminal is wider than the hot terminal (these are known as polarized terminals). The reasons for this arrangement may not be obvious at first, and two questions can be raised: (1) why do we have polarized terminals? and (2) why do we have two ground terminals? The answers will become apparent later in this section.

Household plugs come in various shapes such as the ones shown in Figure 8.18. In Figure 8.18(a), the plug is unpolarized with two equal-width prongs; so either prong can be connected to the neutral terminal of the outlet. In Figure 8.18(b), the plug has two prongs and is polarized, where the neutral prong is wider than the hot prong. This way, the neutral side of the plug can only fit the neutral side of the outlet. Figure 8.18(c), shows three-terminal plugs; one is used in the United States and the other is international. The United States plug has three prongs; two of them are flat and the third is rounded. The ground prong is normally longer than the other two. In this configuration, when the plug is inserted into the outlet, the ground terminal is connected first. Also, the neutral and hot terminals cannot be interchanged. For the international version in Figure 8.18(c), the ground terminal is on the side of the plug.

8.4.1 Neutral versus Ground

One of the most confusing issues in electric safety is the difference between the neutral and the ground terminals. They are both grounded, so why do we sometimes use both of them? To answer this question, let us examine the generic representation of the electric equipment shown in Figure 8.19. The

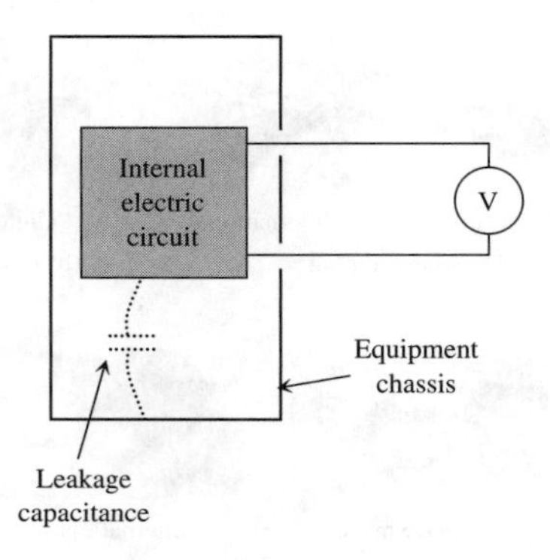

FIGURE 8.19
Generic representation of an appliance.

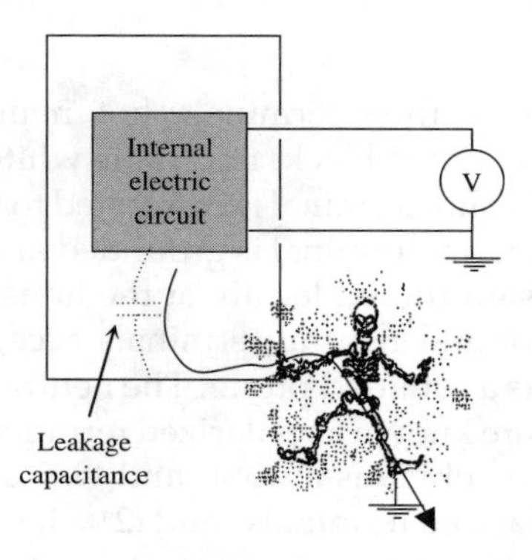

FIGURE 8.20
A person touching a floating chassis.

equipment consists of an internal electric circuit housed inside a chassis. The internal circuit is electrically isolated from the chassis. If the chassis is metallic (conductive), three problems could occur:

1. Any two adjacent conductive elements with different potentials have parasitic (leakage) capacitance between them. In the generic equipment in Figure 8.19, the potential of the internal electric circuit is different from the potential of the chassis. Hence, the capacitive coupling between the internal circuit and the chassis could elevate the potential of the chassis.
2. The current of the internal circuit produces magnetic fields that link the chassis, thus inducing voltage on the chassis.
3. Faults inside the electric circuit could result in the chassis touching circuit components at elevated voltage.

If a person standing on a grounded object touches the chassis as shown in Figure 8.20, the person closes an electrical circuit and a current flows through his body. The current flows from the source to the circuit, then part of it passes through the leakage capacitance to the person, and finally to ground. This current could be small and the person may feel nothing or just skin sensations. However, if the circuit is faulty, the current could reach unsafe levels. To protect people from this hazard, the neutral terminal of the receptacle in some equipment is connected to the chassis as shown in Figure 8.21. This way, the chassis is at the potential of the neutral wire, and no current flows through the person. This, however, requires the polarized plug in Figure 8.18(b), where the neutral is always connected to the chassis, and not the other way!

The system in Figure 8.21 is not always safe. Consider for example the case in Figure 8.22 where the service transformer is far from the equipment. In this case, the cables connecting the transformer to the house are long and have some resistances — the hot wire resistance is R_h and the neutral wire resistance is R_n. The current of the hot wire I starts from the source to the circuit of the electric equipment, then returns to ground through two parallel paths: one through the neutral wire I_n and the other through the man I_{man}.

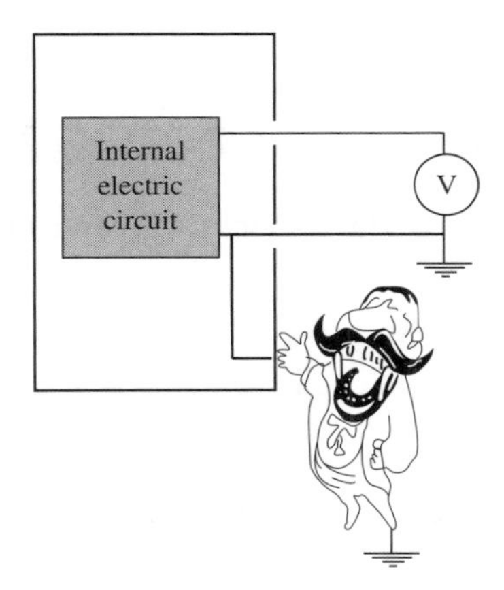

FIGURE 8.21
A person touching a grounded chassis.

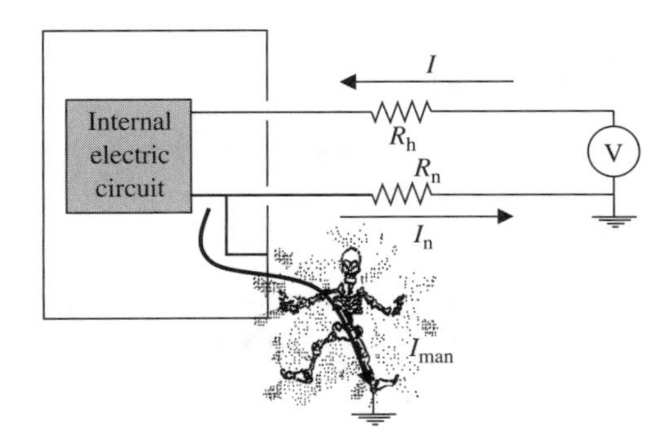

FIGURE 8.22
A person touching a chassis where the neutral wire is long.

The current through the man is determined by the current divider equation

$$I_{man} = I \frac{R_n}{R_n + R_{man}} \tag{8.16}$$

R_{man} is the body resistance of the man plus his ground resistance. As seen in the equation, the current through the man depends on the resistance of the neutral wire and could cause a secondary shock in most cases, and a primary shock under faulty conditions as shown in the following examples.

EXAMPLE 8.8 For the case in Figure 8.22, assume the equipment is powered by 240 V and the resistance of the neutral wire is 0.5 Ω. Assume that the resistance of the man plus his ground resistance is 1500 Ω. If the equipment draws 20 A, compute the current through the man.

Solution

The current through the man is given in Equation (8.16).

$$I_{man} = I \frac{R_n}{R_n + R_{man}} = 20 \frac{0.5}{0.5 + 1500} = 6.7 \text{ mA}$$

This current produces a secondary shock as given in Table 8.1.

EXAMPLE 8.9 For the case in Example 8.8, assume that the equipment has an internal fault and draws 50 A from the line without tripping the circuit breaker. Compute the current through the man.

Solution

The current through the man is given in Equation (8.16).

$$I_{man} = I \frac{R_n}{R_n + R_{man}} = 50 \frac{0.5}{0.5 + 1500} = 16.75 \text{ mA}$$

This current is within the primary shock and may induce respiratory tetanus as given in Table 8.2.

The last two examples show that the neutral wire may not provide protection from electric shocks. To correct this problem, the chassis must be connected directly to the local ground. This can be done by using the three-prong plug in Figure 8.18(c) and the three-terminal outlet in Figure 8.17. Since the local ground is adjacent to the house, its resistance is much smaller than the resistance of the neutral wire. The local ground can be established by a ground

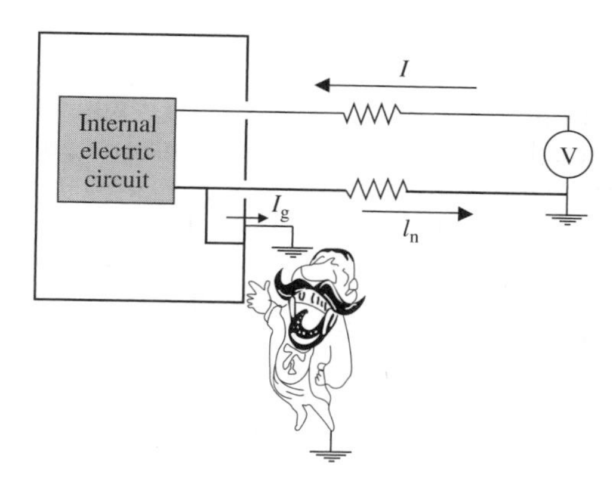

FIGURE 8.23
A person touching a grounded chassis with a long neutral wire—chassis can carry high currents.

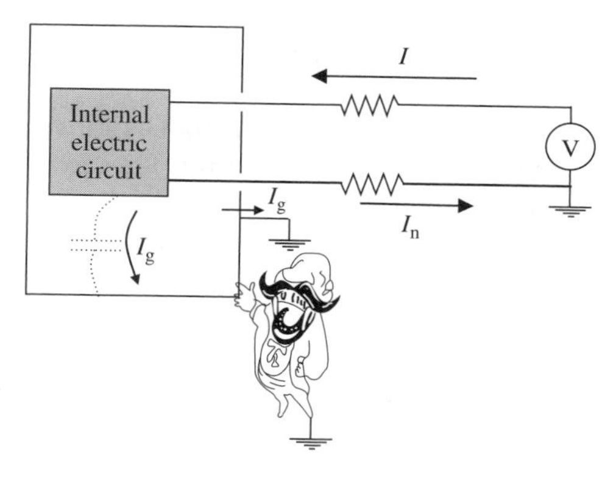

FIGURE 8.24
A person touching a grounded chassis with a long neutral wire—chassis carries a small current.

rod outside the house, or by a wire connected to a well-grounded water pipe. In this case, the current can have two paths as shown in Figure 8.23: one through the neutral wire I_n and the other through the local ground I_g. The resistance of the man is much higher than the resistance of the local ground, and almost no current flows through his body.

The system in Figure 8.23, although safe, could have high current passing through its chassis. A better system is the one shown in Figure 8.24 where the neutral wire is isolated from the chassis. Hence the ground current is just the leakage current of the leakage capacitance. This is the most common system today.

Needless to say, if the chassis is nonconductive, the problems described above vanish. In some cases, however, even if the chassis is made out of isolation material (such as plastic); the knobs, levers, screws, or displays could be made out of metals that are connected together by conductive material inside the chassis. In this case, touching any of these components could have the same effects as touching a conductive chassis.

8.4.2 Ground Fault Circuit Interrupter (GFCI)

One of the most common electric shock scenarios is when electric equipment is mixed with water. Tap water is conductive, and its intrusion allows the circuit components to be electrically connected to the chassis. This is as bad a situation as having the circuit components arcing to the chassis.

Assume that a hair dryer is plugged into an electric outlet. If the hair dryer falls into a bathroom tub filled with water, the high-voltage wire and the circuit components may come in contact with the dryer's chassis. If a person touches the wet dryer while part of his/her body is in contact with a grounded surface, such as a faucet, the person could be electrocuted as depicted in Figure 8.25. The fundamental question is why doesn't the house breaker interrupt the circuit? After all, the breakers are installed to disconnect faulty circuits. In fact, the house breaker may not operate at all if the fault current is below its interruption rating. The current through the man in Figure 8.25, although lethal, could be well below the interruption level of the breaker. For this reason the *Ground Fault Circuit Interrupter* (GFCI) is used in wet areas such as kitchens, bathrooms, gardens, etc.

The GFCI is similar in shape to the regular outlet as shown in Figure 8.26, but has two buttons and a built-in circuit interrupter. The circuit interrupter of the GFCI detects the currents of the hot and neutral terminals. If the two currents are not equal, a leakage current flows through the chassis to the ground. In this case, the GFCI interrupts the circuit by disconnecting the terminal of the hot wire. One of the buttons is used to test the circuit for functionality, and the other is used to reset the circuit after interrupting a fault.

The basic component in the circuit interrupter of the GFCI is the current differential detector shown in Figure 8.27. The detector consists of a core with windings wrapped around it. The hot wire and the neutral wire pass inside the core. If the hot wire current I_{hot} and the neutral current $I_{neutral}$ are equal in magnitude, no voltage is induced across the windings. If there is a difference between the two currents, a voltage is induced across the windings, triggering a relay that switches off the power.

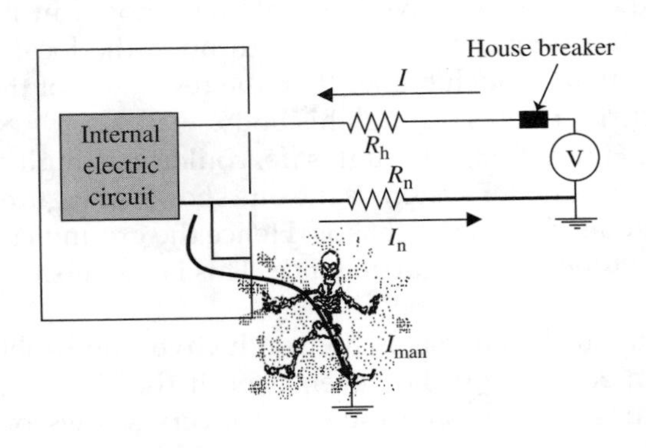

FIGURE 8.25
Water inside a device can create a hazardous condition.

FIGURE 8.26
(see color insert following Page 208) Electric outlet with ground fault circuit interrupter.

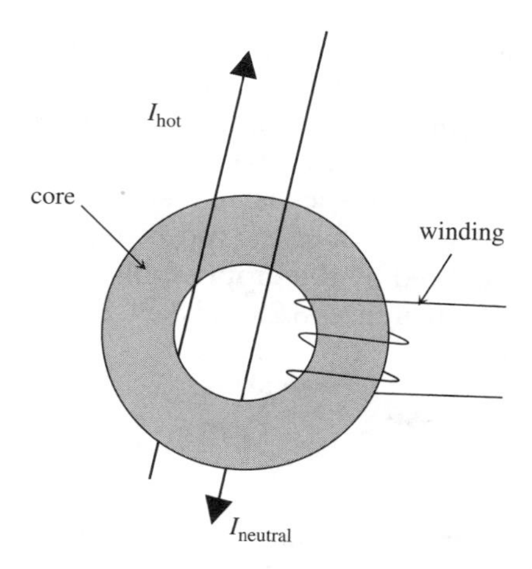

FIGURE 8.27
Current differential detector of the GFCI circuit.

8.5 Safety of Personnel Working on Power Lines

Working on power lines is a daring job that demands extra safety precautions. The United States, through the *Occupational Safety & Health Administration* (OSHA) and the *Institute of Electrical and Electronics Engineers* (IEEE) has established strict safety rules for power line work. Fatalities and severe injuries can and do occur when workers use the wrong equipment, do not follow the established rules, or work without the proper protective gear. A large number of injuries are attributed to workers ignoring the safety rules due to time pressures, lack of knowledge, or poor supervisions.

Power line work can be divided into two categories: live-line work and de-energized line work. Each of these two types requires different sets of rules.

8.5.1 Live-Line Work

When a person is working on a live-line, he/she cannot get closer than an established minimum distance to ensure the worker's safety. This is known as the *Minimum Approach Distance* (MAD), which is established by OSHA and given in Table 8.6. The minimum approach distance is a function of the voltage of the line as shown in the table. The *phase to ground exposure* means that the worker is exposed to only one line. The *phase to phase exposure* is for the work on two different lines. The established rule is that the worker, and all conductive equipment in contact with the worker, must stay outside the minimum approach distance.

The obvious question is how the worker can accomplish his job from a distance. This is done by several methods; the most common ones are by using live-line tools or by using insulated equipment. In either case, the worker must not be part of any electric circuit. Live-line tools include what is known in the industry as *hot sticks*. A hot stick is an insulated rod normally made out of fiberglass material. The length of the rod is dependent on the voltage of the line. The rod can be fitted by the proper tools for the job. Hot sticks with different tools are shown in Figures 8.28 and 8.29. Figure 8.30 shows two more common tools.

TABLE 8.6

Minimum Approach Distance for Live-Line Work (OSHA regulation)

	Distance	
Line Voltage (kV)	**Phase to Ground Exposure (m)**	**Phase to Phase Exposure (m)**
1.1 to 15	0.64	0.66
15.1 to 36	0.72	0.77
36.1 to 46	0.77	0.85
46.1 to 72.5	0.90	1.05
72.6 to 121	0.95	1.29
138 to 145	1.09	1.50
161 to 169	1.22	1.71
230 to 242	159	2.27
345 to 362	2.59	3.8
500 to 550	3.42	5.5
765 to 800	4.53	7.91

FIGURE 8.28
(see color insert following Page 208) Hot stick with fuse puller tool.

FIGURE 8.29
Hot stick with cable handler tool.

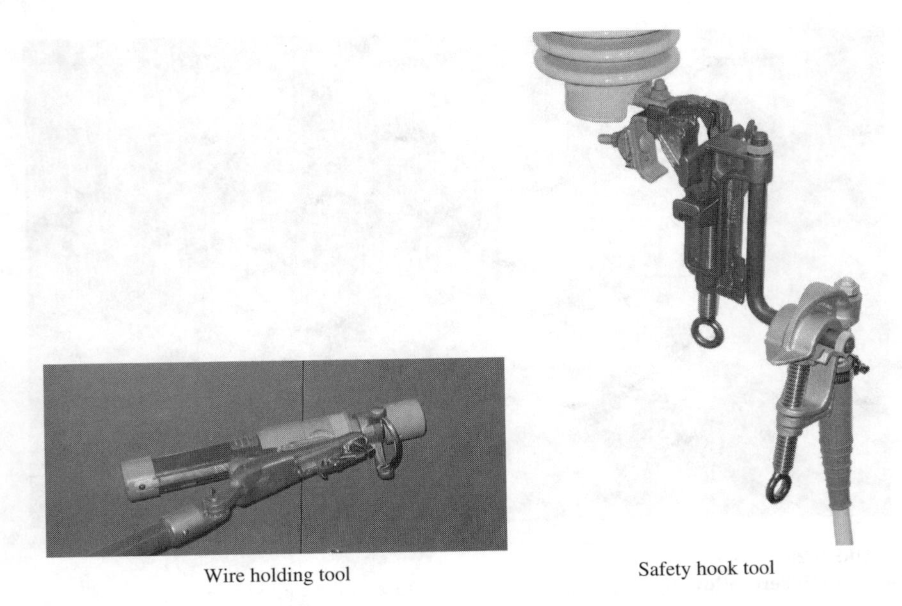

FIGURE 8.30
(see color insert following Page 208) Various tools used on hot sticks.

FIGURE 8.31
(see color insert following Page 208) Worker using a hot stick on an energized line. (Images courtesy of W.H. Salisbury & Co.)

Figure 8.31 shows workers performing maintenance work on energized lines using a hot stick. The worker can be placed inside a basket at the top of a boom mounted on a truck, or can simply climb the pole using special climbing equipment and shoes.

FIGURE 8.32
(see color insert following Page 208) Working on a power line using a helicopter. (Images courtesy of USA Airmobile, Inc.)

A more daring method for live-line maintenance is by using a helicopter, as shown in the left photograph in Figure 8.32. The basic safety rule is to not have the worker's body touching two different potential points at the same time. Hence, the worker can be working on a power line from a platform mounted on the helicopter; the worker in this case is like a bird standing on a high-voltage wire. His body is at the same potential as the line, and no potential difference exists between any two points of his body. The maintenance work must be performed after the helicopter is connected electrically to the wire being worked on, so the helicopter is at the same potential as the line. The worker, who is wearing a highly conductive suite, sits on the platform while performing the work. The pilot must be skilful enough to keep the helicopter in position at all times. The photograph on the right side shows a worker performing maintenance work while being surrounded by high-voltage wires. The four conductors in the figure are actually bundled together to form one phase of a transmission line, and all of the four conductors have the same potential. This is a normal design for high-voltage wires to reduce the corona discharge. The worker inside the bundle is safe even when he/she contacts any two of these conductors. Needless to say, this type of work cannot be performed during bad weather conditions.

8.5.2 De-Energized Line Work

A de-energized line is a line disconnected from both ends. It is a fatal mistake to assume that the de-energized lines are safe to touch, they are not. Power lines always share the right-of-way, and run adjacent to each other. If the de-energized line is in the vicinity of an energized line, the electromagnetic field produced by the energized line induces voltage on the de-energized line as shown in Figure 8.33. The magnitude of the induced voltage depends on several factors, such as the voltage and current of the energized line, the proximity of the two lines, and the length by which the two lines are running

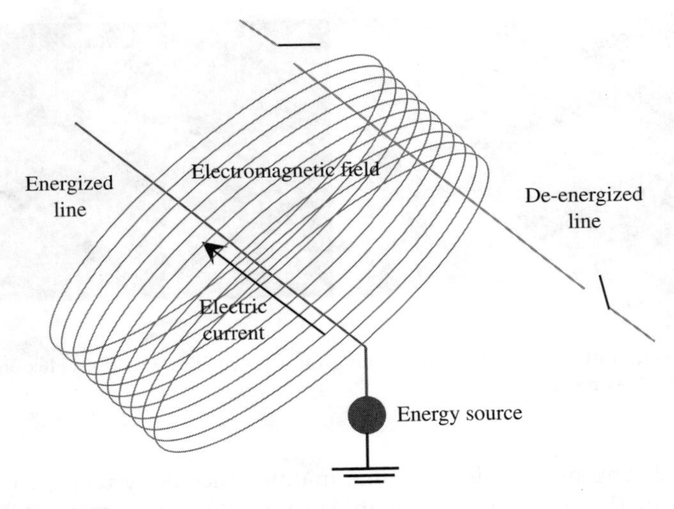

FIGURE 8.33
Electromagnetic coupling between energized and de-energized lines.

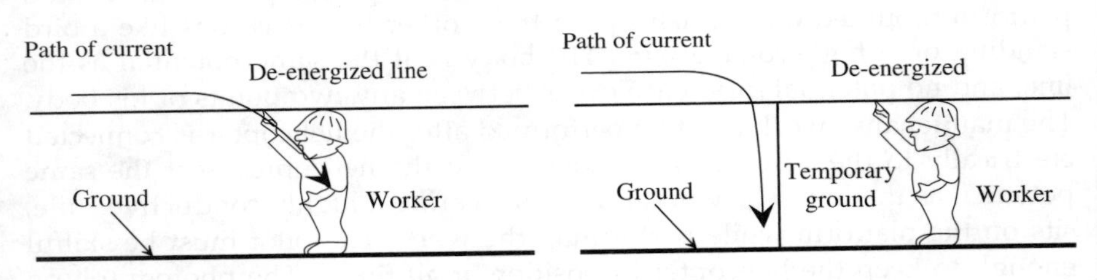

FIGURE 8.34
Worker without and with a temporary ground.

adjacent to each other. The induced voltage on the de-energized line could be as high as several kilovolts. If a person touches the de-energized line while part of his body is in contact with a grounded object, a lethal current could flow through his body.

To protect the linemen from electric shocks, work on the de-energized line must be carried out by one of the follwing two methods: (1) by using live-line tools and maintaining the minimum approach distance; or (2) by establishing an *equipotential zone*.

Before we discuss the equipotential zone, let us recall the basic concept of grounding an object. Assume that a person is touching a de-energized line while his feet are on the ground (or grounded object) as shown on the left side of Figure 8.34. If for some reason a voltage is induced on this line, a current will flow through the person, which could be lethal. This is because the hands and feet of the person are touching two different potential points. However, if before the person touches the line, a temporary ground wire is installed between the line and the ground as shown on the right side of Figure 8.34,

Byron Nuclear Power Plant

ITAIPU Hydropower Plant

COLOR FIGURE 2.2
Power plants. (Images courtesy of the U.S. Department of Energy.)

COLOR FIGURE 4.2
Fox River diversion hydropower plant, Wisconsin. (Image courtesy of the U.S. Army Corps of Engineers.)

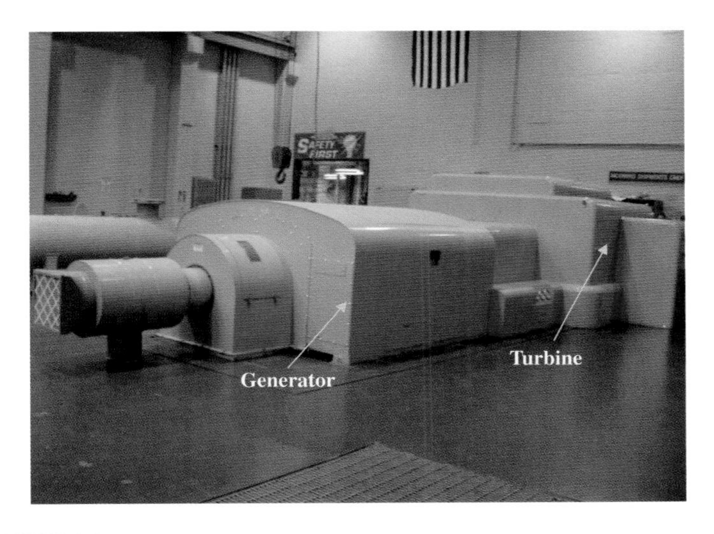

COLOR FIGURE 4.9
Inside a thermal power plant.

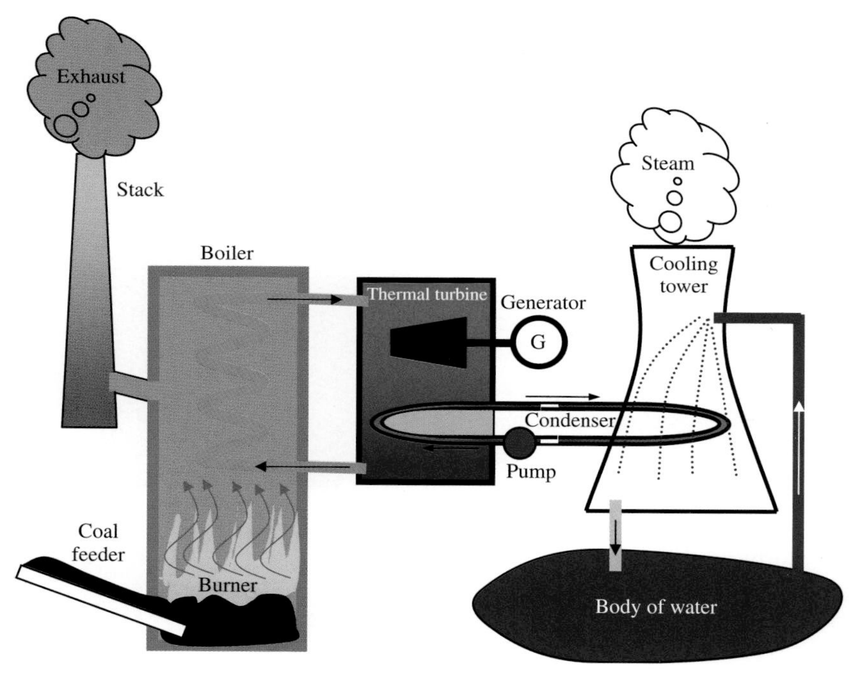

COLOR FIGURE 4.11
Main components of coal-fired power plant.

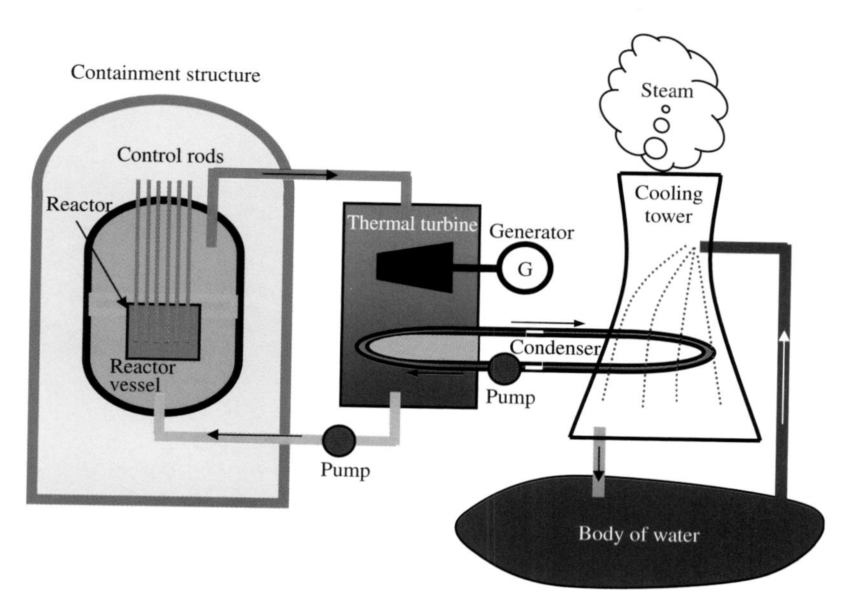

COLOR FIGURE 4.15
Main components of a BWR.

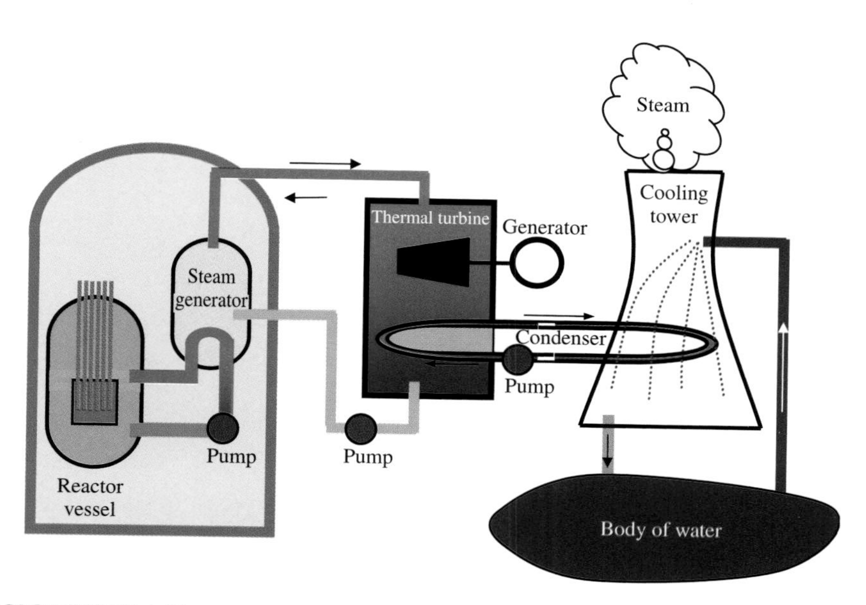

COLOR FIGURE 4.16
Pressurized water reactor.

COLOR FIGURE 4.17
PWR nuclear power plants.

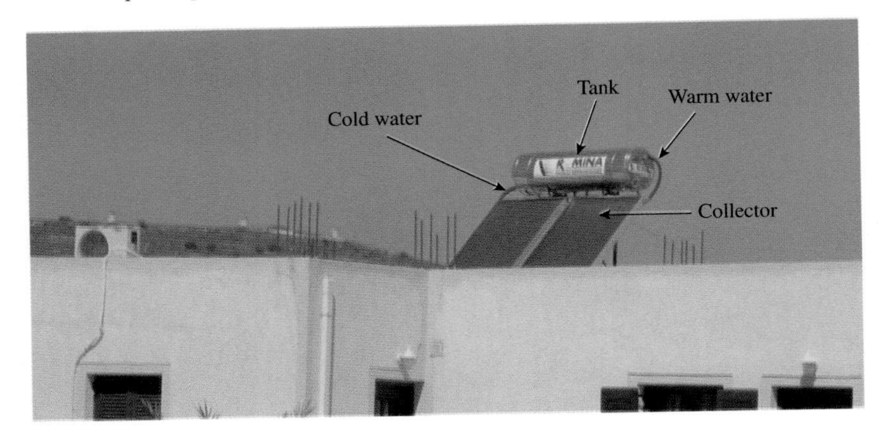

COLOR FIGURE 6.3
Passive thermosiphon hot water solar system.

COLOR FIGURE 6.6
PV module and PV array. (Images courtesy of the U.S. Department of Energy.)

COLOR FIGURE 6.7
Various photovoltaic systems. (Images courtesy of the U.S. Department of Energy.)

COLOR FIGURE 6.11
High-power photovoltiac systems. (Images courtesy of the U.S. Department of Energy.)

COLOR FIGURE 6.14
Wind farm located in California. (Images courtesy of the U.S. Department of Energy.)

COLOR FIGURE 6.15
2 MW off-shore wind turbine farm in Denmark. (Courtesy of LM Glasfiber Group.)

COLOR FIGURE 6.20
PEM fuel cell.

COLOR FIGURE 6.25
Steam generated from rain even in cold environment. (Courtesy of the U.S. Department of Energy.)

COLOR FIGURE 6.28
The first geothermal power plants in the United States (The Geysers) in northern California. (Courtesy of the U.S. Department of Energy.)

COLOR FIGURE 6.29
Free-flow tidal energy system. (Images courtesy of Marine Current Turbines Limited.)

On a distribution pole

To a distribution substation

COLOR FIGURE 7.30
Capacitors for power factor correction.

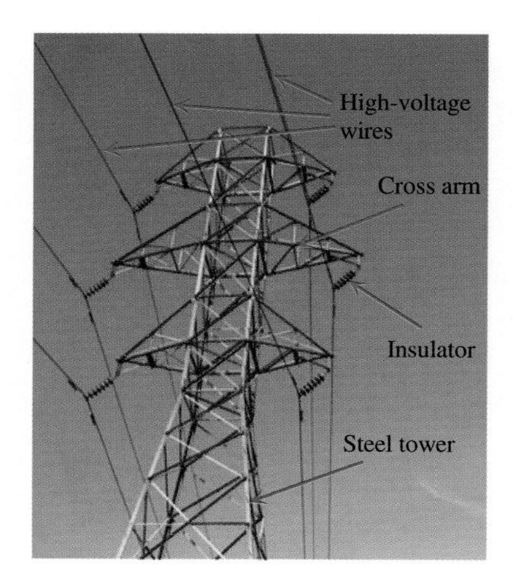

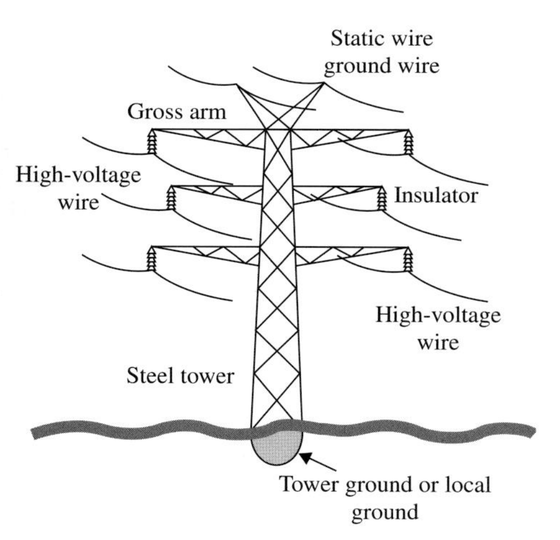

COLOR FIGURE 8.10
Main components of a steel power line.

COLOR FIGURE 8.26
Electric outlet with ground fault circuit interrupter.

COLOR FIGURE 8.28
Hot stick with fuse puller tool.

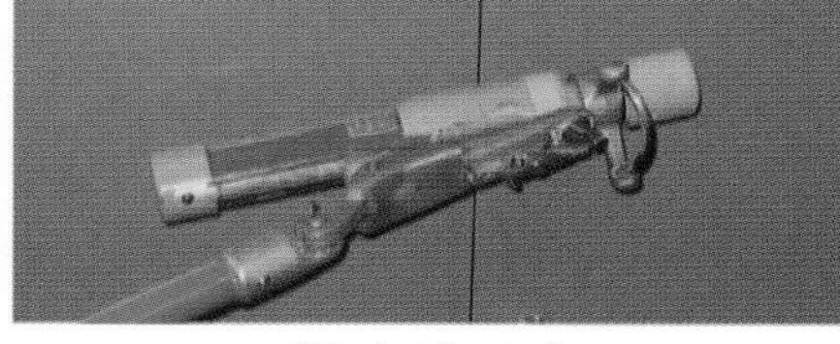

Wire holding tool

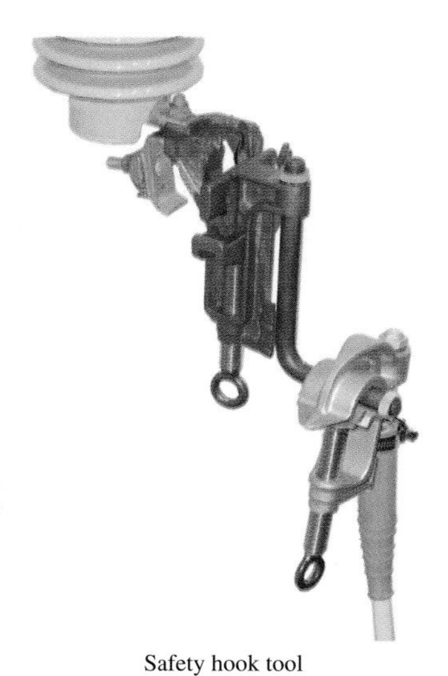

Safety hook tool

COLOR FIGURE 8.30
Various tools used on hot sticks.

COLOR FIGURE 8.31
Worker using a hot stick on an energized line. (Images courtesy of W.H. Salisbury & Co.)

COLOR FIGURE 8.32
Working on a power line using a helicopter. (Images courtesy of USA Airmobile, Inc.)

(a) Transmission transformer

(b) Distribution transformer

(c) Service transformer

(d) Circuit transformer

COLOR FIGURE 11.2
Various types of transformers.

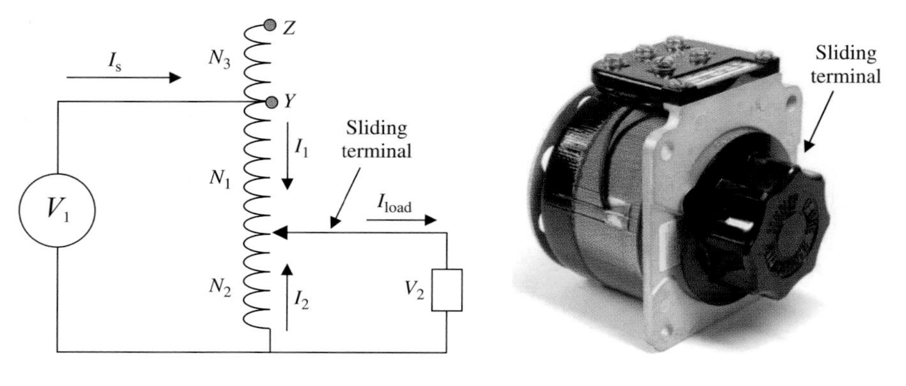

COLOR FIGURE 11.9
Autotransformer with adjustable voltage.

Single-phase transformer

Two transformer bank

Three transformer bank

COLOR FIGURE 11.15
Transformer banks.

MARS rover (Courtesy of NASA)

Underwater unmanned robot

Electric propulsion ferry

Hybrid electric vehicle

City transportation

City electric bus

Maglev train
(Courtesy of transrapid int.)

Electric train

Monorail

COLOR FIGURE 12.6
Induction machines are the prime movers of these vehicles.

Stator

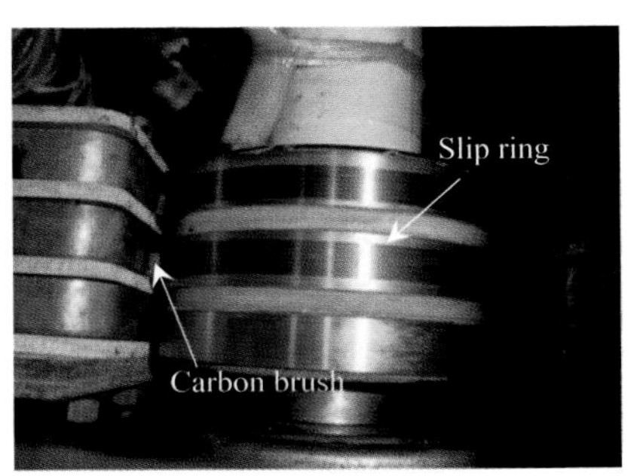

Slip ring rotor

COLOR FIGURE 12.7
Main parts of induction machine.

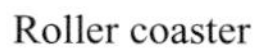

Roller coaster

Maglev launcher
(Courtesy of NASA Marshall Space Flight Center)

Maglev rapid transportation

COLOR FIGURE 12.22
Example of the applications of linear induction motors. (Images courtesy of Transrapid International.)

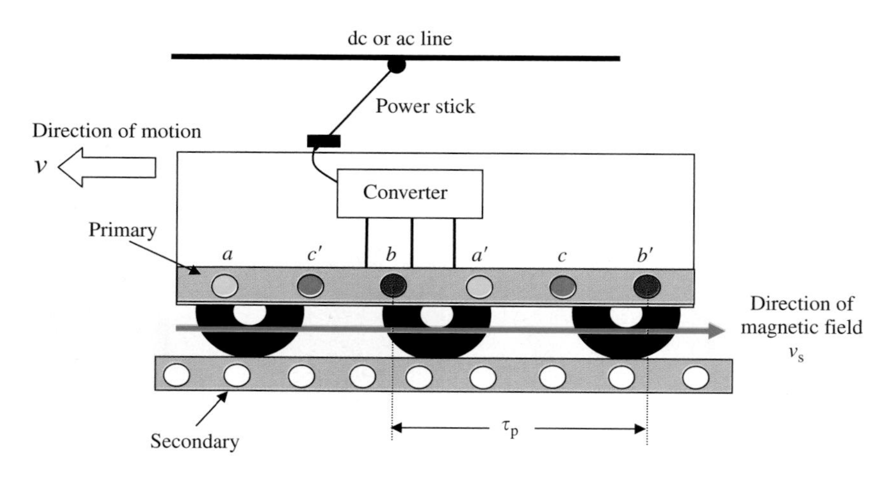

COLOR FIGURE 12.26
Motion of the linear induction motor.

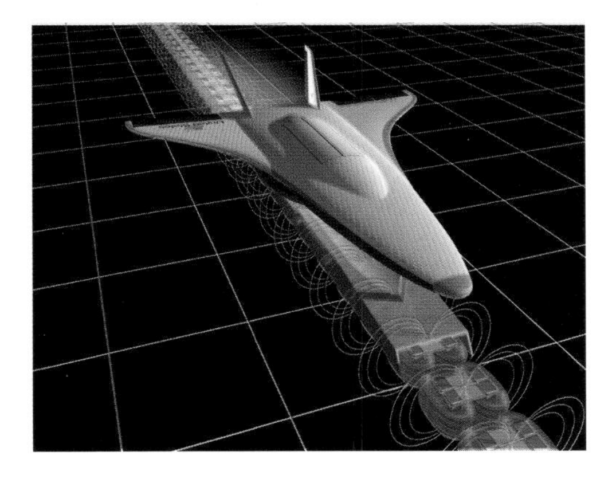

COLOR FIGURE 12.29
Concept of Maglev for a space launcher. (Courtesy of NASA Marshall Space Flight Center.)

COLOR FIGURE 12.33
Machine model showing slip-ring arrangement.

COLOR FIGURE 12.56
Hybrid electric vehicle uses permanent magnet synchronous motor.

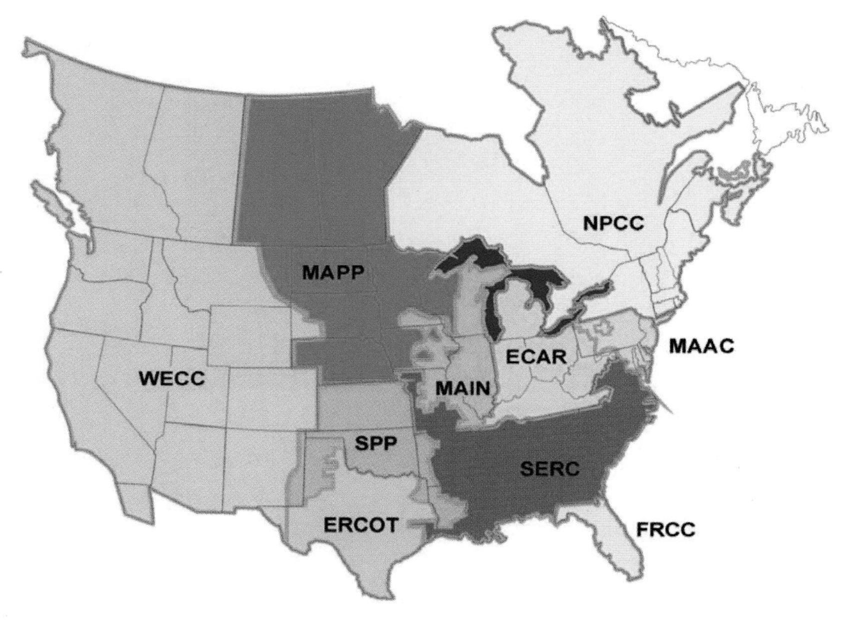

COLOR FIGURE 13.12
U.S. regional reliability councils.

almost all the current will flow through the ground wire, and the person is safe to work on the de-energized line. In this case, the hands and feet of the person are at equal potential points (equipotential).

EXAMPLE 8.10 Assume that a 20 km de-energized line is accidentally energized by a 2 kV source while a lineman is working at the far end of the line. The line resistance is 0.1 Ω/km, and the worker body resistance plus his ground resistance is 3000 Ω.

1. Compute the current through the lineman and the voltage across his body.
2. Assume that a 0.01 Ω temporary ground is installed between the line and the ground in parallel with the worker. Compute the current through the lineman and the voltage across his body.

Solution

1. The equivalent circuit of the system without the temporary ground is shown in Figure 8.35.

   The current through the man is given by

$$I_{\text{man}} = \frac{V}{R_{\text{line}} + R_{\text{man}}} = \frac{2000}{20 \times 0.1 + 3000} \approx 667 \text{ mA}$$

This level of current is lethal and the worker cannot survive for more than 55 ms according to Dalziel's formula.

The voltage across the man is

$$V_{\text{man}} = I_{\text{man}} R_{\text{man}} = 0.667 \times 3000 \approx 2 \text{ kV}$$

The voltage across the lineman is almost equal to the source voltage since the line resistance is small.

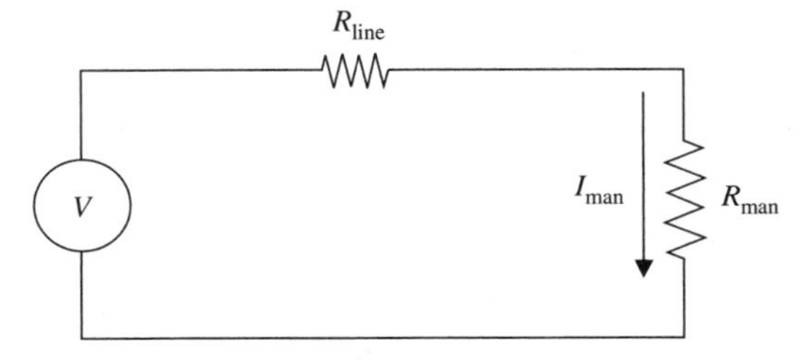

FIGURE 8.35
Equivalent circuit of part 1.

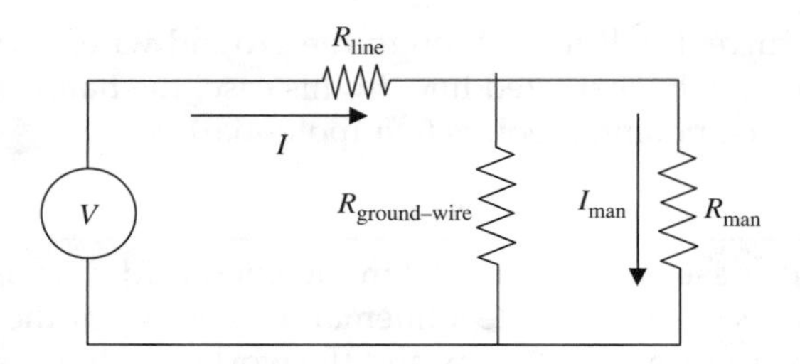

FIGURE 8.36
Equivalent circuit of part 2.

2. Now assume that a ground wire is installed. The equivalent circuit in this case is given in Figure 8.36.

   The parallel resistance of the man and ground wire is

$$R = \frac{R_{man} R_{ground-wire}}{R_{lman} + R_{ground-wire}} = \frac{3000 \times 0.01}{3000.01} \approx 0.01\ \Omega$$

   The current in the line is

$$I_{line} = \frac{V}{R_{line} + R} = \frac{2000}{2 + 0.01} \approx 995\ \text{A}$$

   The current through the man is

$$I_{man} = I_{line} \frac{R_{ground-wire}}{R_{ground-wire} + R_{man}} = 995 \frac{0.01}{0.01 + 3000} \approx 3.32\ \text{mA}$$

   This current is safe.

   The voltage across the man is

$$V_{man} = I_{man} R_{man} = 3.32 \times 3000 \approx 10\ \text{V}$$

Note that with the ground wire, the current through the lineman and the voltage across his body are very small. However, the ground wire protects the worker only if its resistance is very small, it is installed adjacent to the worker and can carry high currents.

The idea of the equipotential zone is to keep all equipment and people at the work site at the same potential. The helicopter maintenance work in Figure 8.32 is an example of an ungrounded equipotential zone. However, when some of the equipment or people are at ground potential level, the equipotential zone must be grounded. This is done by bonding all equipment and hardware at the work site to the best available ground system. The work

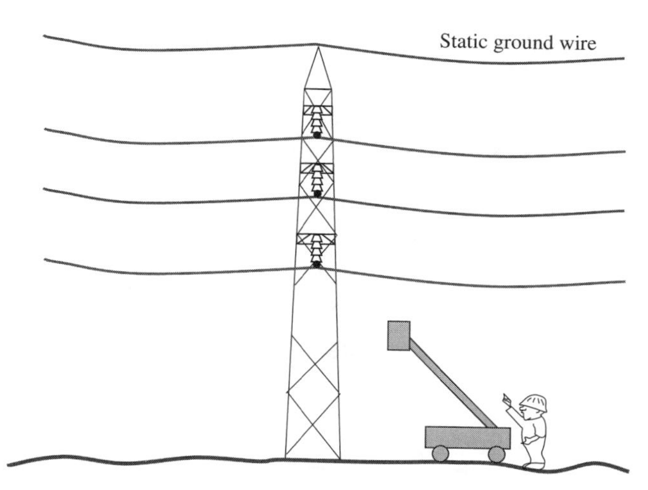

FIGURE 8.37
Work site before establishing a grounded equipotential zone.

site is the area where the workers are performing maintenance or installation jobs. All people at the site must be inside the *grounded equipotential zone*. This includes the people on the ground, on the lift, or on the tower. An effective grounded equipotential zone keeps the step and touch potential of all people within the safe limits, even when the de-energized line is accidentally energized. As seen in Example 8.10, the worker is protected when the voltage across his body is very small.

Now let us assume that a lineman is to work on the lowest conductor in the system shown in Figure 8.37. Assume that all three-phase wires are de-energized. Before the lineman approaches the de-energized line, he must establish a grounded equipotential zone. The zone must have the following characteristics:

1. The equipotential zone must encompass the entire work site (people on the truck and on the ground).
2. Any worker inside the equipotential zone must have his/her body bypassed by very low resistances, much lower than the body resistance of the worker.
3. The equipotential zone must protect the workers even during faults and lightning.
4. The ground wires must withstand the fault currents without being damaged, and must have very low resistances.
5. The ground wires must be bonded to clean surfaces to ensure that the contact resistance is very small.
6. All grounds must be bonded to the best possible ground available at the site.

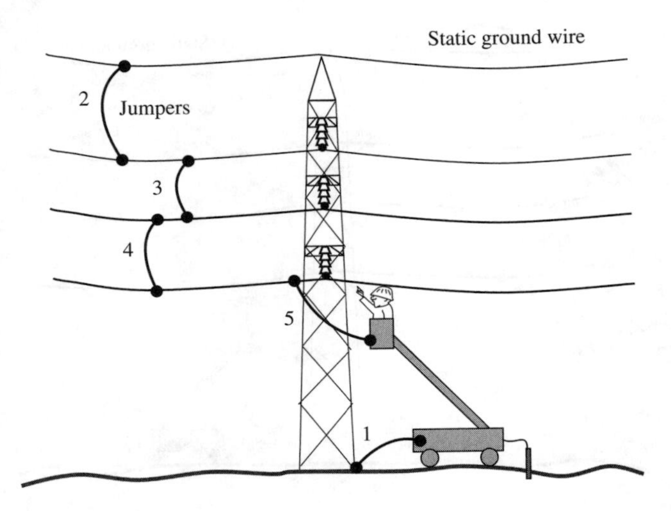

FIGURE 8.38
Grounded equipotential zone.

Based on the above conditions, the worker selects the static ground wire to be the reference ground since it is the best ground path at the site. The static wire is bonded to the tower, making the tower hardware at the static wire potential. Assume that the boom truck is conductive. The workers can establish the grounded equipotential zone by following the subsequent steps:

1. Bond the truck to a ground rod. This way the touch potential of the truck can be reduced, but not eliminated. The effectiveness of the ground rod is substantially reduced if the resistivity of the soil is high, or the rod is not driven deep enough into the soil.
2. Use a ground wire to bond the truck to the base of the tower. This is line 1 in Figure 8.38. This ground wire ensures that the touch potential of the truck is further reduced. Keep in mind that the tower is bonded to the static wire, which is the lowest ground resistance at the site.
3. Use live-line tools and connect a ground jumper wire to the static wire, then to the top de-energized line. This is line 2 in Figure 8.38. Now the top wire is grounded.
4. Use the live-line tools to connect the top conductor to the middle one. This is line 3 in Figure 8.38. The two top conductors are now grounded. If the distance between the middle and the static wires is small, a better method is to connect the middle conductor to the static wire directly to reduce the resistance between the middle and the static wire.
5. Use the live-line tools to connect the middle conductor to the lowest one. This is line 4 in Figure 8.38. The three de-energized conductors are now grounded.

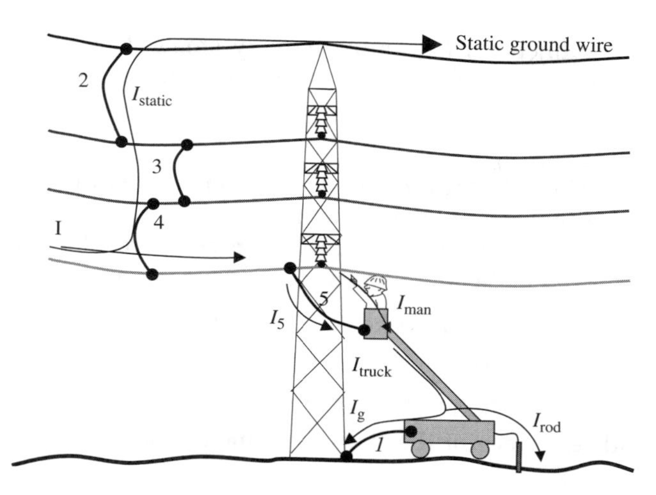

FIGURE 8.39
Fault current distribution in grounded equipotential zone.

6. Connect a ground wire between the basket of the truck and the conductor being worked on. This is line 5 in Figure 8.38. The worker in the basket is then bypassed by the low resistance of the ground wire.

The linemen, after all these steps, can work safely on any of the three conductors, and on the ground around the truck or tower. In the case when the ground rod does not provide adequate protection for the ground personnel around the truck, two extra protection methods can be used. The first is to isolate the area around the truck by using barricades to prevent people from touching the truck while standing on ground. The second is to have the truck parked over a ground mat that is large enough to accommodate the truck and any person working in its immediate vicinity. The ground mat is connected to the truck and is grounded.

By the above measures, the linemen are protected even if the line being worked on is accidentally energized as shown in Figure 8.39. Assume that a surge of current I flows through the line that is being touched by the lineman. This current is split into three components: one goes through the low-impedance path to the static wire I_{static}; the second I_5 goes through the ground wire 5 to the truck; and the third goes through the lineman I_{man}, then to the truck. The truck current I_{truck} is split into two components: one I_g goes to the ground through ground wire 1, and the other I_{rod} goes to the ground through the ground rod. The relationships of all these currents can be represented by

$$\begin{aligned} I &= I_{static} + I_5 + I_{man} \\ I_{truck} &= I_5 + I_{man} \\ I_{truck} &= I_g + I_{rod} \end{aligned} \tag{8.17}$$

To protect the lineman, I_{man} must be very small. This can be achieved by using ground wires with very low resistances adjacent to the worker as explained in Example 8.11.

EXAMPLE 8.11 Consider the grounded equipotential zone of the de-energized lines in Figure 8.39. Assume that an accident occurs when the de-energized line comes in contact with another high-voltage circuit resulting in a fault current of 5 kA flowing in the line being maintained. Assume the following data:

The ground resistance of the rod (R_{rod}) = 30 Ω.

The ground resistance of the tower (R_{gt}) = 15 Ω.

The resistance of the tower structure (R_{tower}) = 0.001 Ω.

The ground resistance of the static wire including the wire resistance (R_{static}) = 0.01 Ω.

The resistance of any ground wire (R_{gw}) = 0.002 Ω.

The body resistance of the man (R_{man}) = 3000 Ω.

1. Draw the equivalent circuit.
2. Compute the current through the lineman and the voltage across his body.
3. Compute the potential of the truck.
4. Assume that the temporary ground wire 5 is not present; repeat parts 2 and 3.
5. Assume that the temporary ground wires 1 and 5 are not present; repeat parts 2 and 3.

Solution

1. The equivalent circuit is shown in Figure 8.40.
2. The currents and voltages across the lineman, and the voltage of the truck, can be found by circuit analysis using the model in Figure 8.40.

$$I_{man} = 1.82 \text{ mA}$$

The voltage across the lineman is

$$V_{man} = I_{man} R_{man} = 5.46 \text{ V}$$

3. The voltage of the truck is

$$V_{truck} = I_{rod} R_{rod} = 58.1 \text{ V}$$

With these values the lineman and any person touching the truck are safe.

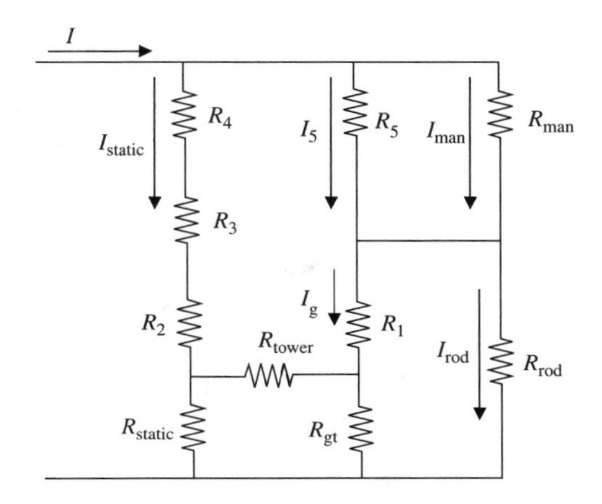

FIGURE 8.40
Equivalent circuit of Figure 8.39.

4. If the temporary ground wire 5 is not present, the values of the currents and voltages can be computed by using the circuit in Figure 8.40 without R_5.

$$I_{man} = 10 \text{ mA}$$

The voltage across the lineman is

$$V_{man} = I_{man} R_{man} = 30 \text{ V}$$

The voltage of the truck is

$$V_{truck} = I_{rod} R_{rod} = 50 \text{ V}$$

Note that all workers are still safe even if one ground fails.

5. If the temporary ground wires 1 and 5 are not present, the values of the currents and voltages are

$$I_{man} = 26.4 \text{ mA}$$

The voltage across the lineman is

$$V_{man} = I_{man} R_{man} = 79.2 \text{ V}$$

The voltage of the truck is

$$V_{truck} = I_{rod} R_{rod} = 0.8 \text{ V}$$

Note that the lineman inside the bucket is exposed to hazardous current.

EXAMPLE 8.12 Assume that a person is touching the truck in Example 8.11. Assume that the body resistance of the person is 3000 Ω and his foot resistance is 1000 Ω. Compute the voltage across the man, and the current through him.

Solution

The worst case scenario is when the truck potential is high. This is the case for part 3 of Example 8.11 when the truck voltage is 58.1 V. You can solve this problem by using Thevenin's theorem as explained earlier. Thevenin's voltage is 58.1 V. Examining the circuit in Figure 8.40 one would conclude that Thevenin's resistance is very small.

$$R_{th} \approx 0.001\ \Omega$$

Hence, the current through the man is

$$I_{\text{man}} = \frac{V_{\text{th}}}{R_{\text{th}} + R_{\text{man}} + 0.5R_f} = \frac{58.1}{0.001 + 3000 + 500} = 16.6 \text{ mA}$$

The current through the worker touching the truck is high, and the worker has two options: (1) to create a barricade around the truck where no person is allowed to enter the barricaded area, or (2) have a conductive mat under the truck where the worker stands on the mat and the mat is connected to the truck by a low-resistance ground wire.

The level of protection when an equipotential zone is used depends on many factors; the most important one is the use of ground wires with very low resistance. Repeat any of the above examples with higher resistance wires and you will find out that all workers could be exposed to primary shocks.

Exercise

8.1. A person is standing on a wet organic soil; compute his/her ground resistance.

8.2. Compute the resistance of a 10 m wide stretch of soil that is 2 m away from the edge of a grounded hemisphere. The radius of the hemisphere is 0.5 m. Assume that the soil is dry.

8.3. A person is working on a steel structure while standing on the ground. An accident occurrs where 5 A pass through the structure to the ground. The structure is grounded by a metal rod 6 cm in diameter. The rod is dug 2 m into the ground. The surrounding soil is of a dry type. Assume that the resistance of the man's body is 2000 Ω. Compute the current through the man.

8.4. Compute the survival time of the man in the previous problem by using Dalziel's formula. Assume that the weight of the man is 80 kg.

8.5. Repeat Problems 8.3 and 8.4 for wet organic soil.

8.6. During a weather storm, an atmospheric discharge hits a lightning pole. The pole is grounded through a hemisphere. The maximum value of the lightning current through the pole is 10 kA. The soil of the area is moist. A man who is walking 20 m away from the center of the hemisphere experiences an excessive step potential. The man's body resistance is 1500 Ω. Compute the current through his legs and his step potential.

8.7. During a weather storm, an atmospheric discharge hits a lightning pole that is grounded through a hemisphere. The maximum value of the lightning current through the rod is 20 kA. The soil of the area is moist. A man is playing golf 50 m away from the center of the hemisphere. At the moment of the lightning strike, the distance between his two feet is 0.4 m. Compute the current through the person's body assuming that his resistance is 1500 Ω.

8.8. Repeat the previous problem and assume that the person is 5 m away from the center of the hemisphere. What is the effect of the proximity of the man to the grounding hemisphere?

8.9. A power line insulator partially fails and 10 A pass through the structure to the tower's ground. The tower's ground is a hemisphere with a radius of 0.5 m. The soil resistivity is 100 Ωm. Assume that a man touches the tower while standing on the ground. Compute the current going through the man, assuming that his body resistance is 3000 Ω.

8.10. An electric circuit is powered by an unpolarized 120 V, 60 Hz outlet. The circuit is inside an ungrounded chassis. A 100 nf capacitance exists between the circuit

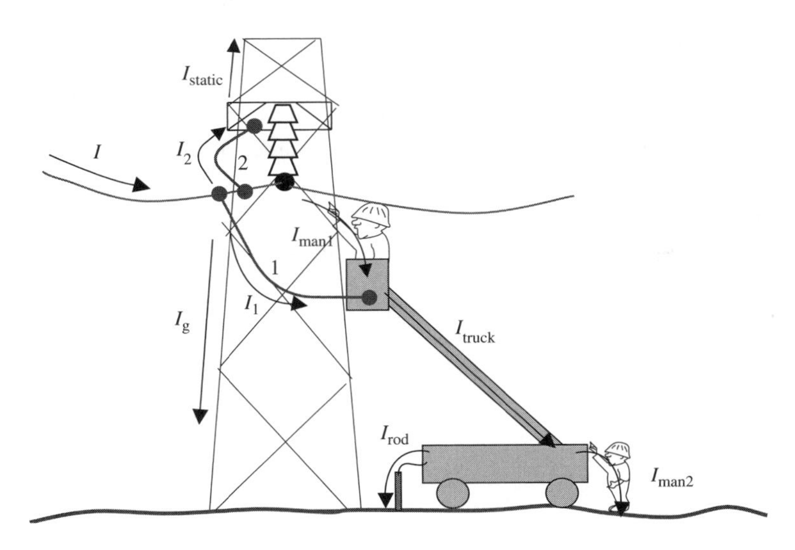

FIGURE 8.41
Hazard to workers on the ground due to a poor grounding method.

and the chassis. If a man touches the chassis, compute the current through his body. Assume that the body resistance of the man plus his ground resistance is 3000 Ω.

8.11. An electric circuit is powered by a two-prong polarized outlet through a feeder. The resistance of the neutral wire is 0.2 Ω. The chassis of the circuit is metallic and is connected to the neutral terminal of the outlet. A current of 200 A is drawn from the outlet through the hot wire. If a person with 2 kΩ resistance (including ground resistance) touches the chassis, compute the current through his body. Also state the type of hazard the person is exposed to.

8.12. State 6 factors that determine the severity of the electric shock when a person comes in contact with an energized conductor.

8.13. Compute the survival time of a heavy man receiving an electric shock of 100 mA.

8.14. A lineman working on a de-energized line uses only two ground wires as shown in Figure 8.41. One of the ground wires is connected between the tower and the conductor and the other between the bucket and the conductor. Assume an accident occurs, where the de-energized line comes in contact with another high-voltage circuit. This results in a fault current of 5 kA. Assume the following data:

The ground resistance of the rod (R_{rod}) = 30 Ω.

The ground resistance of the tower (R_{gt}) = 15 Ω.

The ground resistance of the static wire including the wire resistance (R_{static}) = 0.01 Ω.

The resistance of any ground wire (R_{gw}) = 0.002 Ω.

The body resistance of either man (R_{man}) = 3000 Ω.

The ground resistance (foot resistance) of the man on the ground (R_f) = 3000 Ω

Assume that the lineman fails to remove the high-resistance oxide layer at the connection point where ground wire 2 is attached to the tower. This results in a 1 Ω contact resistance. Ignore the resistance of the tower structure and evaluate the grounding system.

8.15. Repeat the previous problem assuming that the man on the ground is standing on a ground mat attached to the truck through a ground cable of 0.02 Ω.

9

Three-Phase Systems

High-power devices such as generators, transformers, and transmission lines are built as three-phase equipment. The three-phase system has many advantages over the single-phase system; the most important ones are:

1. The three-phase system produces rotating magnetic fields inside the ac motors, and therefore causes the motors to rotate without the need for extra controls. Since electric motors constitute the majority of the electric energy consumed worldwide, having a rotating field is a very important advantage of three-phase systems. The theory of the rotating fields is given in Chapter 12.
2. A three-phase generator produces more power than a single-phase generator of equivalent volume.
3. A three-phase transmission line transmits three times the power of a single-phase line.
4. The three-phase system is more reliable — when one phase is lost, the other two phases can still deliver some power to the loads.

9.1 Generation of Three-Phase Voltages

According to Faraday's law, if a conductor cuts magnetic field lines, a voltage is induced across the conductor. The synchronous generator uses this phenomenon to produce three-phase voltage. A simple schematic of the generator is shown in Figure 9.1 and a more detailed analysis of the machine is given in Chapter 12. The generator consists of an outer frame called a *stator*, and a rotating magnet called a *rotor*. At the inner perimeter of the stator, coils are placed inside slots. Each of the two slots separated by 180° houses a single coil (a–a', b–b', or c–c'). The coil is built by placing a wire inside a slot in one direction (e.g., a), and winding it back inside the opposite slot (a'). In Figure 9.1, we have three coils; each is separated by 120° from the other coils.

Now let us assume that the magnet is spinning clockwise inside the machine by an external prime mover. The magnetic field then cuts all coils and, therefore, induces voltage across each of them. If each coil is connected to load

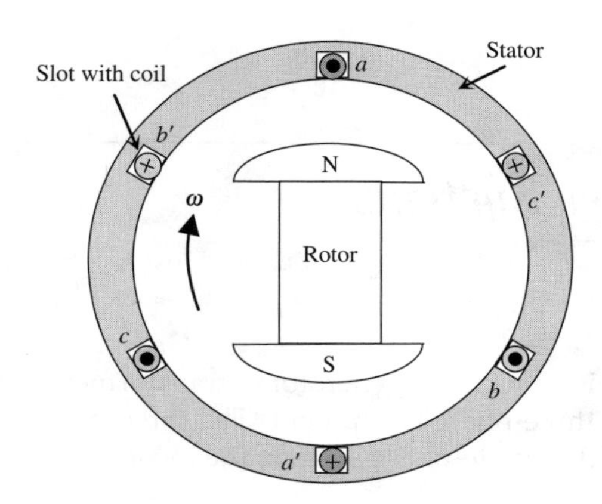

FIGURE 9.1
A three-phase generator.

impedance, a current would flow into the load and the generator would produce energy that is consumed by the load. The dot inside the coil indicates a current direction toward the reader, and the cross indicates a current in the opposite direction.

The voltage across any of the three coils can be expressed by Faraday's law.

$$e = NBl\omega \tag{9.1}$$

where e is the voltage induced across the coil, N is the number of turns in the coil, B is the flux density of the field, l is the length of the slot, and ω is the angular speed of the rotor.

The induced voltage on a conductor is proportional to the perpendicular component of the field with respect to the conductor. At the rotor position in Figure 9.1, coil a–a' has the maximum perpendicular flux, and, hence, has the maximum induced voltage. Coil b–b' will have its maximum voltage when the rotor moves clockwise by 120°, and coil c–c' will have its maximum voltage when the rotor is at 240°. If the rotation is continuous, the voltage across each coil is sinusoidal, all coils have the same magnitude of maximum voltage, and the induced voltages are shifted by 120° from each other as shown in Figure 9.2. These characteristics form what is known as a balanced three-phase system.

To express the waveforms of a balanced system mathematically, we need to select one of the waveforms as a reference and express all other waveforms in relation to it. Assuming that $v_{aa'}$ is our reference voltage, the waveforms in Figure 9.2 can be expressed by the following equations:

$$\begin{aligned} v_{aa'} &= V_{\max}\cos(\omega t) \\ v_{bb'} &= V_{\max}\cos(\omega t - 120°) \\ v_{cc'} &= V_{\max}\cos(\omega t - 240°) = V_{\max}\cos(\omega t + 120°) \end{aligned} \tag{9.2}$$

where $v_{aa'}$, $v_{bb'}$, and $v_{cc'}$ are the instantaneous voltages of the three coils; these voltages are known as *phase voltages*. The phase in this case is the coil; so

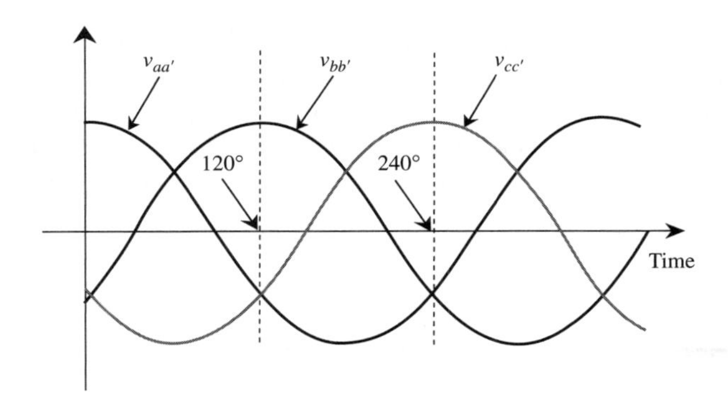

FIGURE 9.2
Waveforms of the three phases.

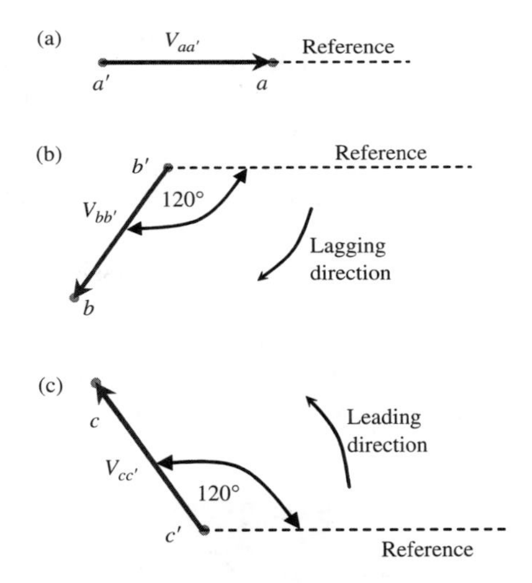

FIGURE 9.3
Phasor diagram of balanced three phases: (a) phasor diagram of phase $\overline{V}_{aa'}$; (b) phasor diagram of phase $\overline{V}_{bb'}$; and (c) phasor diagram of phase $\overline{V}_{cc'}$.

phase a is coil aa', phase b is coil bb', and phase c is coil cc'. Note that $v_{bb'}$ in Figure 9.2 lags $v_{aa'}$ by 120°, and $v_{cc'}$ leads $v_{aa'}$ by 120°. The maximum values of the phase voltages $V_{\max}$ are equal. As given in Chapter 7, the waveforms in Equation (9.2) can be written in the polar form.

$$
\begin{aligned}
\overline{V}_{aa'} &= \frac{V_{\max}}{\sqrt{2}} \angle 0^\circ = V \angle 0^\circ \\
\overline{V}_{bb'} &= \frac{V_{\max}}{\sqrt{2}} \angle -120^\circ = V \angle -120^\circ \\
\overline{V}_{cc'} &= \frac{V_{\max}}{\sqrt{2}} \angle 120^\circ = V \angle 120^\circ
\end{aligned}
\tag{9.3}
$$

where $\overline{V}$ is the phasor voltage in complex number and V is the magnitude of the rms voltage. As explained in Chapter 7, the phasor diagram of $\overline{V}_{aa'}$ (the reference voltage) can be drawn as shown in Figure 9.3(a). The direction of the arrow is from a' to a. The length of the arrow is equal to the rms value of $\overline{V}_{aa'}$. Similarly, the phasor diagrams of $\overline{V}_{bb'}$ and $\overline{V}_{cc'}$ are shown in Figure 9.3(b) and (c), respectively.

9.2 Connections of Three-Phase Circuits

The three-phase generator shown in Figure 9.1 has three independent coils and each coil represents a phase. Since each coil has two terminals, the generator has six terminals. To transmit the generated power from the power plant to the load centers, six wires seem to be needed. Since the transmission lines are often very long (hundreds of miles), the cost of the six wires is very high. To reduce the cost of the transmission system, the three coils are often connected in *wye* or *delta* configuration. In each of these configurations, only three wires are needed to transmit the power. These three wires are considered one circuit.

Figure 9.4 shows a photo of a 500 kV transmission line that consists of two circuits. Each circuit is a single three-phase configuration. The photo in Figure 9.5 shows a distribution pole with one circuit. The voltage level of this circuit is 15 kV. The photo also shows transformers mounted on the pole to step down the voltage from 15 kV to the household level.

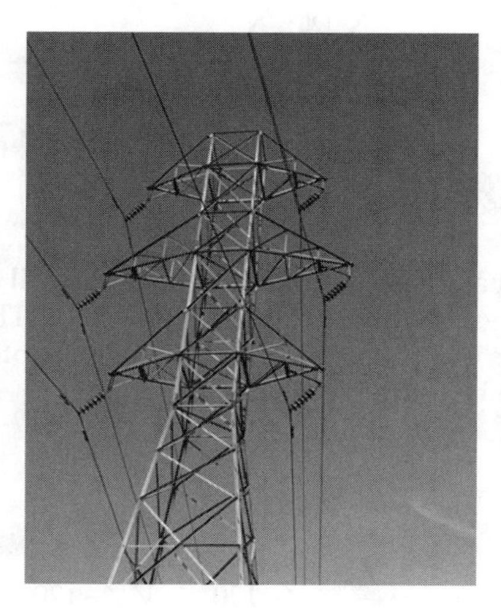

FIGURE 9.4
Transmission line with double circuits.

FIGURE 9.5
Single-circuit distribution line.

9.2.1 Wye-Connected Source

The wye connection is also referred to as "Y" or "star." For the generator in Figure 9.1, assume that the induced voltages in the three coils are represented by three voltage sources. To connect the coils in wye configuration, terminals a', b', and c' of the three coils are bonded into a common point called neutral or n as shown on the left side of Figure 9.6. Since the coils are mechanically separated by 120° with respect to each other, we can also draw the three-phase wye configuration in the convenient diagram on the right side of Figure 9.6.

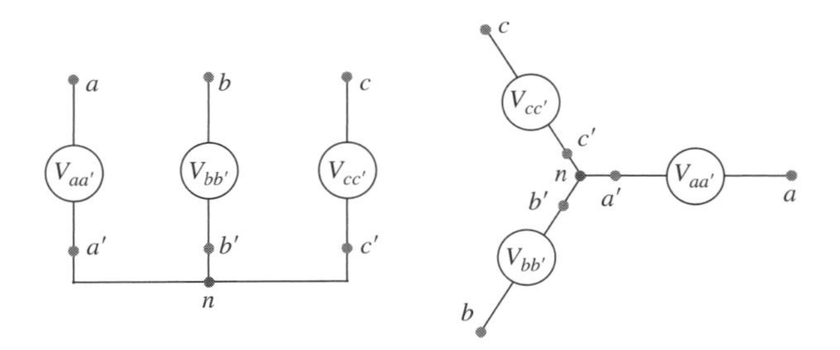

FIGURE 9.6
The connection of the three coils in wye.

The phasor diagram of the three-phase voltages in wye configuration can be obtained by grouping the three phasors in Figure 9.3 as shown in Figure 9.7(a). Since the common point is n, the phasors can be labeled $\overline{V}_{an}$, $\overline{V}_{bn}$, and $\overline{V}_{cn}$ as shown in Figure 9.7(b). Keep in mind that $\overline{V}_{an}$, $\overline{V}_{bn}$, and $\overline{V}_{cn}$ are the phase voltages.

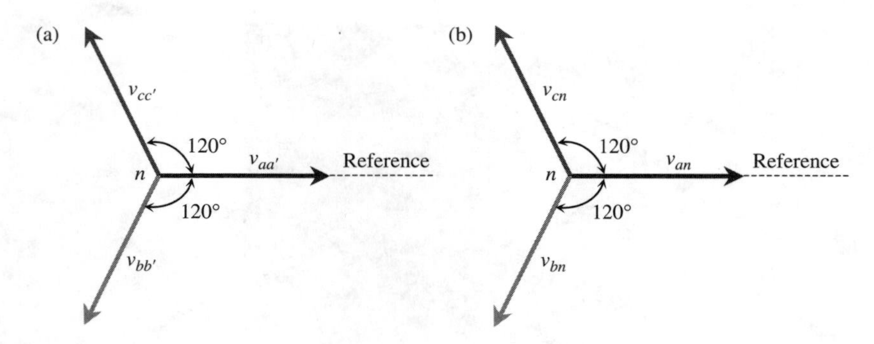

FIGURE 9.7
Phasor diagram of the three phases.

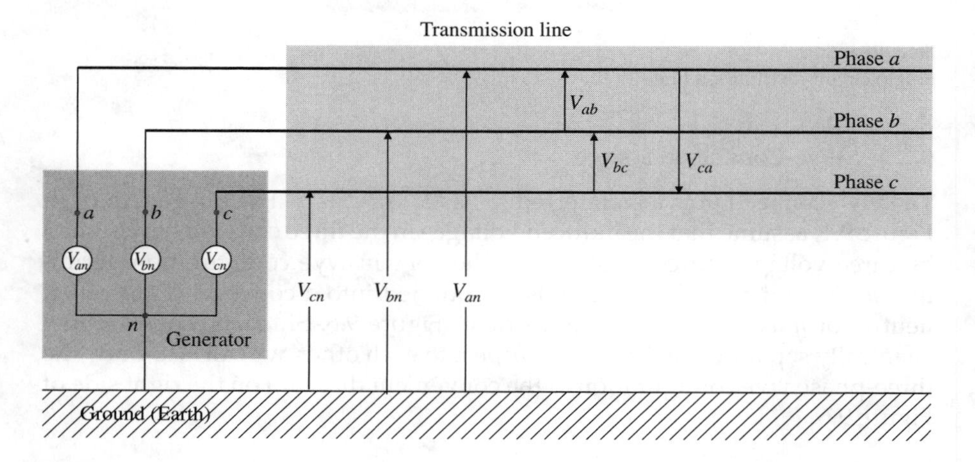

FIGURE 9.8
Three-phase wye generator connected to three-phase transmission line.

For balanced systems, the magnitudes of the rms voltages for all phases are equal, that is,

$$V_{an} = V_{bn} = V_{cn} \tag{9.4}$$

By this wye connection, the three terminals of the generator (*a*, *b*, and *c*) in Figure 9.6 are the only ones connected to the three-wire transmission line as shown in Figure 9.8. The neutral point is often connected to the ground and no additional wire is needed. The phase voltage of this system is the voltage between any line and the ground (or neutral). Hence,

$$V_{\text{ph}} = V_{an} = V_{bn} = V_{cn} \tag{9.5}$$

where V_{ph} is the magnitude of the phase voltage. Equation (9.3) can then be rewritten as

$$\begin{aligned} \overline{V}_{an} &= V_{\text{ph}} \angle 0^\circ \\ \overline{V}_{bn} &= V_{\text{ph}} \angle -120^\circ \\ \overline{V}_{cn} &= V_{\text{ph}} \angle 120^\circ \end{aligned} \tag{9.6}$$

The voltage between any two lines is known as the *line-to-line* voltage. The line-to-line voltage between a and b is $\overline{V}_{ab}$; it is the potential of phase a minus the potential of phase b.

$$\overline{V}_{ab} = \overline{V}_{an} - \overline{V}_{bn} \tag{9.7}$$

Substituting the values of $\overline{V}_{an}$ and $\overline{V}_{bn}$ of Equation (9.6) into Equation (9.7) yields

$$\overline{V}_{ab} = V_{\text{ph}} \angle 0^\circ - V_{\text{ph}} \angle -120^\circ = \sqrt{3} V_{\text{ph}} \angle 30^\circ \tag{9.8}$$

Two important observations can be made from Equation (9.8):

1. The magnitude of the line-to-line voltage is larger than the magnitude of the phase voltage by $\sqrt{3}$.
2. The line-to-line voltage $\overline{V}_{ab}$ leads the phase voltage $\overline{V}_{an}$ by 30°.

Similarly, the line-to-line voltages $\overline{V}_{bc}$ and $\overline{V}_{ca}$ can be computed as given in Equation (9.9).

$$\begin{aligned} \overline{V}_{bc} &= \overline{V}_{bn} - \overline{V}_{cn} = V_{\text{ph}} \angle -120^\circ - V_{\text{ph}} \angle 120^\circ = \sqrt{3} V_{\text{ph}} \angle -90^\circ \\ \overline{V}_{ca} &= \overline{V}_{cn} - \overline{V}_{an} = V_{\text{ph}} \angle 120^\circ - V_{\text{ph}} \angle 0^\circ = \sqrt{3} V_{\text{ph}} \angle 150^\circ \end{aligned} \tag{9.9}$$

Note that phasor $\overline{V}_{bc}$ leads $\overline{V}_{bn}$ by 30°, and phasor $\overline{V}_{ca}$ leads $\overline{V}_{cn}$ by 30°. The phasor diagram of all phase and line-to-line voltages is shown in Figure 9.9. Keep in mind that the direction of any phasor is determined by the subscript of the phasor; for example, the direction of $\overline{V}_{ca}$ is from point a to c.

From Equations (9.8) and (9.9), we can write the line-to-line voltages as

$$\begin{aligned} \overline{V}_{bc} &= \overline{V}_{ab} \angle -120^\circ \\ \overline{V}_{ca} &= \overline{V}_{ab} \angle 120^\circ \end{aligned} \tag{9.10}$$

Hence, the line-to-line voltages are also balanced, equal in magnitude, and shifted by 120° from each other. In a generic term, we can relate the magnitude of the line-to-line voltage V_{ll} to the phase voltage V_{ph} by

$$V_{\text{ll}} = \sqrt{3} V_{\text{ph}} \tag{9.11}$$

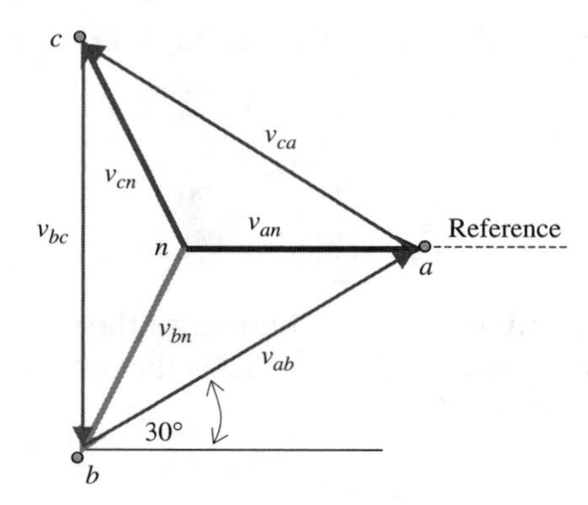

FIGURE 9.9
Phasor diagram of the phase and line-to-line voltages.

EXAMPLE 9.1 A balanced three-phase system has its phase voltage $\overline{V}_{an} = 120\ \angle 40°$ V. Compute the line-to-line voltage $\overline{V}_{bc}$.

Solution

Note that $\overline{V}_{an}$ is not in phase with the reference. Since $\overline{V}_{bc}$ lags $\overline{V}_{an}$ by 120°, and $\overline{V}_{cn}$ leads $\overline{V}_{an}$ by 120°,

$$\overline{V}_{bn} = \overline{V}_{an}\angle{-120°} = (120\angle 40°)\angle{-120°} = 120\angle{-80°}\ \text{V}$$
$$\overline{V}_{cn} = \overline{V}_{an}\angle 120° = (120\angle 40°)\angle 120° = 120\angle 160°\ \text{V}$$

The line-to-line voltages can be computed as follows:

$$\overline{V}_{bc} = \overline{V}_{bn} - \overline{V}_{cn} = (120\angle{-80°}) - (120\angle 160°)$$
$$\overline{V}_{bc} = 133.59 - j159.22 = 207.84\angle{-50°}\ \text{V}$$

Note that if we use the basic relationships between the phase and line-to-line voltages given in Equations (9.8) through (9.10), we can solve this problem without the need for complex number computations. First, we know that the magnitude of the line-to-line voltage is $\sqrt{3}V_{\text{ph}}$. Then

$$V_{ab} = \sqrt{3}V_{\text{ph}} = \sqrt{3} \times 120 = 207.84\ \text{V}$$

Second, we know that the angle of $\overline{V}_{ab}$ is leading $\overline{V}_{an}$ by 30° as given in Equation (9.8). Then

$$\overline{V}_{ab} = 207.84\ \angle(40° + 30°) = 207.84\angle 70°\ \text{V}$$

From Equation (9.10), $\overline{V}_{bc}$ lags $\overline{V}_{ab}$ by 120°.

$$\overline{V}_{bc} = \overline{V}_{ab}\angle{-120°} = 207.84\angle(70° - 120°) = 207.84\angle{-50°}\ \text{V}$$

9.2.2 Delta-Connected Source

The delta (Δ) configuration can be made by cascading the coils of the generator in Figure 9.1; terminal a' is connected to b, b' is connected to c, and c' is connected to a as shown on the left side of Figure 9.10. The voltage sources of the delta connection can also be drawn in the convenient diagram on the right side of Figure 9.10. Since point a' is the same as b, b' is the same as c, and c' is the same as a; hence

$$\begin{aligned}\overline{V}_{aa'} &= \overline{V}_{ab}\\ \overline{V}_{bb'} &= \overline{V}_{bc}\\ \overline{V}_{cc'} &= \overline{V}_{ca}\end{aligned} \tag{9.12}$$

The wiring connection and the phasor diagram of the delta configuration are shown in Figure 9.11. In the wiring connection, a', b', and c' are removed as they are the same points as b, c, and a, respectively. Since the reference for any phasor diagram can be arbitrarily selected, the phasor diagram in Figure 9.11 has $\overline{V}_{ab}$ as a selected reference.

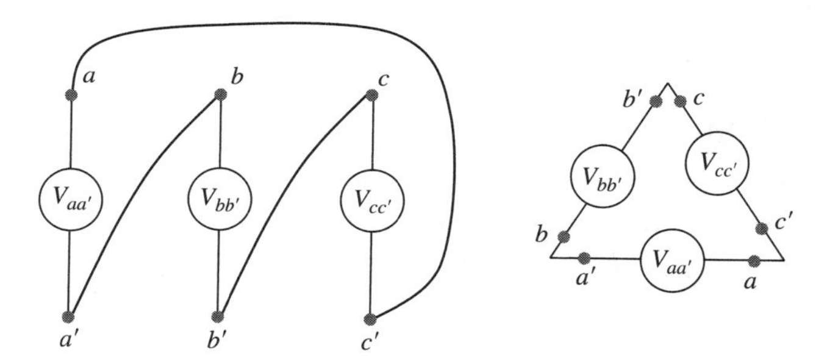

FIGURE 9.10
Delta connection.

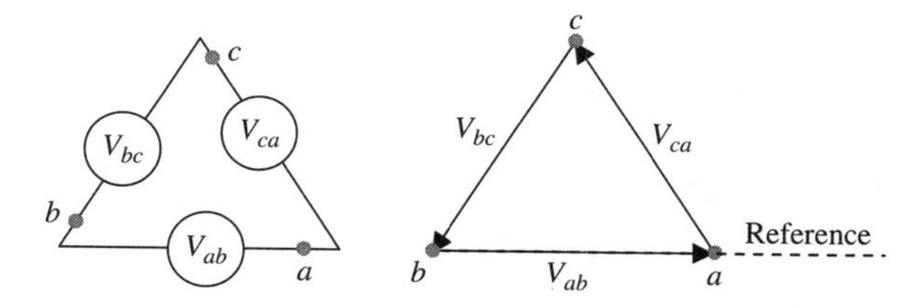

FIGURE 9.11
Delta connection and its phasor diagram.

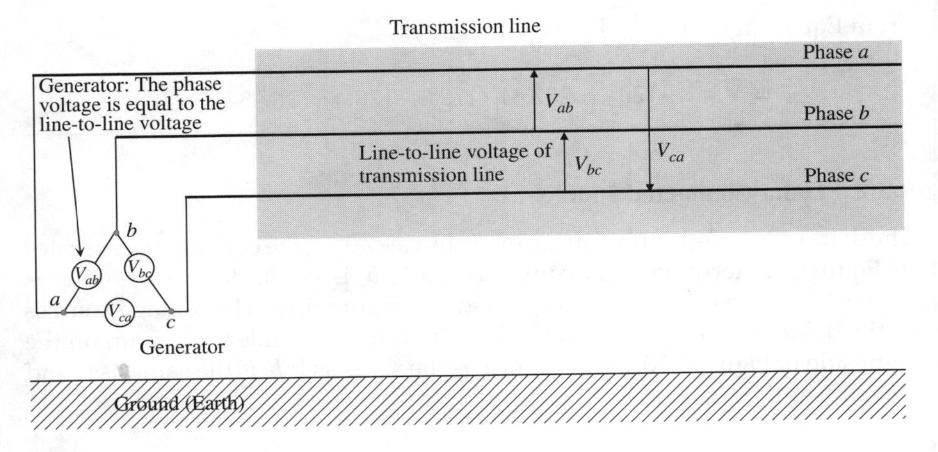

FIGURE 9.12
A three-phase generator connected to a three-phase transmission line.

As we stated earlier, $\overline{V}_{aa'}$ is the voltage of coil aa', which is defined as the phase voltage of the generator. Also, $\overline{V}_{ab}$ is defined as the line-to-line voltage of the generator. Since $\overline{V}_{aa'} = \overline{V}_{ab}$ in a delta-connected generator, the phase voltage is equal to the line-to-line voltage.

A delta-connected generator is attached to a three-phase transmission line as shown in Figure 9.12. Note that the delta connection has only three wires with no neutral terminal.

9.2.3 Wye-Connected Load

Residential loads in the United States are normally single-phase loads at 120 or 240 V. Small loads that draw small currents such as lights, televisions, and radios are powered at 120 V. The heavier loads such as furnaces, stoves, and dryers are powered at 240 V. The industrial and commercial loads are mostly three-phase loads at various line-to-line voltage levels (208 V, 480 V, 5 kV, 15 kV). As a general rule, the larger the power demand, the higher is the voltage.

Although household loads are single-phase loads, clustered residential areas are powered by three phases. Each group of houses is powered by one phase while other groups are powered by the other phases. This is done to balance the loads among the three phases.

The three-phase loads are connected in various configurations with the most common ones being the wye and the delta. All residential loads and most commercial and industrial loads are connected in wyes to ensure that the voltage across the load is constant regardless of any fluctuation in the load currents. Figure 9.13 shows a wye-connected load. Each of Z_{an}, Z_{bn}, and Z_{cn}

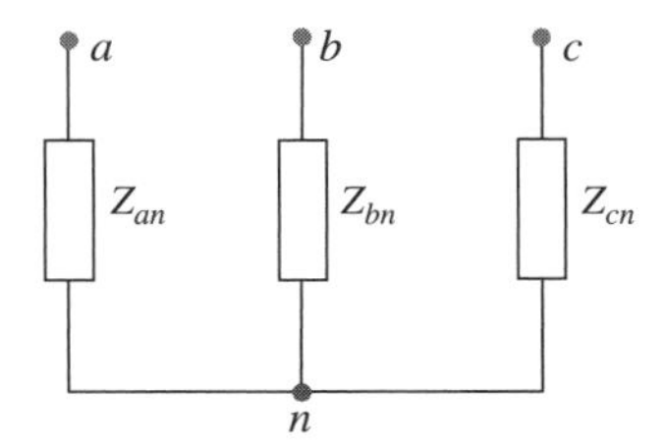

FIGURE 9.13
Three loads connected in wye.

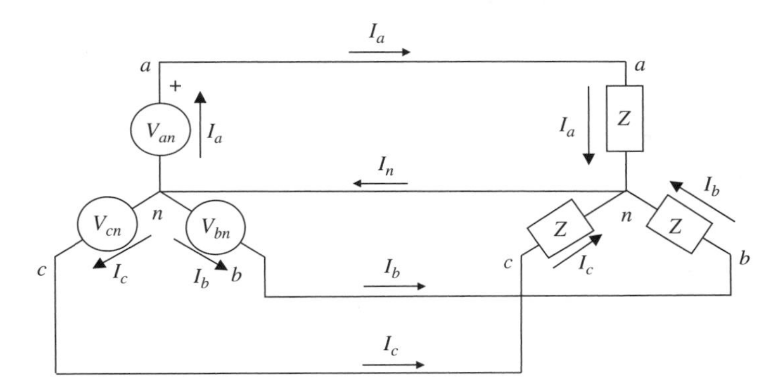

FIGURE 9.14
Wye connected load powered by wye connected source.

is known as the *phase impedance*. For balanced systems, the phase impedance of each phase is equal to that of the other phases, that is,

$$\overline{Z}_{an} = \overline{Z}_{bn} = \overline{Z}_{cn} = \overline{Z} \tag{9.13}$$

For residential loads, each Z_{an}, Z_{bn}, and Z_{cn} represents a group of houses. Although the demands in any neighborhood differ from house to house due to the various energy consumption habits of the customers, the three-phase loads are almost balanced over a wide residential area.

Figure 9.14 shows a wye-connected load powered by a wye-connected source. Four lines connect the load to the source. The neutral points of the load and that for the source are connected either directly by wire or indirectly through the ground.

Assume that the load is balanced and the voltage $\overline{V}_{an}$ is shifted from an arbitrary reference by an angle θ. The load currents, which flow from the source to the load, can then be computed as

$$\begin{aligned}
\bar{I}_a &= \frac{\bar{V}_{an}}{\bar{Z}_{an}} = \frac{\bar{V}_{an}}{\bar{Z}} = \frac{V_{\text{ph}}\angle\theta}{Z\angle\phi} = \frac{V_{\text{ph}}}{Z}\angle(\theta-\phi) \\
\bar{I}_b &= \frac{\bar{V}_{bn}}{\bar{Z}_{bn}} = \frac{\bar{V}_{bn}}{\bar{Z}} = \frac{V_{\text{ph}}\angle(\theta-120^\circ)}{Z\angle\phi} = \frac{V_{\text{ph}}}{Z}\angle(\theta-\phi-120^\circ) \\
\bar{I}_c &= \frac{\bar{V}_{cn}}{\bar{Z}_{cn}} = \frac{\bar{V}_{cn}}{\bar{Z}} = \frac{V_{\text{ph}}\angle(\theta+120^\circ)}{Z\angle\phi} = \frac{V_{\text{ph}}}{Z}\angle(\theta-\phi+120^\circ)
\end{aligned} \tag{9.14}$$

Four observations can be made by examining Equation (9.14):

1. The magnitudes of $\bar{I}_a$, $\bar{I}_b$, and $\bar{I}_c$ are equal.
2. The current $\bar{I}_a$ leads $\bar{I}_b$ by 120°, and $\bar{I}_b$ leads $\bar{I}_c$ by 120°.
3. The currents of the source (known as the *phase currents* of the source) are equal to the currents flowing in the lines connecting the source and the load (known as the *line currents*).
4. The line currents are equal to the currents flowing into the phase impedances (also known as the *phase current* of the load).

The currents and voltages of the circuit in Figure 9.14 are shown in the phasor diagram in Figure 9.15. Using Kirkhoff's current rule, the sum of the three currents $\bar{I}_a$, $\bar{I}_b$, and $\bar{I}_c$ entering the neutral node n of the load is equal to the current exiting the node $\bar{I}_n$. Hence,

$$\bar{I}_n = \bar{I}_a + \bar{I}_b + \bar{I}_c \tag{9.15}$$

If the load is balanced, the sum of all load currents in Equation (9.14) is zero.

$$\bar{I}_n = \bar{I}_a + \bar{I}_a\angle{-120^\circ} + \bar{I}_a\angle 120^\circ = \bar{I}_a(1 + 1\angle{-120^\circ} + 1\angle 120^\circ) = 0 \tag{9.16}$$

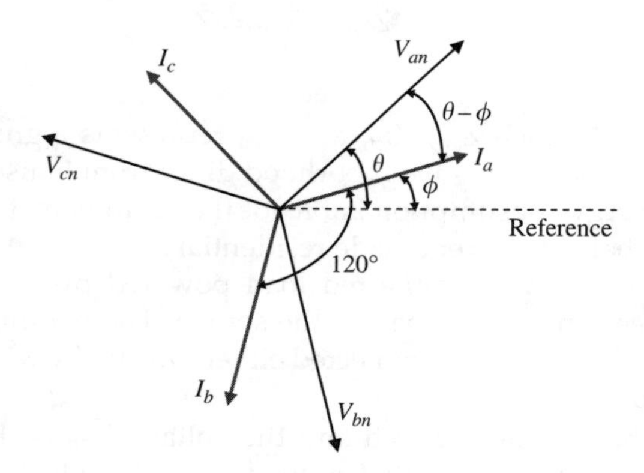

FIGURE 9.15
Phasor diagram of loads connected in wye.

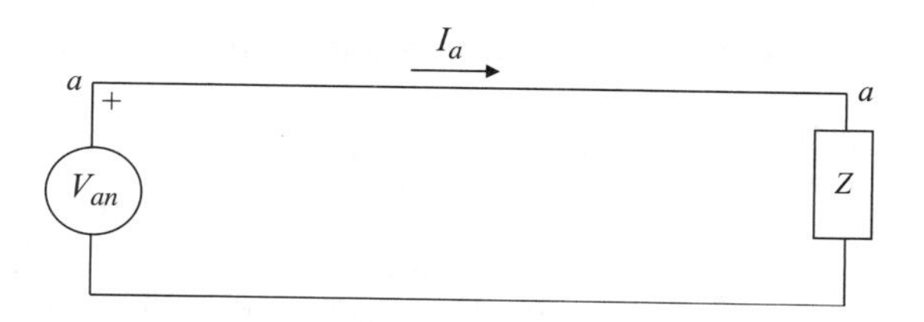

FIGURE 9.16
Single-phase representation of the circuit in Figure 9.14.

Since $\overline{I}_n = 0$, there is no need to have a wire connecting the neutral points of the wye-connected generator and the wye-connected load.

For the three-phase system in Figure 9.14, there are three independent loops with the neutral wire as a common branch. If the system is balanced, we need to analyze one phase only; the voltages and currents of the other two phases can be obtained by the relationships in Equations (9.10) and (9.14). Figure 9.16 shows a single-phase representation of a balanced three-phase system.

EXAMPLE 9.2 The phase voltage of a balanced three-phase source is $\overline{V}_{an} = 120\angle{-40^\circ}$ V. The source is energizing a balanced three-phase load connected in a wye. The phase impedance of the load is $\overline{Z} = 20\angle 30^\circ\ \Omega$. Compute the following:

1. The load current of each phase.
2. The neutral current.
3. The line-to-line voltages of the load.

Solution

Use the single-phase representation of the balanced three-phase system shown in Figure 9.16.

1. $\overline{I}_a = \dfrac{\overline{V}_{an}}{\overline{Z}} = \dfrac{120\angle{-40^\circ}}{20\angle 30^\circ} = 6\angle{-70^\circ}$ A

   The currents of the other phases can be computed by the relationships of the balanced system.

$$\overline{I}_b = \overline{I}_a\angle{-120^\circ} = 6\angle{-190^\circ}\ \text{A}$$
$$\overline{I}_c = \overline{I}_a\angle 120^\circ = 6\angle 50^\circ\ \text{A}$$

2. $\overline{I}_n = \overline{I}_a + \overline{I}_b + \overline{I}_c = 6\angle{-70^\circ} + 6\angle{-190^\circ} + 6\angle 50^\circ = 0$

3. $\overline{V}_{ab} = \sqrt{3}\ \overline{V}_{an}\angle 30^\circ = \sqrt{3}(120\angle{-40^\circ})\angle 30^\circ = 207.84\angle{-10^\circ}$ V
   $\overline{V}_{bc} = \overline{V}_{ab}\angle{-120^\circ} = (207.84\angle{-10^\circ})\angle{-120^\circ} = 207.84\angle{-130^\circ}$ V
   $\overline{V}_{ca} = \overline{V}_{ab}\ \angle 120^\circ = 207.84\ \angle 110^\circ$ V

The main conclusions of the wye-connected load are the following:

1. The magnitude of the line-to-line voltage is greater than the phase voltage by $\sqrt{3}$.
2. The line-to-line voltage $\overline{V}_{ab}$ leads the phase voltage of the load $\overline{V}_{an}$ by 30°. Similar conclusions can be made for the other two phases.
3. The line currents are equal to the phase currents of the load.
4. No current flows through the neutral wire, so there is no need for a wire between the neutral points in Figure 9.14.

9.2.4 Delta-Connected Load

A balanced three-phase load connected in delta is shown in Figure 9.17. The load impedance of each phase is connected between two lines. Hence, the voltage across any load impedance is the line-to-line voltage. Although the load impedance is connected between lines, it is still called *phase impedance*. There is no line-to-line impedance.

The line currents feeding the load are $\overline{I}_a$, $\overline{I}_b$, and $\overline{I}_c$. The currents of the load impedances are $\overline{I}_{ab}$, $\overline{I}_{bc}$, and $\overline{I}_{ca}$, which are known as the *phase currents* (they are the currents of the phase impedances). The relationship between the line and phase currents can be obtained by examining Kirkhoff's nodal equations at a, b, and c.

$$\begin{aligned}\overline{I}_{ab} &= \overline{I}_a + \overline{I}_{ca} \\ \overline{I}_{bc} &= \overline{I}_b + \overline{I}_{ab} \\ \overline{I}_{ca} &= \overline{I}_c + \overline{I}_{bc}\end{aligned} \tag{9.17}$$

The phase currents can be computed by dividing the phase voltage (voltage across the phase impedance) by the phase impedance. However, the voltage

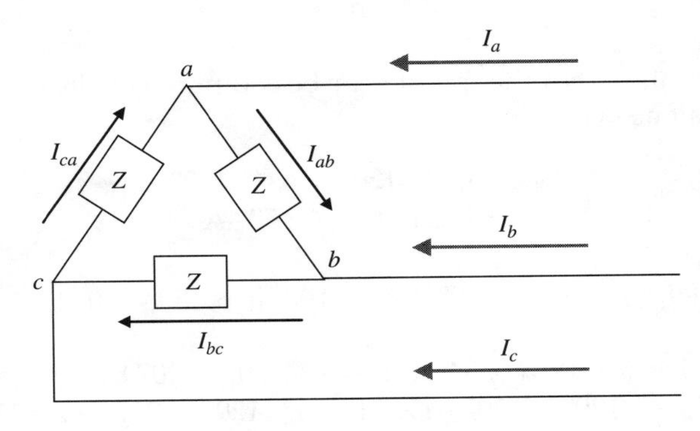

FIGURE 9.17
Three-phase load connected in delta.

across the phase impedance is equal to the line-to-line voltage. Thus,

$$\begin{aligned} \overline{I}_{ab} &= \frac{\overline{V}_{ab}}{\overline{Z}} \\ \overline{I}_{bc} &= \frac{\overline{V}_{bc}}{\overline{Z}} \\ \overline{I}_{ca} &= \frac{\overline{V}_{ca}}{\overline{Z}} \end{aligned} \tag{9.18}$$

Since the impedances of all phases are equal and the line-to-line voltages are balanced, the phase currents must also be balanced.

EXAMPLE 9.3 The phase voltage of a balanced three-phase source is $\overline{V}_{an} = 120\angle{-40^\circ}$ V. The source energizes a three-phase load connected in delta. The phase impedance of the load is $\overline{Z} = 10\angle{-30^\circ}\ \Omega$. Compute the phase currents of the load.

Solution

The first step is to compute the phase voltages of the load, which are the same as the line-to-line voltages of the source.

$$\begin{aligned} \overline{V}_{ab} &= \sqrt{3}\ \overline{V}_{an}\angle 30^\circ = \sqrt{3}(120\angle{-40^\circ})\angle 30^\circ = 207.84\ \angle{-10^\circ}\ \text{V} \\ \overline{V}_{bc} &= \overline{V}_{ab}\angle{-120^\circ} = 207.84\ \angle{-130^\circ}\ \text{V} \\ \overline{V}_{ca} &= \overline{V}_{ab}\angle{+120^\circ} = 207.84\ \angle 110^\circ\ \text{V} \end{aligned}$$

Now compute the phase currents.

$$\begin{aligned} \overline{I}_{ab} &= \frac{\overline{V}_{ab}}{\overline{Z}} = \frac{207.84\angle{-10^\circ}}{10\angle{-30^\circ}} = 20.784\ \angle 20^\circ\ \text{A} \\ \overline{I}_{bc} &= \frac{\overline{V}_{bc}}{\overline{Z}} = \frac{207.84\angle{-130^\circ}}{10\angle{-30^\circ}} = 20.784\ \angle{-100^\circ}\ \text{A} \\ \overline{I}_{ca} &= \frac{\overline{V}_{ca}}{\overline{Z}} = \frac{207.84\angle 110^\circ}{10\angle{-30^\circ}} = 20.784\ \angle 140^\circ\ \text{A} \end{aligned}$$

Note that the phase currents are equal in magnitude and are separated by 120° with respect to each other.

As seen in Example 9.3, the phase currents are balanced. Let us rewrite Equation (9.17) as given below.

$$\begin{aligned} \overline{I}_a &= \overline{I}_{ab} - \overline{I}_{ca} \\ \overline{I}_b &= \overline{I}_{bc} - \overline{I}_{ab} \\ \overline{I}_c &= \overline{I}_{ca} - \overline{I}_{bc} \end{aligned} \tag{9.19}$$

Let us assume that the phase current $\bar{I}_{ab}$ is the chosen reference. Then,

$$\begin{aligned} \bar{I}_{ab} &= I_{\text{ph}} \angle 0^\circ \\ \bar{I}_{bc} &= \bar{I}_{ab} \angle{-120^\circ} = I_{\text{ph}} \angle{-120^\circ} \\ \bar{I}_{ca} &= \bar{I}_{ab} \angle 120^\circ = I_{\text{ph}} \angle 120^\circ \end{aligned} \tag{9.20}$$

where I_{ph} is the magnitude of the phase current. Now substitute the values of the phase currents of Equation (9.20) into Equation (9.19).

$$\begin{aligned} \bar{I}_a &= I_{\text{ph}} \angle 0 - I_{\text{ph}} \angle 120^\circ = \sqrt{3}\, I_{\text{ph}} \angle{-30^\circ} = \sqrt{3}\, \bar{I}_{ab} \angle{-30^\circ} \\ \bar{I}_b &= I_{\text{ph}} \angle{-120^\circ} - I_{\text{ph}} \angle 0 = \sqrt{3}\, I_{\text{ph}} \angle{-150^\circ} = \sqrt{3}\, \bar{I}_{bc} \angle{-30^\circ} \\ \bar{I}_c &= I_{\text{ph}} \angle 120^\circ - I_{\text{ph}} \angle{-120^\circ} = \sqrt{3}\, I_{\text{ph}} \angle 90^\circ = \sqrt{3}\, \bar{I}_{ca} \angle{-30^\circ} \end{aligned} \tag{9.21}$$

As seen in Equation (9.21), the line current $\bar{I}_a$ is greater than the phase current $\bar{I}_{ab}$ by $\sqrt{3}$, and lags the phase current by 30°. Similar relationships can be obtained for the other two phases.

EXAMPLE 9.4 Calculate the line currents in Example 9.3. Also sketch the phasor diagram showing the line-to-line voltages, phase currents, and line currents.

Solution

$$\begin{aligned} \bar{I}_a &= \sqrt{3}\, \bar{I}_{ab} \angle{-30^\circ} = \sqrt{3}(20.784 \angle 20^\circ) \angle{-30^\circ} = 36 \angle{-10^\circ} \text{ A} \\ \bar{I}_b &= \sqrt{3}\, \bar{I}_{bc} \angle{-30^\circ} = \sqrt{3}(20.784 \angle{-100^\circ}) \angle{-30^\circ} = 36 \angle{-130^\circ} \text{ A} \\ \bar{I}_c &= \sqrt{3}\, \bar{I}_{ca} \angle{-30^\circ} = \sqrt{3}(20.784 \angle 140^\circ) \angle{-30^\circ} = 36 \angle 110^\circ \text{ A} \end{aligned}$$

The phasor diagram is sketched in Figure 9.18.

The main conclusions of the delta-connected circuits are the following:

1. The magnitude of the line-to-line voltage is equal to the phase voltage across the load.
2. The line current $\bar{I}_a$ lags the phase current $\bar{I}_{ab}$ by 30°. Similar conclusions can be made for the other two phases.
3. The line current is greater than the phase current by $\sqrt{3}$.
4. The delta connection has no neutral terminal.

9.2.5 Circuits with Mixed Connections

The source and load are not always connected in the same fashion; the load could be connected in delta and the source in wye, or vise versa. In either

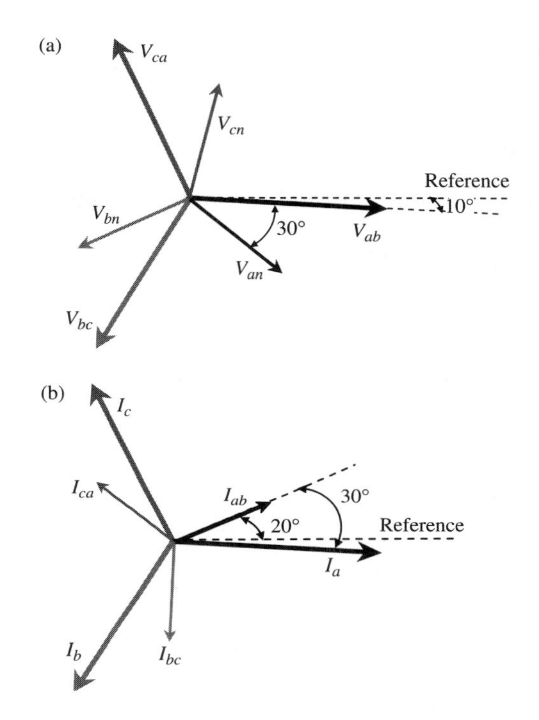

FIGURE 9.18
Phasor diagram of Example 9.4: (a) phasor diagram of voltages and, (b) of currents.

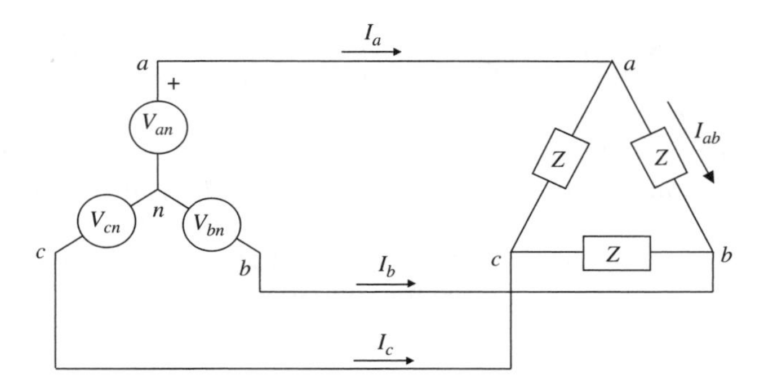

FIGURE 9.19
Circuit with wye source and delta load.

case, careful attention must be given to the line and phase quantities. Examine Example 9.5 carefully.

EXAMPLE 9.5 Figure 9.19 shows a balanced delta-connected load energized by a balanced wye-connected source. The phase voltage of the source is

$\overline{V}_{an} = 120\angle 0°$ V, and the line current is $\overline{I}_c = 10\angle 75°$ A. Compute the load impedance.

Solution

To compute the impedance, we need to compute the voltage across the load and the current of the load.

The voltage across the load is

$$\overline{V}_{ab} = \sqrt{3}\,\overline{V}_{an}\angle 30° = \sqrt{3}(120\angle 0°)\angle 30° = 207.84\ \angle 30°\text{ V}$$

The current of the load can be computed by using the relationship

$$\overline{I}_a = \sqrt{3}\,\overline{I}_{ab}\angle -30°$$

where $\overline{I}_a = \overline{I}_c\angle -120° = (10\angle 75°)\angle -120° = 10\angle -45°$ A.

$$\overline{I}_{ab} = \frac{10\angle -45°}{\sqrt{3}\angle -30°} = 5.774\angle -15°\text{ A}$$

Hence,

$$\overline{Z} = \frac{\overline{V}_{ab}}{\overline{I}_{ab}} = \frac{207.84\angle 30°}{5.774\angle -15°} = 36\angle 45°\ \Omega$$

The loads in three-phase circuits can be connected in parallel even if their configurations are different. An example is shown in Figure 9.20 where the source is wye-connected, one load is delta-connected, and the other is wye-connected. This circuit can be easily analyzed by using the superposition theorem. For example, to solve for the line current $\overline{I}_a$, you can first compute the contribution of each load to the line current independently. Then sum up the two contributions to find the line current due to both loads. Hence, the circuit in Figure 9.20 can be divided into the two subcircuits in Figure 9.21 and

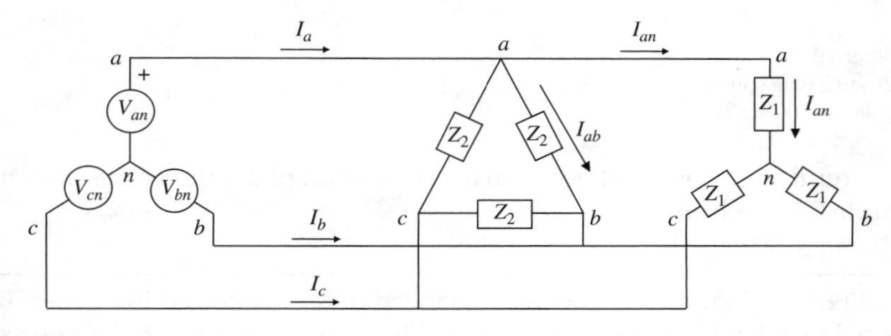

FIGURE 9.20
Circuit with wye and delta loads.

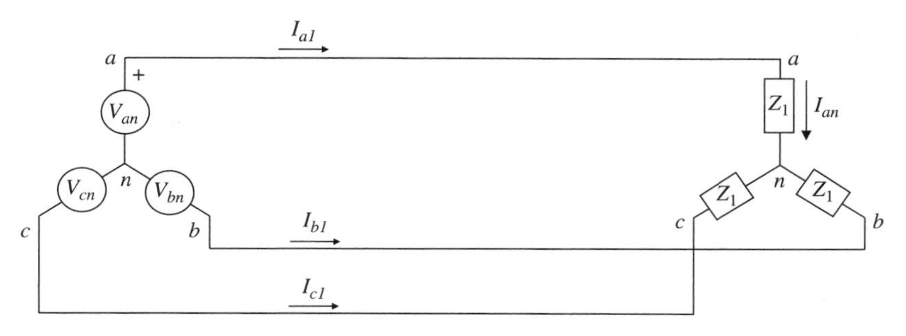

FIGURE 9.21
Contribution of the wye load.

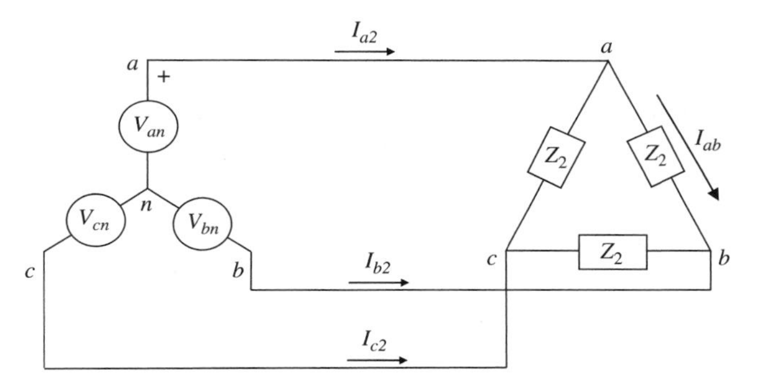

FIGURE 9.22
Contribution of the delta load.

Figure 9.22. In Figure 9.21, the contribution of the wye-connected load to the total line current is $\overline{I}_{a1}$. In Figure 9.22, the contribution of the delta-connected load to the total line current is $\overline{I}_{a2}$. The line current due to both loads is just the sum of $\overline{I}_{a1}$ and $\overline{I}_{a2}$.

$$\overline{I}_a = \overline{I}_{a1} + \overline{I}_{a2} \tag{9.22}$$

EXAMPLE 9.6 Assume that the source voltage in Figure 9.20 is $\overline{V}_{an} = 120\angle 0^\circ$ V, the phase impedance of the wye load is $\overline{Z}_1 = 10\angle 20^\circ\ \Omega$, and the phase impedance of the delta load is $\overline{Z}_2 = 15\angle{-30^\circ}\ \Omega$. Compute the line currents of the source.

Solution

First find the contribution of the wye load to the line current $\overline{I}_{a1}$.

$$\overline{I}_{a1} = \overline{I}_{an} = \frac{\overline{V}_{an}}{\overline{Z}_1} = \frac{120\angle 0}{10\angle 20^\circ} = 12\angle{-20^\circ}\ \text{A}$$

Now let us compute the contribution of the delta load $\overline{I}_{a2}$.

$$\overline{I}_{ab} = \frac{\overline{V}_{ab}}{\overline{Z}_2} = \frac{\sqrt{3}\,\overline{V}_{an}\angle 30^\circ}{15\angle -30^\circ} = \frac{\sqrt{3}(120\angle 0)\angle 30^\circ}{15\angle -30^\circ} = 13.85\ \angle 60^\circ\ \text{A}$$

$$\overline{I}_{a2} = \sqrt{3}\,\overline{I}_{ab}\angle -30^\circ = \sqrt{3}(13.85\angle 60^\circ)\angle -30^\circ = 24\ \angle 30^\circ\ \text{A}$$

The total line current at the source side is

$$\overline{I}_a = \overline{I}_{a1} + \overline{I}_{a2} = 12\angle -20^\circ + 24\angle 30^\circ = 33\ \angle 13.84^\circ\ \text{A}$$

9.2.6 Wye–Delta Transformation

When the source and the load are both in wye or delta, the analysis of the circuit is simple. However, when the connections are different, the process is more involved as you must be attentive to the phase shifts and the changes in magnitude between the line and phase quantities. An alternative method is to find an equivalent load connection that matches that of the source. For example, if the source is wye-connected and the load is delta-connected, it would be easier to analyze the circuit if the delta load is replaced by an equivalent wye load. This is known as the Y–Δ transformation.

Let us consider the delta load in Figure 9.23 where the phase impedance of the load is labeled $\overline{Z}_\Delta$. Our objective is to find its equivalent wye load shown in Figure 9.24. The phase impedance of the equivalent wye is $\overline{Z}_Y$. But how does one find the value of $\overline{Z}_Y$ so that the two circuits are equivalent? The simplest way is to treat both circuits as an impedance box with three terminals. The two circuits are equivalent if the voltage applied between any two terminals in both circuits produces the same terminal currents. This means the impedance between any two terminals in the delta circuit is equal to the impedance between the same terminals in the wye circuit. For example, the impedance measured between terminals a and b, $\overline{Z}_{ab}$, must be the same for both circuits.

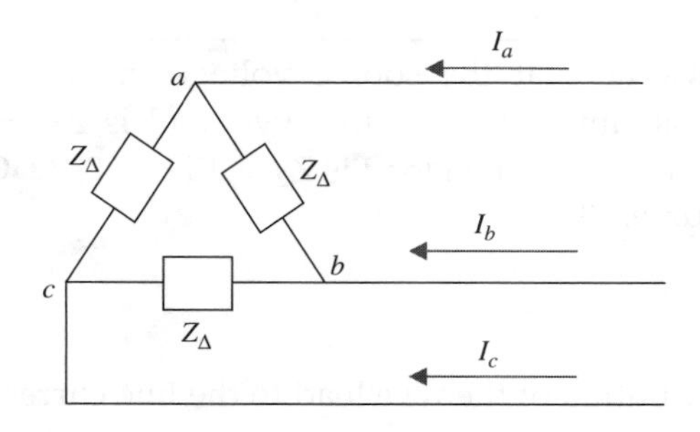

FIGURE 9.23
Delta load.

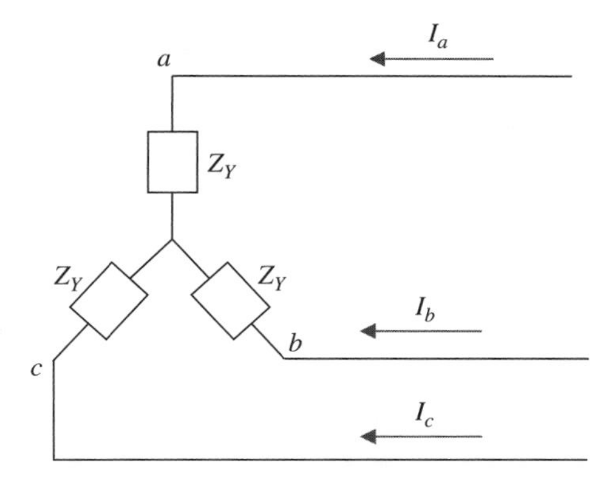

FIGURE 9.24
An equivalent wye load to the delta load in Figure 9.23.

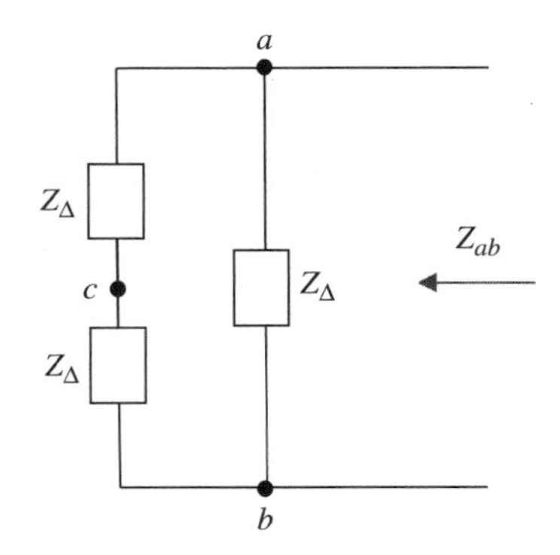

FIGURE 9.25
Terminal impedance of delta load.

$\overline{Z}_{ab}$ of the delta-connected load can be quickly computed by rearranging the impedances in Figure 9.23 as shown in Figure 9.25. It is easy to see that the impedance measured between terminals a and b is a parallel combination of $\overline{Z}_\Delta$ and $2\overline{Z}_\Delta$ as given in Equation (9.23). Keep in mind that the third terminal c is unconnected.

$$\overline{Z}_{ab} = \frac{\overline{Z}_\Delta(2\overline{Z}_\Delta)}{3\overline{Z}_\Delta} = \frac{2}{3}\overline{Z}_\Delta \tag{9.23}$$

Now let us do the same with the equivalent wye load. Rearrange the wye load as shown in Figure 9.26. Since the third terminal c is floating, the impedance between terminals a and b is a series combination of two $\overline{Z}_Y$ as given in Equation (9.24).

$$\overline{Z}_{ab} = 2\overline{Z}_Y \tag{9.24}$$

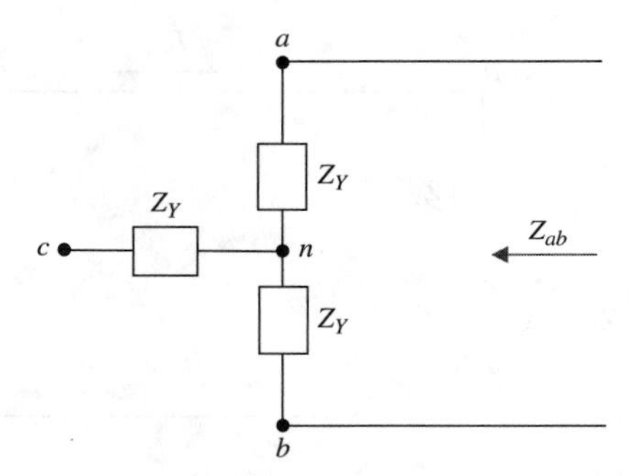

FIGURE 9.26
Terminal impedance of wye load.

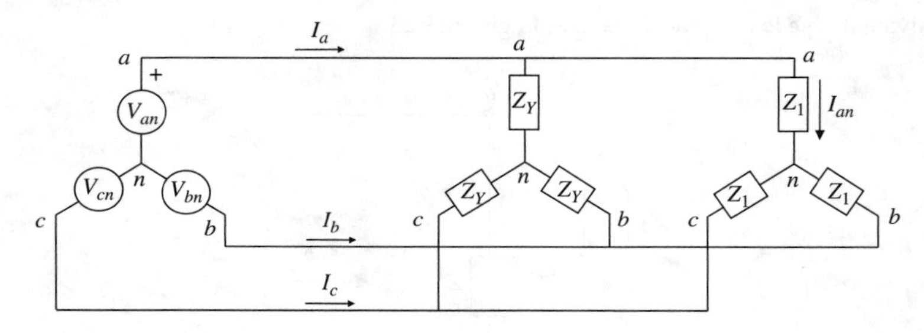

FIGURE 9.27
Equivalent circuit of the system in Figure 9.20.

To make the wye load equivalent to the original delta load, the terminal impedances of both circuits must be equal; that is, Equations (9.23) and (9.24) are equal. Hence,

$$\bar{Z}_Y = \frac{\bar{Z}_\Delta}{3} \tag{9.25}$$

The relationship in Equation (9.25) is valid for transforming wye into delta or delta into wye.

EXAMPLE 9.7 Repeat Example 9.6 by using the wye–delta transformation.

Solution

Convert the delta load in Figure 9.20 to wye.

$$\bar{Z}_Y = \frac{\bar{Z}_\Delta}{3} = \frac{\bar{Z}_2}{3} = \frac{15\angle -30°}{3} = 5\ \angle -30°\ \Omega$$

After the delta is converted into wye, we have two wye-connected loads in parallel as shown in Figure 9.27. These two loads can be replaced by one

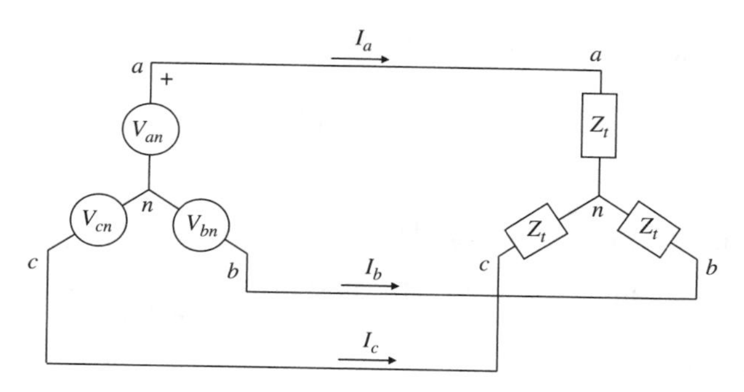

FIGURE 9.28
Equivalent wye load.

equivalent wye load as shown in Figure 9.28. The new value of the equivalent load impedance is

$$\overline{Z}_t = \frac{\overline{Z}_Y \overline{Z}_1}{\overline{Z}_Y + \overline{Z}_1} = \frac{(5\angle -30°)(10\angle 20°)}{(5\angle -30°) + (10\angle 20°)} = 3.6343\ \angle -13.84°\ \Omega$$

Now, the line current can be directly computed as

$$\overline{I}_a = \frac{\overline{V}_{an}}{\overline{Z}_t} = \frac{120\angle 0}{3.6343\angle -13.84°} = 33\ \angle 13.84°\ \text{A}$$

This process is obviously much simpler than that in Example 9.6.

9.3 Power Calculations of Three-Phase Circuits

The power in a three-phase circuit is the sum of the powers of each phase. The power per phase is

$$\begin{aligned} P_{\text{ph}} &= V_{\text{ph}} I_{\text{ph}} \cos\theta \\ Q_{\text{ph}} &= V_{\text{ph}} I_{\text{ph}} \sin\theta \end{aligned} \tag{9.26}$$

where V_{ph} is the phase voltage (the voltage across the phase impedance), I_{ph} is the phase current (the current in the phase impedance), and θ is the power factor angle (the angle between the phase voltage and the phase current). θ is also the impedance angle.

Since the system is balanced, the three-phase power is just three times the power of a single-phase.

$$P = 3P_{\text{ph}} = 3V_{\text{ph}} I_{\text{ph}} \cos\theta \tag{9.27}$$

A similar equation can be obtained for the reactive power.

$$Q = 3Q_{ph} = 3V_{ph}I_{ph}\sin\theta \tag{9.28}$$

9.3.1 Three-Phase Power of Wye Loads

When the load is connected in wye, the line current I_l is the same as the phase current I_{ph}, and the line-to-line voltage V_{ll} is greater than the phase voltage V_{ph} by $\sqrt{3}$, hence

$$P = 3P_{ph} = 3V_{ph}I_{ph}\cos\theta = 3\frac{V_{ll}}{\sqrt{3}}\,I_l\cos\theta \tag{9.29}$$

Rewriting Equation (9.29) yields

$$P = \sqrt{3}\,V_{ll}I_l\cos\theta \tag{9.30}$$

Similarly, the reactive power is

$$Q = \sqrt{3}\,V_{ll}I_l\sin\theta \tag{9.31}$$

Keep in mind that θ is the angle between the *phase voltage* and the *phase current*.

9.3.2 Three-Phase Power of Delta Loads

The phase voltage across the impedance of the delta load is the same as the line-to-line voltage, and the line current is equal to the phase current multiplied by $\sqrt{3}$. Hence,

$$P = 3P_{ph} = 3V_{ph}I_{ph}\cos\theta = 3V_{ll}\frac{I_l}{\sqrt{3}}\cos\theta \tag{9.32}$$

Equation (9.32) can be rewritten as

$$P = \sqrt{3}\,V_{ll}I_l\cos\theta \tag{9.33}$$

Similarly, the three-phase reactive power is

$$Q = \sqrt{3}\,V_{ll}I_l\sin\theta \tag{9.34}$$

Note that Equations (9.30) and (9.33) are identical. Also, Equations (9.31) and (9.34) are identical. Therefore the power of a three-phase circuit can be computed by the same equation regardless of the load connection. Keep in mind that θ is the angle between the phase voltage and the phase current in both cases. In other words, θ is the angle of the load impedance.

EXAMPLE 9.8 A three-phase source of $\overline{V}_{an} = 120\angle 0^\circ$ V is powering a delta-connected load of $\overline{Z} = 15\angle 20^\circ$. Compute the real and reactive powers consumed by the load.

Solution

Compute the magnitude of the line-to-line voltage V_{ll}.

$$V_{ll} = \sqrt{3}V_{an} = 207.84 \text{ V}$$

Now compute the line current, but first calculate the magnitude of the phase current I_{ab}.

$$I_{ab} = \frac{V_{ab}}{Z} = \frac{V_{ll}}{Z} = \frac{207.84}{15} = 13.85\ A$$
$$I_l = \sqrt{3}I_{ab} = \sqrt{3}13.85 = 24 \text{ A}$$

The phase angle of the load is given as 20°. Hence the real and reactive powers are

$$P = \sqrt{3}V_{ll}I_l\cos\theta = \sqrt{3}\times 207.84\times 24\times\cos 20^\circ = 8.12 \text{ kW}$$
$$Q = \sqrt{3}V_{ll}I_l\sin\theta = \sqrt{3}\times 207.84\times 24\times\sin 20^\circ = 2.95 \text{ kVAr}$$

EXAMPLE 9.9 A three-phase source of $\overline{V}_{an} = 120\angle 0^\circ$ V delivers 10 kW and 5 kVAr to a wye-connected load. Compute the load impedance.

Solution

Compute the power factor angle θ.

$$\theta = \tan^{-1}\left(\frac{Q}{P}\right) = \tan^{-1}\left(\frac{5}{10}\right) = 26.56^\circ$$

Compute the line-to-line voltage.

$$V_{ll} = \sqrt{3}\ V_{an} = 207.84 \text{ V}$$

Use the power equation to compute the line current.

$$P = \sqrt{3}\ V_{ll}I_l\cos\theta$$
$$10{,}000 = \sqrt{3}\times 207.84\times I_l\times\cos 26.56^\circ$$
$$I_l = 31.05 \text{ A}$$

The line current is the same as the phase current in wye loads. Now compute the magnitude of the load impedance.

$$Z = \frac{V_{ph}}{I_{ph}} = \frac{120}{31.05} = 3.86\ \Omega$$

The complex value of the load impedance is

$$\overline{Z} = Z(\cos\theta + j\sin\theta) = 3.86(\cos 26.56 + j\sin 26.56) = 3.45 + j1.73\ \Omega$$

Exercise

9.1. A wye-connected balanced three-phase source is feeding a balanced three-phase load. The phase voltage and phase current of the source are

$$v(t) = 340\sin(377t + 0.5236)\text{V}$$
$$i(t) = 100\sin(377t + 0.87266)\text{A}$$

Calculate the following:

a. The rms phase voltage.
b. The rms line-to-line voltage.
c. The rms phase current.
d. The rms line current.
e. The frequency of the supply.
f. The power factor at the source side; state leading or lagging.
g. The three-phase real power delivered to the load.
h. The three-phase reactive power delivered to the load.
i. If the load is connected in delta configuration, calculate the load impedance.

9.2. The current and voltage of a wye-connected load are: $\overline{V}_{ca} = 480\angle{-60°}$ V, $\overline{I}_b = 20\angle 120°$ A.

a. Compute $\overline{V}_{an}$.
b. Compute $\overline{I}_a$.
c. Compute the power factor angle.
d. Compute the real power of the load.

9.3. A three-phase, 480-volt system is connected to a balanced three-phase load. The line current I_a is 10 A, and is in phase with the line-to-line voltage V_{bc}. Calculate the impedance of the load for the following cases:

a. If the load is wye-connected.
b. If the load is delta-connected.

9.4. The waveforms of the line-to-line voltage and line current of a three-phase delta-connected load are

$$v_{ll} = V_{max} \sin \omega t$$
$$i_l = I_{max} \sin(\omega t - 50°)$$

Calculate the power factor angle.

9.5. A balanced wye-connected load with a per-phase impedance of $(4 + j3)\Omega$ is connected across a three-phase source of 173 V (line-to-line).

a. Find the line current, the power factor, the complex power, the real power, and the reactive power consumed by the load.

b. With V_{ab} as the reference, sketch the phasor diagram that shows all voltages and currents.

9.6. Two three-phase wye-connected loads are in parallel across a three-phase supply. The first load draws a phase current of 20 A at 0.9 power factor leading, and the second load draws a phase current of 30 A at 0.8 power factor lagging. Calculate the following:

a. The line current from the source side and its power factor.

b. The real power supplied by the source if the supply voltage is 400 V.

9.7. The line-to-line voltage of a three-phase system is $\overline{V}_{bc} = 340 \angle 20°$ V.

a. Calculate the phase voltage $\overline{V}_{an}$.

b. If the load impedance is connected in wye, and the impedance per phase is $Z = 10 \angle 60°\ \Omega$, calculate the current $\overline{I}_b$.

c. Calculate the current in the neutral line $\overline{I}_n$.

9.8. A three-phase motor is rated at 5.0 Hp. What is the power of the motor per phase in kW?

9.9. The waveform of an ac voltage can be expressed by $V = 180 \sin(300t + 3)$V.

a. Calculate the rms value of the voltage.

b. Calculate the frequency of the supply.

c. Calculate the phase shift in degrees.

9.10. The following are the voltage and current measured for a wye-connected load.

$$\overline{V}_{ab} = 200 \angle 50°\ \text{V}$$
$$\overline{I}_c = 10 \angle 140°\ \text{A}$$

a. Calculate the power factor angle.

b. Calculate the real power consumed by the load.

9.11. A delta-connected source energizes two parallel loads. One of the loads is connected in delta and the other in wye. The line-to-line voltage of the source is 208 V. The delta load has a phase impedance of $\overline{Z}_\Delta = 10 \angle -25°\ \Omega$. The wye load has a phase impedance of $\overline{Z}_Y = 5\angle 40°\ \Omega$. Compute the line current of phase a.

9.12. A three-phase wye-connected source energizes a delta-connected load. The phase voltage of the source is $\overline{V}_{bn} = 120\angle 0°$ V and the phase impedance of the load is $\overline{Z} = 9\angle 30°\ \Omega$. Compute $\overline{I}_a$ and the power consumed by the delta load.

10

Power Electronics

The electric power grid must operate at fixed voltage and fixed frequency to ensure that all generators are synchronized and the entire system is stable. If any generator produces voltage or frequency different from the rest of the system, blackouts can occur, as discussed in Chapter 13. Because of these limitations, the early productions of electrical equipment were directly energized from the utility grid without any conditioning for the voltage or frequency. Therefore, they were limited in performance, heavy in weight, and inefficient in operation.

With the development of power electronic devices and circuits, newer equipment often have converters that allow the user to change the voltage and frequency applied to the equipment. The new washing machine, for example, allows the user to change the speed and torque of the washing cycle by adjusting the voltage and frequency applied to the electric motor of the washer. The new air conditioners operate at continuous and variable speeds to maintain the environment at the users' settings with minimum deviation, rather than the old inefficient switch-in switch-out operation.

Besides the household usage, power electronics provides the industry with effective methods to save energy and improve performance. Among the main advantages of using power electronics are:

- Electric loads that demand different waveforms from that provided by the utility can still be energized from the utility supply through a proper converter. For example, the motors used in all printers, scanners, and hard disks cannot operate at the utility voltage or frequency, but can still be energized from the utility by using the proper power electronic converters.
- Power electronic circuits can increase the efficiency of the system for a wide range of operating conditions.
- Equipment operating through power electronic devices is more reliable and last longer.
- Equipment designed to operate through power electronic converter is lighter in weight and smaller in volume from the ones designed to operate at the utility's fixed voltage and frequency.

- The performance of the electric load can be greatly enhanced, and the load can perform functions that are difficult to implement without power electronics. For example, the smooth speed control and braking of the electric car cannot be achieved without the use of power electronics circuits.
- Electrical equipment can be easily and effectively controlled by power electronic converters.

The history of power electronics started with the vacuum tube invented in 1904 by the British scientist John Fleming. His vacuum tube was quite revolutionary because it allowed the electric current to flow in one direction only (similar to the diode). Later developments created tubes that amplified signals and changed the shape of the electric waveforms. After the end of World War II, the brilliant scientists Bill Shockley, John Bardeen, and Walter Brattain invented the first transistor in 1948. Their invention opened the door wide to the use of solid-state devices in almost all electrical equipment manufactured today. The research team named their device "transistor" because it has "trans-resistance," that is, variable resistance. In 1956, these scientists received the Nobel Prize for their marvelous transistor.

Since the invention of the transistors, scientists worked diligently to invent other semiconductor devices with broader characteristics and higher power ratings. The early transistors were rated at a small fraction of an ampere and at very low voltages. Nowadays, solid-state devices can be built to withstand about 5 kV and can carry thousands of amperes.

10.1 Power Electronic Devices

Power electronic devices can be divided into two categories: single component and hybrid component. The single component is composed of one solid-state device, and the hybrid component is a combination of several single components manufactured as a single block. The single devices are loosely divided into three types: (1) the two-layer devices such as the diodes, (2) the three-layer devices such as the transistors, and (3) the four-layer devices such as the thyristors. The hybrid components include a wide range of devices such as the Insulated Gate Bipolar Transistor (IGBT), Static Induction Transistor (SIT), and Darlington Transistor (DT).

Solid-state devices are the main building components of any converter. Their function is mainly to mimic the mechanical switches by connecting and disconnecting electric loads, but at very high speeds. When the mechanical switch is open (off), the current through the switch is zero and the voltage across its terminals is equal to the source voltage. When the switch is closed (on), its terminal voltage is zero and its current is determined by the load impedance. The characteristics of the ideal mechanical switch are

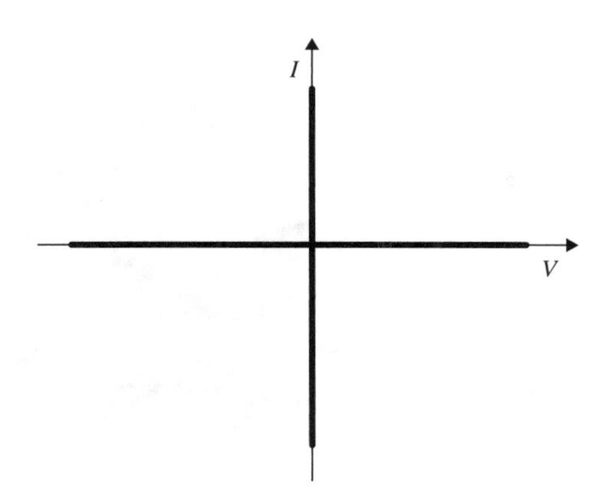

FIGURE 10.1
Current–voltage characteristics of a mechanical switch.

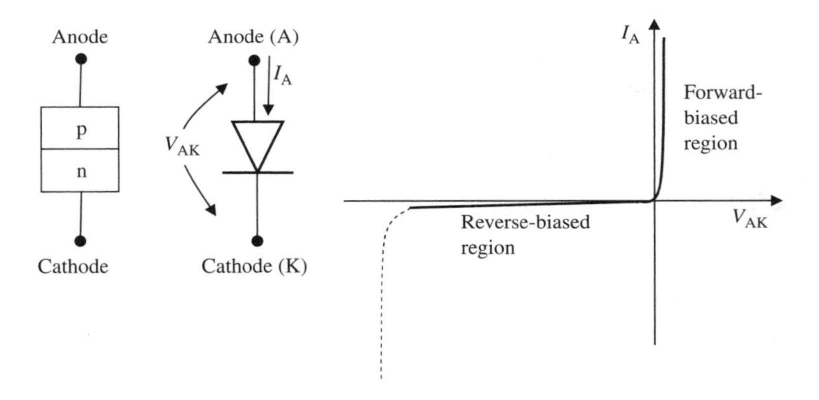

FIGURE 10.2
Solid-state diode.

shown in Figure 10.1. A major part of the ongoing research and development in power electronic devices is devoted to improving the characteristics of solid-state devices to make them as close to the mechanical switch as possible.

10.1.1 Solid-State Diodes

The diode, which is the simplest power electronic device, allows the current to flow in one direction only. It is built out of two semiconductor materials p-type and n-type, placed in contact with each other as shown on the left side of Figure 10.2. The symbol of the diode is shown in the middle of the figure. The p-junction is the *anode* of the device and the n-junction is the *cathode*. The current of the diode flows only from the anode to the cathode when the anode-to-cathode voltage V_{AK} is positive (forward biased) as shown at the right side of the figure. When the voltage is reversed (reverse biased), almost no current

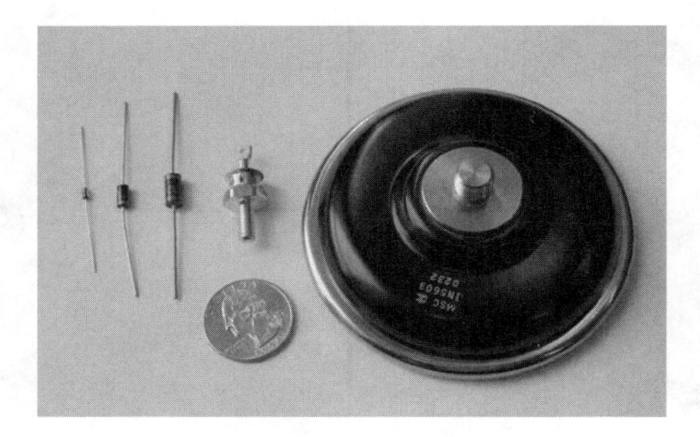

FIGURE 10.3
Several diodes.

flows through the diode. However, if the reverse-biased voltage is very high, the diode is destroyed by the current avalanche shown by the dashed section of the characteristics. When the diode is forward biased, the voltage drop across the diode V_{AK} is very small (about 0.7 V), and the diode in this case resembles a closed mechanical switch. When the diode is reverse biased (the thick portion of the line), the diode's current is very small and the diode resembles an open mechanical switch.

Figure 10.3 shows a picture of several diodes ranging from 6 V, 10 mA to the large one rated at 5 kV, 5 A. Higher rating diodes can be found for thousands of kiloamperes. For higher voltage diodes, several single component diodes are often connected in series to achieve a voltage level of as high as 1 MV.

10.1.2 Transistors

In power electronic applications, the transistors are mainly used as high-frequency electronic controlled switches. Two main types of transistors are commonly used: the Bipolar Junction Transistor (BJT) and the Field Effect Transistor (FET).

10.1.2.1 Bipolar junction transistor

The BJT consists of three layers of semiconductor material in the n-p-n or p-n-p arrangements; the n-p-n shown on the left side of Figure 10.4 is the most common configuration in power electronic applications. The middle layer of the transistor is very thin compared with the other two layers. The three layers of the transistor are called *collector* (C), *base* (B), and *emitter* (E). The bipolar transistor can be viewed as a pair of solid-state diodes joined back-to-back. But because the middle layer is fairly thin, the transistor behaves differently

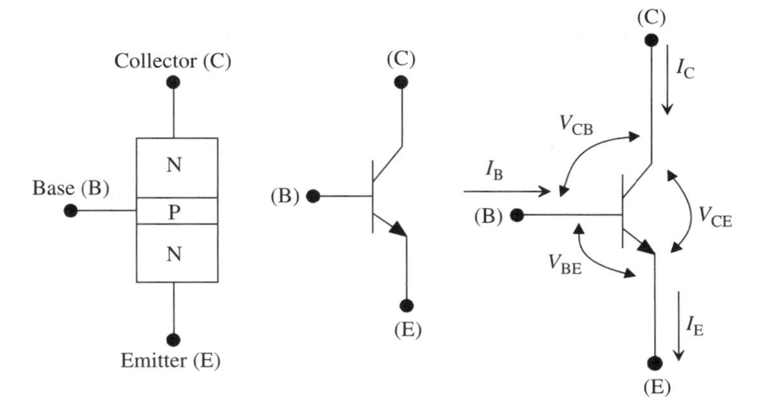

FIGURE 10.4
Bipolar junction transistor.

FIGURE 10.5
Various size transistors.

from just two back-to-back diodes. If we apply a positive voltage between the collector and emitter V_{CE}, one of the back-to-back diodes is reverse biased. If we apply a small positive voltage between the base and emitter V_{BE}, the base–emitter junction is forward biased, and a current flows between the base and emitter as if the junction is a forward-biased diode. In this case, electrons move from the emitter to the base. Once the electrons reach the base layer, they are quickly attracted to the collector layer because of its high positive bias voltage. Hence, a current I_C flows through the collector terminal. This collector current is a function of the voltage we apply between the base and emitter.

Figure 10.5 shows bipolar power transistors of various sizes. The rating of the largest transistor is 100 A.

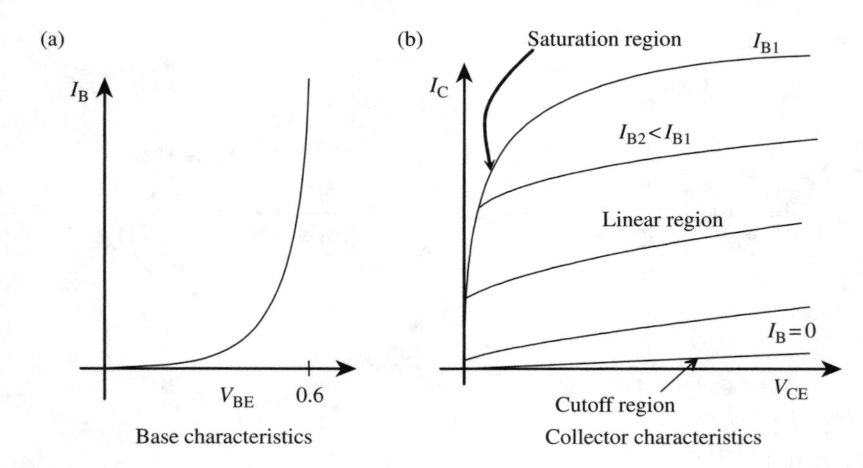

FIGURE 10.6
Characteristics of the BJT.

The basic equations of the BJT are

$$I_C = \beta I_B \tag{10.1}$$

$$I_E = I_B + I_C \tag{10.2}$$

$$V_{CE} = V_{CB} + V_{BE} \tag{10.3}$$

where β is the current gain (the ratio of collector to base currents), I_B is the base current, I_E is the emitter current, V_{CE} is the collector–emitter voltage, V_{CB} is the collector–base voltage, and V_{BE} is the base–emitter voltage.

The characteristics of the transistor are shown in Figure 10.6. The base characteristic (I_B versus V_{BE}) is very similar to the characteristic of the diode in the forward-biased mode. In the forward direction the base–emitter voltage V_{BE} is below 0.7 V. The collector characteristic is I_C versus V_{CE} for various values of base currents I_B. This characteristic can be divided into three regions: linear, cutoff, and saturation. In the linear region, the transistor operates as an amplifier, where β is almost constant and in the order of a few hundreds. Hence, any base current is amplified a few hundred times in the collector circuit. This is the region where most audio amplifiers operate.

The cutoff region is the area of the characteristic where the base current is zero. In this case, the collector current is negligibly small regardless of the value of the collector–emitter voltage. In the saturation region, the collector–emitter voltage is very small compared to the transistor ratings, and the collector current is determined by the load. β in the saturation region is often less than 30.

When a transistor is used to emulate a mechanical switch, it operates in the cutoff and saturation regions. The power transistors cannot operate in the linear region for long periods because V_{CE} in the linear region is high and the

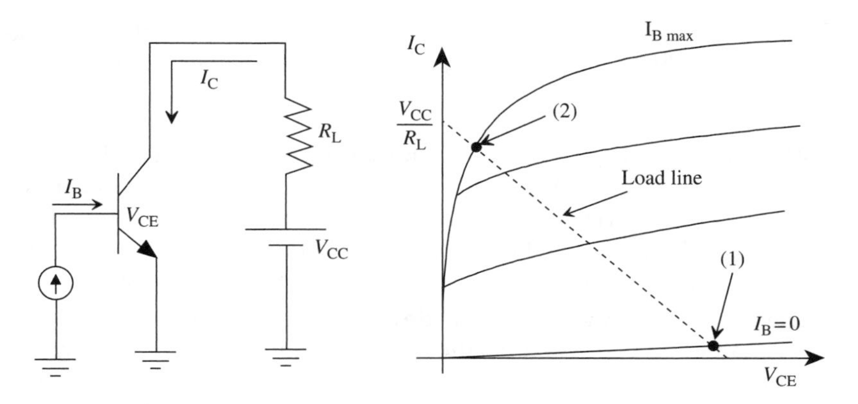

FIGURE 10.7
Switching of a transistor.

current of the power transistor is often high. Therefore, the power losses of the transistor ($V_{CE}I_C$) are excessive and eventually damage the transistor due to the excessive thermal heat.

The circuit in Figure 10.7 explains how the transistor operates as a switch. The transistor is connected to an external circuit consisting of a dc source V_{CC} and a load resistance R_L. The base circuit of the transistor is connected to a separate current source to produce the base current I_B. The loop equation of the collector circuit is

$$V_{CC} = V_{CE} + R_L I_C \tag{10.4}$$

The collector equation, which is known as the *load line equation*, shows a linear relationship between I_C and V_{CE}. The equation has a negative slope and intersects the x-axis at V_{CC}, and the y-axis at V_{CC}/R_L. If the base current is zero, the operating point of the circuit is in the cutoff region at point (1). The collector current in this case is very small and the collector–emitter voltage of the transistor is almost equal to the source voltage V_{CC}. Hence, the operation of the transistor in the cutoff region resembles an open mechanical switch. If the base current is set to the maximum value, the transistor is in the saturation region at point (2). At this operating point, the voltage drop across the collector–emitter is very small and the collector (or load) current is almost equal to V_{CC}/R_L. The transistor in this case is equivalent to a closed mechanical switch.

The bipolar transistor emulates a normally open mechanical switch where a control current is needed to close the switch and keep it latched in the closed position. The control current of the transistor is the base current I_B. To open the transistor, the base current is set to zero.

The bipolar junction transistor has several advantages; the best among them are its high switching speeds and high reliability. However, the BJT has two main drawbacks:

1. When the transistor is closed in the saturation region, V_{CE} is about 0.1 V, which is very low for high-voltage circuits. However, if the transistor carries high currents while closed, the losses could be excessive, leading to high internal heat. This heat must be dissipated as quickly as it is generated to prevent the transistor from being damaged.
2. The BJT is a current-controlled device; the transistor is closed and maintained in the closed position only if the maximum base current is present. Since β is small in the saturation region, the base current is a large part of the collector current. This high base current causes high losses in the base circuit. Furthermore, the driving circuit that generates the base current must be capable of producing a large current for as long as the transistor is closed. Such a circuit is large in size and complex to build.

EXAMPLE 10.1 A bipolar junction transistor has a current gain of 10 in the saturation region. When the transistor is closed, and the collector current is 100 A, compute the following:

1. The base current to keep the transistor closed.
2. The losses of the transistor.

Solution

1. $I_B = I_C/\beta = \frac{100}{10} = 10$ A
   Note that the base current is very high and may require an elaborate driving circuit.
2. The losses inside the transistor are mainly in the base–emitter loops and the collector–emitter loop.
   The base–emitter losses are

$$P_{BE} = I_B V_{BE}$$

Since the base–emitter junction is just a diode in the forward bias, we can assume that V_{BE} is about 0.7 V.

$$P_{BE} = I_B V_{BE} = 10 \times 0.7 = 7 \text{ W}$$

The collector–emitter losses are

$$P_{CE} = I_C V_{CE}$$

Assume V_{CE} is about 0.1 V when the transistor is fully closed.

$$P_{CE} = I_C V_{CE} = 100 \times 0.1 = 10 \text{ W}$$

Total losses = 17 W.

EXAMPLE 10.2 In the saturation region, the BJT circuit shown in Figure 10.7 has a current gain of 20 and $V_{CE} = 0.5$ V. The source voltage is 100 V and the load resistance is 5 Ω.

1. Calculate the minimum base current that maintains the transistor closed.
2. Calculate the losses in the collector–emitter loop.
3. Assume the base current is reduced to half the value computed in part 1, and the transistor is pulled out of the saturation region. Assume the current gain increased to 30; compute the losses in the collector–emitter circuit.

Solution

1. First, compute the collector current by using Equation (10.4).

$$I_C = \frac{V_{CC} - V_{CE}}{R_L} = \frac{100 - 0.5}{5} = 19.9 \text{ A}$$

Hence, the minimum base current is

$$I_B = \frac{I_C}{\beta} = \frac{19.9}{20} = 0.995 \text{ A}$$

2. $P_{CE} = I_C V_{CE} = 19.9 \times 0.5 = 9.95$ W
3. When the base current is reduced, V_{CE} increases as given in Equation (10.4). But first, let us compute the new collector current assuming the current gain is unchanged.

$$I_C = \beta I_B = 30 \times \frac{0.995}{2} = 14.925 \text{ A}$$

$$V_{CE} = V_{CC} - R_L I_C = 100 - 5 \times 14.925 = 25.375 \text{ V}$$

The new losses are

$$P_{CE} = I_C V_{CE} = 14.925 \times 25.375 = 378.72 \text{ W}$$

Note that the losses in the collector–emitter loop are much higher when the transistor is outside the saturation region. This is why the power transistor must always be closed in the saturation region.

10.1.2.2 Metal Oxide Semiconductor Field Effect Transistor (MOSFET)

The MOSFET, unlike the BJT, is a voltage-controlled device. This is a great advantage over the BJT as the MOSFET is a more efficient device that requires a simpler driving circuit. The MOSFET is widely used in digital circuits as well as high-power electronic circuits.

The MOSFET is based on the principle of the Field Effect Transistor (FET); the current near the surface of a semiconductor material can be altered by electric fields. Keep in mind that the current of the BJT is altered by a current in the base. The MOSFET is structured as a three-layer semiconductor material where the middle layer is subjected to an external electric field that controls the flow of current between the top and bottom layer. The strength of the electric field depends on the applied voltage (not current). The three terminals of the MOSFET in Figure 10.8 are the *drain* (D), *source* (S), and *gate* (G). The control signal V_{GS} is applied between the gate and the source, and the load current is the drain current I_D. There are two main types of MOSFETs, the enhanced-mode and the enhanced-depletion-mode. The characteristics of the enhanced-mode are shown in Figure 10.8 where V_{GS} is always positive. The enhanced-depletion-mode MOSFET has its V_{GS} either positive or negative as shown in Figure 10.9.

The power MOSFET has several advantages. Among them are:

- The MOSFET behaves as a voltage-controlled resistance; the value of the gate voltage determines the on-resistance of the MOSFET (V_{DS}/I_D). When fully closed, the on-resistance is just a few milliohms.
- Its input resistance is high since the gate current is almost zero.
- Because of its low on-resistance and the lack of current in the gate loop, the losses of the MOSFET are lower than the BJT losses.
- The MOSFET can be connected in parallel for heavier loads. This is not possible with the BJT.

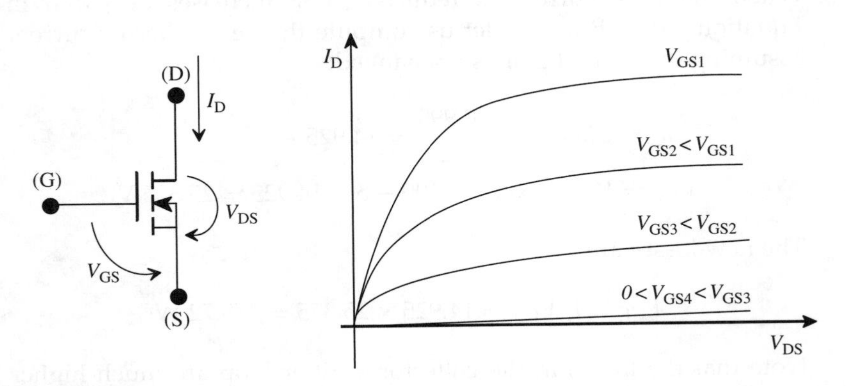

FIGURE 10.8
Enhanced-mode MOSFET.

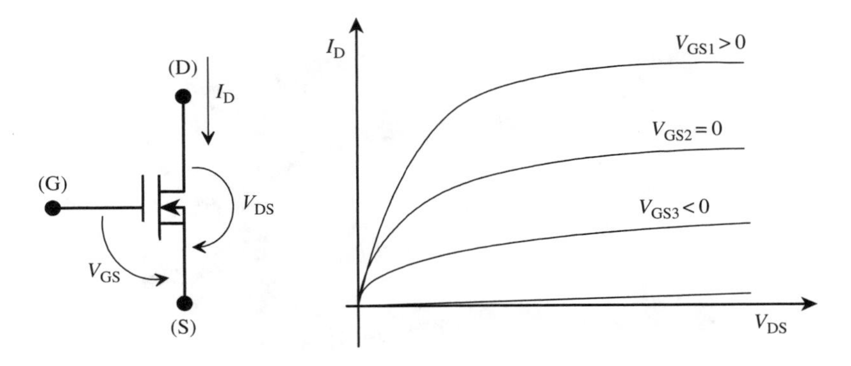

FIGURE 10.9
Enhanced-depletion-mode MOSFET.

The main disadvantage of the MOSFET is its high gate–source capacitance. This is due to the difference in charges between the gate and the source, and the high input resistance of the gate. Therefore, the switching frequency of the MOSFET is limited by this input capacitance. High switching frequency may result in excessive capacitive current in the gate circuit.

10.1.3 Thyristors

The thyristor's family includes several devices such as the silicon-controlled rectifier (SCR), bi-directional switch (Triac), and gate turn-off SCR (GTO). These devices are widely used in high-power applications as they can handle thousand of amperes and can be cascaded to withstand hundreds of kilovolts. Figure 10.10 shows various high-current SCRs with the large one rated at 300 A. The thyristor's family has also several devices in the low power range such as the silicon diode for alternating current (SIDAC), silicon unilateral switch (SUS), and the bilateral diode (Diac).

10.1.3.1 Silicon Controlled Rectifier

The SCR is built out of four layers of semiconductor material as shown in Figure 10.11. The device has three terminals: *anode* (A), *cathode* (K), and *gate* (G). When the anode to cathode voltage is negative, the SCR is reverse-biased and very small leakage current flows from the cathode to the anode; the device in this case is open. If the reverse-biased voltage reaches the *reverse breakdown* limit V_{RB} of the device, the SCR is destroyed. When the anode to cathode voltage is positive, the SCR is forward-biased. If no current pulse is applied at the gate, no current flows in the anode loop as long as V_{AK} is below the *breakover* voltage V_{BO} of the SCR. In this case, the SCR is open. However, if a gate current pulse I_G is applied when $V_{AK} < V_{BO}$, the SCR can be closed. The voltage V_{AK} just before the SCR is closed is called *turn-on*

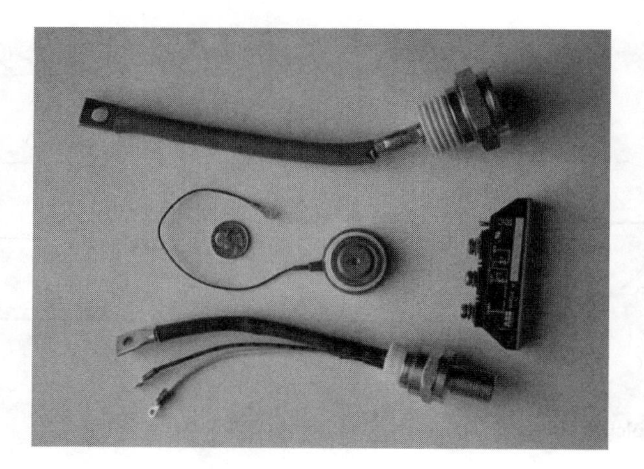

FIGURE 10.10
High-power SCRs.

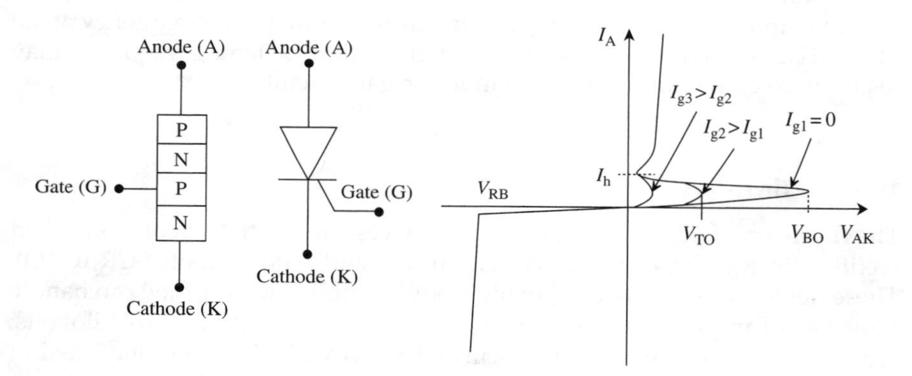

FIGURE 10.11
SCR structure, symbol, and characteristics.

voltage V_{TO}. The higher the gate current, the lower is the turn-on voltage. If it is desired to close the SCR when the turn-on voltage is near zero, the gate pulse must be the maximum allowed by the SCR's specification. If the gate signal is removed after the SCR is closed, the SCR remains latched closed unless the anode current falls below the holding value I_h. At or below the holding current, the SCR is opened.

The SCR is highly popular in heavy power applications due to several reasons. Among them are:

- The SCRs are much cheaper than BJTs and MOSFETs.
- The SCR is triggered by a single pulse, instead of the continuous current needed by the BJT. Thus, the input losses are much less than those of the BJT.

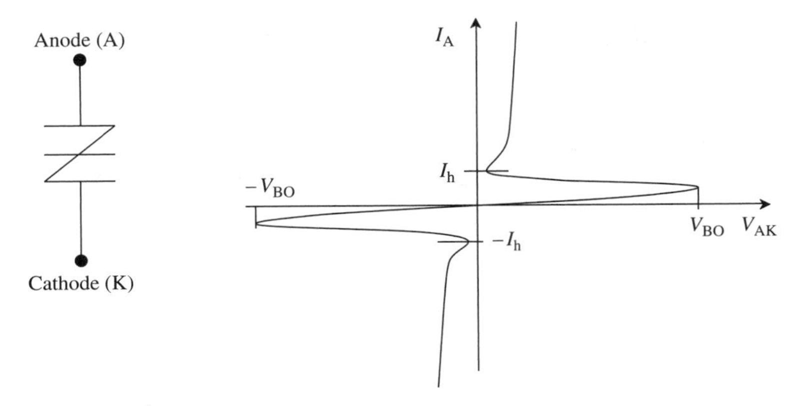

FIGURE 10.12
Characteristics of the SIDAC.

- In ac circuits, the SCR can be self-commutated (opened) without the need for external commutation circuits.
- The SCR can be built with much larger current and voltage ratings than the BJT or MOSFET.

10.1.3.2 SIDAC

The SIDAC is a silicon bilateral voltage-triggered switch used mainly in low power control circuits. As seen in the characteristics of the SIDAC in Figure 10.12, when the anode to cathode voltage exceeds the breakover voltage of the SIDAC, the device switches on (closes) and the anode current flows in the circuit. The conduction is continuous until the current drops below the minimum holding current of the device. The turn-on voltage of the SIDAC is a fixed value and cannot be controlled.

10.1.4 Hybrid Power Electronic Devices

There are various hybrid devices used in power electronic circuits. They are mainly composed of several components such as BJT and MOSFET, SCR and FET, and cascaded BJTs. The hybrid structures offer very desirable features such as higher switching speeds, higher currents, higher current gains, or higher efficiencies. The hybrid devices area is fast growing where newer designs are always developed. In this section, the Darlington Transistor and the Insulated Gate Bipolar Transistor are discussed as they are very common in power applications.

10.1.4.1 Darlington Transistor

One of the main problems with the BJT is its low current gain, which demands high-base current. This problem is solved by cascading two bipolar transistors

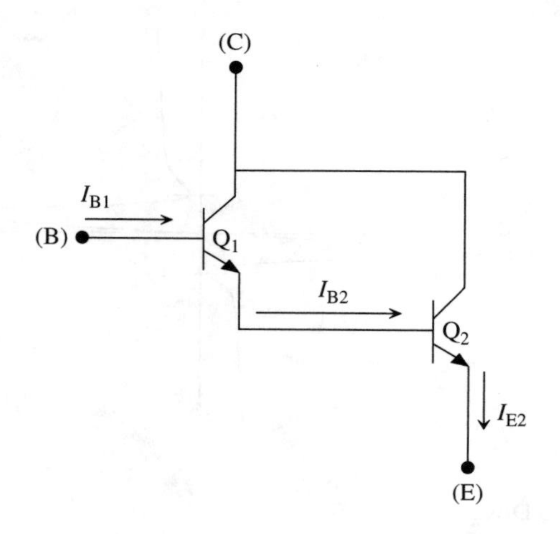

FIGURE 10.13
Darlington transistor.

as in the configuration shown in Figure 10.13. The collector current of Q_1 is $\beta_1 I_{B1}$, and its emitter current is

$$I_{E1} = I_{B1} + I_{C1} = (1 + \beta_1) I_{B1} \tag{10.5}$$

This emitter current is also equal to the base current of transistor Q_2, and the emitter current of Q_2 is

$$I_{E2} = (1 + \beta_2) I_{B2} = (1 + \beta_2)(1 + \beta_1) I_{B1} \tag{10.6}$$

Hence the ratio of the emitter current of Q_2 (i.e., the load current) and the base current of Q_1 (i.e., the control current) is

$$\frac{I_{E2}}{I_{B1}} = (1 + \beta_1)(1 + \beta_2) \tag{10.7}$$

EXAMPLE 10.3 A bipolar junction transistor has $\beta = 9$ in the saturation region. When the transistor is closed, and the emitter current is 100 A, compute the following:

1. The base current to keep the transistor closed.
2. If two of these transistors are connected in Darlington configuration, compute the base current.

Solution

1. $I_B = I_E/(1+\beta) = \frac{100}{10} = 10$ A.
2. With Darlington connection,

$$I_B = \frac{I_{E2}}{(1+\beta_1)(1+\beta_2)} = \frac{100}{10 \times 10} = 1 \text{ A}$$

Note that the base current of the Darlington transistor is 10% of the base current of a single BJT.

10.1.4.2 Insulated Gate Bipolar Transistor (IGBT)

BJTs are very reliable devices that can operate at high switching frequencies. However, they are low-current gains switching devices, which demand high-base currents to keep the devices latched in the closed position. This requires triggering circuits that are bulky, expensive, and of low efficiency. The MOSFETs, on the other hand, are voltage-controlled devices and their triggering circuits are much simpler and less expensive. For these reasons, the power electronic engineers have merged these two devices in the configuration shown in Figure 10.14 to produce a high-current voltage triggered device. This device is called an Insulated Gate Bipolar Transistor or IGBT. The symbol and characteristics of an IGBT are shown in Figure 10.15.

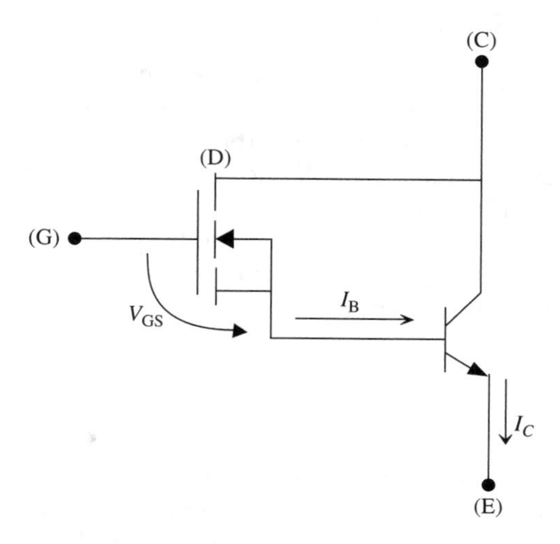

FIGURE 10.14
MOSFET and bipolar circuit.

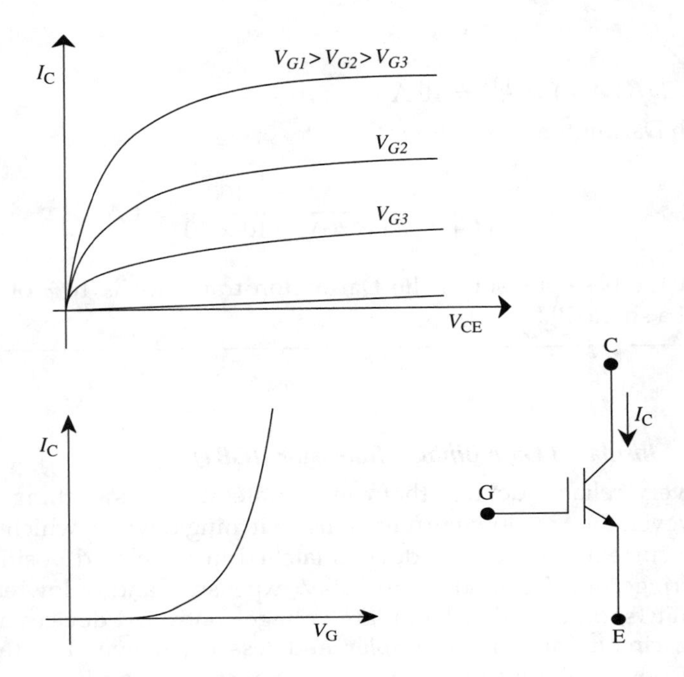

FIGURE 10.15
Characteristics of an IGBT.

10.2 Solid-State Switching Circuits

Solid-state switching circuits have various configurations and are often called converters. There are four types of converters, which are shown in Figure 10.16. The ac/dc converter converts any ac waveform into a dc waveform with adjustable voltage. The dc/dc converter converts a dc waveform into an adjustable voltage dc waveform. The dc/ac converter converts a dc waveform into ac with adjustable voltage and frequency. The ac/ac converter converts the fixed voltage, fixed frequency waveform into waveforms with adjustable voltage and frequency.

10.2.1 ac/dc Converters

These are very common converters used as power supplies or to drive direct current equipment using ac sources. There are three main types of ac/dc converters: the fixed voltage, the variable voltage, and the fixed current. The fixed voltage circuits produce constant voltage across the load and the circuit is known as a rectifier circuit. For variable voltage across the load, a switching device such as the BJT, MOSFET, or SCR is used to regulate the flow of current

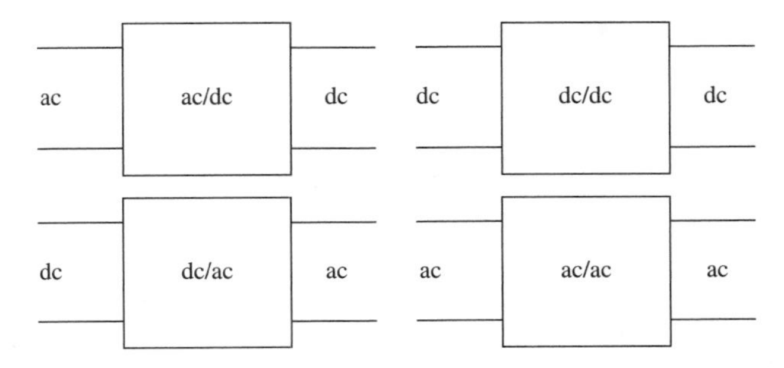

FIGURE 10.16
Four types of converters.

in the external circuit, thus controlling the voltage across the circuit. The fixed current circuit is used in applications such as battery chargers and constant torque of electromechanical devices.

10.2.1.1 Fixed-voltage Circuits

The simplest form of this ac/dc converter is the half-wave diode rectifier circuit shown in Figure 10.17. The circuit consists of an alternating current source of potential v_S, a load resistance R, and a diode connected between the source and the load. Because of the diode, the current flows only in one direction when the voltage source is in the positive half of its cycle; the diode is then forward biased. Hence, the load voltage v_t is only present when the current flows in the circuit as shown in Figure 10.18.

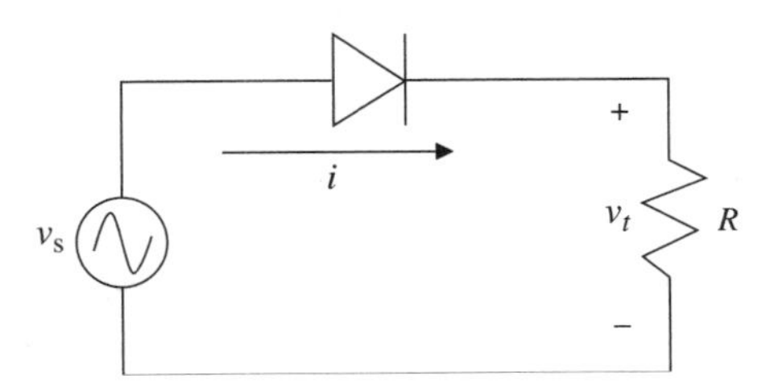

FIGURE 10.17
Half-wave rectifier circuit.

A full-wave rectifier circuit is shown in Figure 10.19. Because the diodes are connected in the bridge configuration shown in the figure, the current of the load flows in the same direction whether the source voltage is in the positive or negative half of its ac cycle. In the positive half cycle, point A has higher potential than point B. Hence, diodes D_1 and D_2 are forward biased and the current flows as shown by the solid arrows. During the negative half, point B has higher potential than point A. Hence, D_3 and D_4 are forward biased and the current flows as shown by the dashed arrows. In either half of the ac cycle,

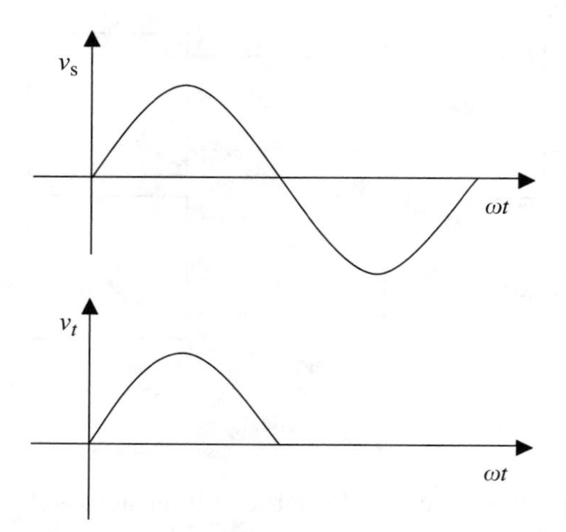

FIGURE 10.18
Waveforms of half-wave rectifier circuit.

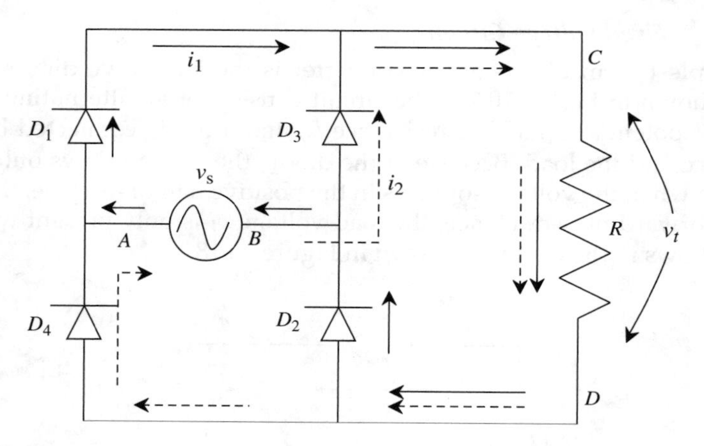

FIGURE 10.19
Full-wave rectifier circuit.

the current in the load is unidirectional; that is, it is dc current. The waveforms of the circuit are shown in Figure 10.20.

Let us assume that the source voltage is sinusoidal, expressed by the equation

$$v_s = V_{max} \sin \omega t \tag{10.8}$$

The dc voltage across the load can be expressed by the average value of v_t. For the half-wave rectifier circuit in Figure 10.17, the average voltage across

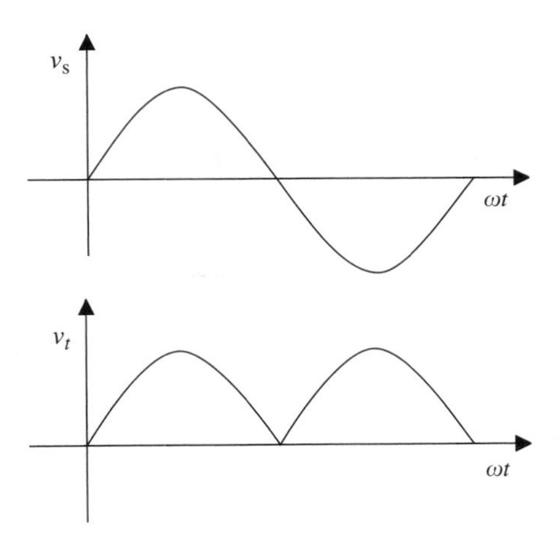

FIGURE 10.20
Waveforms of full-wave rectifier circuit.

the load is

$$V_{\text{ave-hw}} = \frac{1}{2\pi}\int_0^{2\pi} v_t \, \mathrm{d}\omega t = \frac{1}{2\pi}\int_0^{\pi} v_s \, \mathrm{d}\omega t = \frac{1}{2\pi}\int_0^{\pi} V_{\max}\sin\omega t \, \mathrm{d}\omega t = \frac{V_{\max}}{\pi} \tag{10.9}$$

For the full-wave rectifier circuit in Figure 10.19, the average voltage is

$$V_{\text{ave-fw}} = \frac{1}{2\pi}\int_0^{2\pi} v_t \, \mathrm{d}\omega t = \frac{1}{\pi}\int_0^{\pi} v_s \, \mathrm{d}\omega t = \frac{1}{\pi}\int_0^{\pi} V_{\max}\sin\omega t \, \mathrm{d}\omega t = \frac{2V_{\max}}{\pi} \tag{10.10}$$

Note that the average voltage across the load for the full-wave circuit is double that for the half-wave circuit.

The current and voltage are often expressed by their root mean square (rms) values. As given in Chapter 7, the rms voltage is defined as

$$V_{\text{rms}} = \sqrt{\frac{1}{2\pi}\int_0^{2\pi} [v(t)]^2 \, \mathrm{d}\omega t} \tag{10.11}$$

Hence, the rms voltage across the load for the half-wave rectifier circuit is

$$\begin{aligned} V_{\text{rms-hw}} &= \sqrt{\frac{1}{2\pi}\int_0^{2\pi} v_t^2 \, \mathrm{d}\omega t} = \sqrt{\frac{1}{2\pi}\int_0^{\pi} (V_{\max}\sin\omega t)^2 \, \mathrm{d}\omega t} \\ V_{\text{rms-hw}} &= \sqrt{\frac{V_{\max}^2}{4\pi}\int_0^{\pi} (1-\cos 2\omega t)\mathrm{d}\omega t} = \frac{V_{\max}}{2} \end{aligned} \tag{10.12}$$

For the full-wave rectifier circuit, the rms voltage across the load is

$$V_{rms-fw} = \sqrt{\frac{1}{2\pi}\int_0^{2\pi} v_t^2 \, d\omega t} = \sqrt{\frac{1}{\pi}\int_0^{\pi} (V_{max}\sin\omega t)^2 \, d\omega t}$$
$$V_{rms-fw} = \sqrt{\frac{V_{max}^2}{2\pi}\int_0^{\pi} (1-\cos 2\omega t) \, d\omega t} = \frac{V_{max}}{\sqrt{2}} \tag{10.13}$$

Note that the rms voltage across the load for the full-wave rectifier circuit is the same as that for the source voltage. The rms current of the load in either circuit can be computed as

$$I_{rms} = \frac{V_{rms}}{R} \tag{10.14}$$

The electric power consumed by the resistive load is

$$P = \frac{V_{rms}^2}{R} = I_{rms}^2 R \tag{10.15}$$

Hence, for half- and full-wave rectifier circuits, the load power is

$$P_{hw} = \frac{V_{rms-hw}^2}{R} = \frac{V_{max}^2}{4R}$$
$$P_{fw} = \frac{V_{rms-fw}^2}{R} = \frac{V_{max}^2}{2R} \tag{10.16}$$

EXAMPLE 10.4 A full-wave rectifier circuit converts a 120 V (rms) source into dc. The load of the circuit is 10 Ω resistance. Compute the following:

1. The average voltage across the load.
2. The average voltage of the source.
3. The rms voltage of the load.
4. The rms current of the load.
5. The power consumed by the load.

Solution

1. $V_{ave-fw} = 2V_{max}/\pi = 2(\sqrt{2}\,120)/\pi = 108$ V
2. The average voltage of the source is zero since the source waveform is symmetrical across the x-axis.
3. $V_{rms-fw} = V_{max}/\sqrt{2} = 120$ V
4. $I_{rms-fw} = V_{rms-fw}/R = \frac{120}{10} = 12$ A
5. $P_{fw} = V_{rms-fw}^2/R = 1.44$ kW

10.2.1.2 Voltage-Controlled Circuits

The simple half-wave SCR converter in Figure 10.21 provides some control for the load voltage. The circuit consists of an alternating current source of potential v_s, a load resistance R, and an SCR connected between the source and the load. The triggering circuit of the SCR is not shown in the figure.

The waveforms of the circuit are shown in Figure 10.22. When the SCR is open, the current of the circuit is zero and the load voltage is also zero. When the SCR is triggered at $\omega t = \alpha$, the SCR is closed and the voltage across the load resistance is equal to the source voltage. Since the source is a sinusoidal waveform, the current of the circuit is zero at π. Hence, the SCR is opened at π since its current is below the holding value of the SCR. The period when the current is flowing in the circuit is called the conduction period γ. Hence,

$$\gamma = \pi - \alpha \tag{10.17}$$

The full-wave SCR circuit is shown in Figure 10.23 and its waveforms are shown in Figure 10.24. The current flow in the full-wave SCR circuit is similar to that explained for the full-wave rectifier circuit, except that the current starts flowing after the triggering pulse is applied on S_1 and S_2 at α during the positive half of the ac cycle, and on S_3 and S_4 after $\alpha + 180°$ during the negative half of the cycle.

The average voltage across the load for the half-wave SCR circuit is

$$\begin{aligned} V_{\text{ave-hw}} &= \frac{1}{2\pi}\int_0^{2\pi} v_t \, d\omega t = \frac{1}{2\pi}\int_\alpha^{\pi} v_s \, d\omega t = \frac{1}{2\pi}\int_\alpha^{\pi} V_{\max}\sin\omega t \, d\omega t \\ V_{\text{ave-hw}} &= \frac{V_{\max}}{2\pi}(1+\cos\alpha) \end{aligned} \tag{10.18}$$

For full-wave SCR circuits, the average voltage across the load resistance is

$$\begin{aligned} V_{\text{ave-fw}} &= \frac{1}{2\pi}\int_0^{2\pi} v_t \, d\omega t = \frac{1}{\pi}\int_\alpha^{\pi} v_s \, d\omega t = \frac{1}{\pi}\int_\alpha^{\pi} V_{\max}\sin\omega t \, d\omega t \\ V_{\text{ave-fw}} &= \frac{V_{\max}}{\pi}(1+\cos\alpha) \end{aligned} \tag{10.19}$$

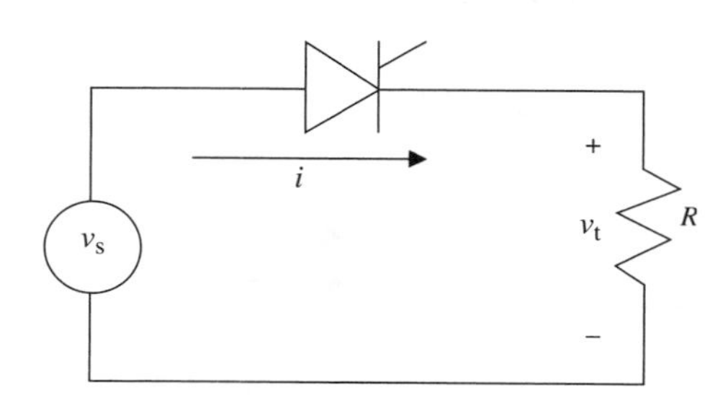

FIGURE 10.21
Half-wave SCR circuit.

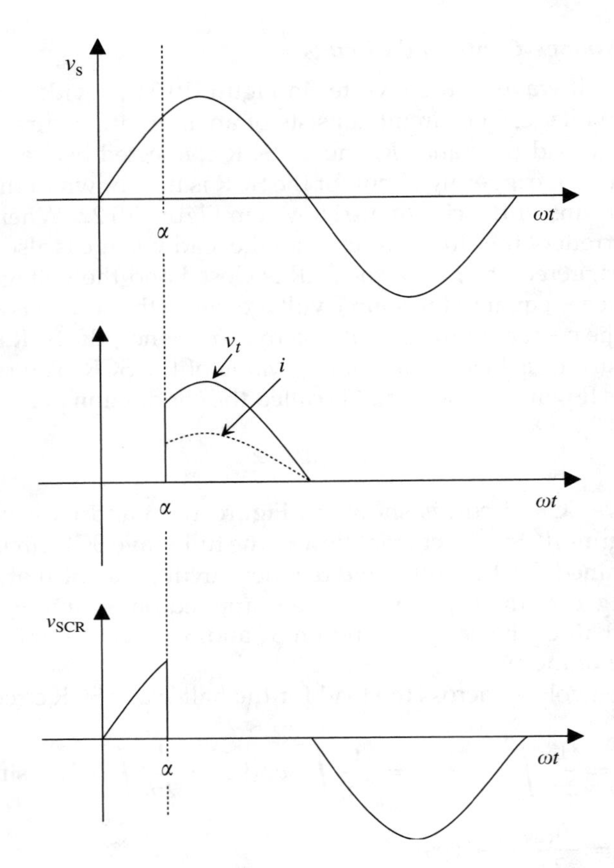

FIGURE 10.22
Waveforms of half-wave SCR circuit.

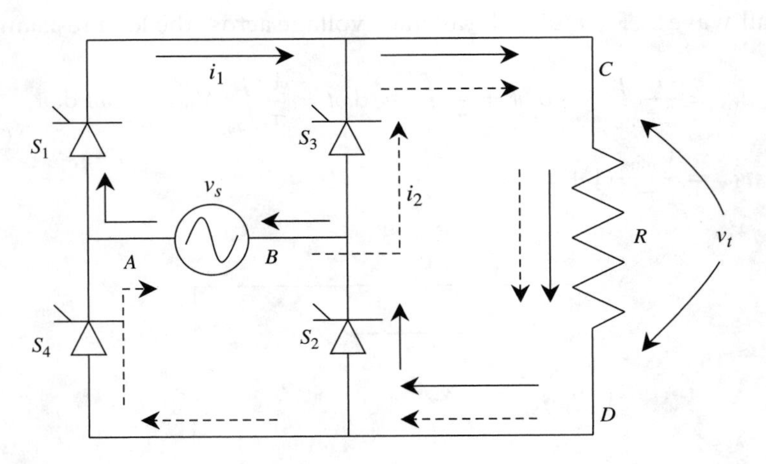

FIGURE 10.23
Full-wave SCR circuit.

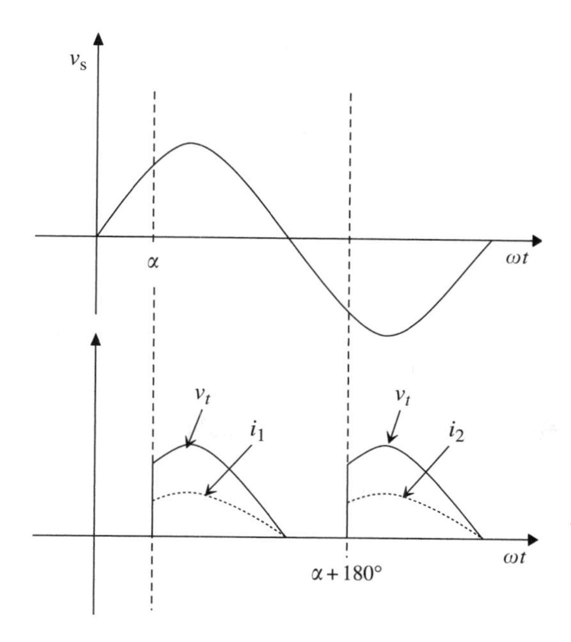

FIGURE 10.24
Waveforms of full-wave SCR circuit.

The rms voltage across the load for the half-wave SCR circuit can be computed as

$$V_{\text{rms-hw}} = \sqrt{\frac{1}{2\pi}\int_0^{2\pi} v_t^2 \, d\omega t} = \sqrt{\frac{1}{2\pi}\int_\alpha^{\pi} (V_{\max}\sin\omega t)^2 \, d\omega t}$$

$$V_{\text{rms-hw}} = \sqrt{\frac{V_{\max}^2}{4\pi}\int_\alpha^{\pi} (1-\cos 2\omega t) \, d\omega t} = \frac{V_{\max}}{2}\sqrt{\left(1-\frac{\alpha}{\pi}+\frac{\sin 2\alpha}{2\pi}\right)} \tag{10.20}$$

A similar process can be used to compute the rms voltage across the load for the full-wave SCR circuit.

$$V_{\text{rms-fw}} = \sqrt{\frac{1}{2\pi}\int_0^{2\pi} v_t^2 \, d\omega t} = \sqrt{\frac{1}{\pi}\int_\alpha^{\pi} (V_{\max}\sin\omega t)^2 \, d\omega t}$$
$$V_{\text{rms-fw}} = \frac{V_{\max}}{\sqrt{2}}\sqrt{\left(1-\frac{\alpha}{\pi}+\frac{\sin 2\alpha}{2\pi}\right)} \tag{10.21}$$

The rms voltage across the load for the full-wave SCR circuit is shown in Figure 10.25. As you can see, the load voltage is controlled from zero to $V_{\max}/\sqrt{2}$ by adjusting the triggering angle.

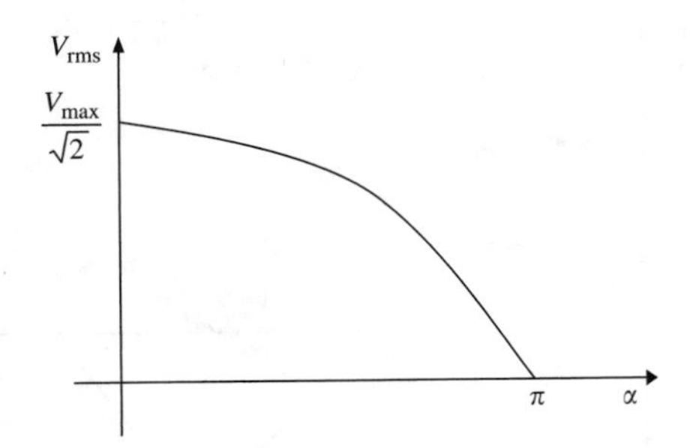

FIGURE 10.25
Rms voltage of the load for the full-wave SCR circuit.

The power across the load can be computed as

$$P = \frac{V_{rms}^2}{R} = I_{rms}^2 R$$

$$P_{hw} = \frac{V_{max}^2}{8\pi R}(2(\pi - \alpha) + \sin 2\alpha) \tag{10.22}$$

$$P_{fw} = \frac{V_{max}^2}{4\pi R}(2(\pi - \alpha) + \sin 2\alpha)$$

EXAMPLE 10.5 A full-wave SCR converter circuit is used to regulate the power across a 10 Ω resistance. The voltage source is 120 V (rms). At what triggering angle is the power across the load 1 kW? Compute the average and rms currents of the load.

Solution

The power expression in Equation (10.22) is nonlinear with respect to the triggering angle α. Therefore, the solution is numerical.

$$P_{fw} = \frac{V_{max}^2}{4\pi R}[2(\pi - \alpha) + \sin(2\alpha)] = \frac{(\sqrt{2} \times 120)^2}{4\pi \times 10}[2(\pi - \alpha) + \sin(2\alpha)] = 1000$$

$$2\alpha - \sin(2\alpha) = 1.92$$

The solution of the above nonlinear equation is iterative.

$$\alpha = 71.9^\circ$$

To compute the average current of the load, we need to compute the average voltage across the load as given by Equation (10.19).

$$V_{ave-fw} = \frac{V_{max}}{\pi}(1 + \cos\alpha) = \frac{\sqrt{2} \times 120}{\pi}(1 + \cos 71.9^\circ) = 70.8 \text{ V}$$

$$I_{ave-fw} = \frac{V_{ave-fw}}{R} = 7.08 \text{ A}$$

For the rms current, we can compute the rms voltage first, and then divide the voltage by the resistance of the load. Another, simpler, method is to use the power formula in Equation (10.15).

$$I_{\text{rms}} = \sqrt{\frac{P_{\text{fw}}}{R}} = 10 \text{ A}$$

10.2.1.3 Constant-Current Circuits

The constant current circuits are used mainly to charge batteries as well as to drive electric motors in constant torque applications. The current during the charging process of the battery must be controlled to prevent the battery from being damaged by excessive charging currents. The simplest charging circuit is the full-wave shown in Figure 10.26. The load of this circuit is the battery and the resistance R represents the internal resistance of the battery.

The range of the triggering angle of the charger circuit is less than the range for the full-wave SCR circuit in Figure 10.23. This is because the minimum triggering angle and the conduction period depend on the voltage of the battery. Examine the circuit in Figure 10.26 and its waveforms in Figure 10.27. During the first half of the ac cycle, S_1 and S_2 can only close when the source voltage is higher than the battery voltage, that is, when $\beta > \alpha \geq \alpha_{\text{min}}$, where α_{min} is the angle at which the source voltage is equal to the battery voltage. If the triggering angle of the SCRs $\alpha < \alpha_{\text{min}}$, the voltage across S_1 or S_2 is negative and the SCRs cannot close. At α_{min},

$$V_{\text{b}} = V_{\text{max}} \sin \alpha_{\text{min}} \tag{10.23}$$

The current to the battery will flow as long as $V_b < v_{\text{s}}$. Hence, the current is commutated at β, where

$$\beta = 180° - \alpha_{\text{min}} \tag{10.24}$$

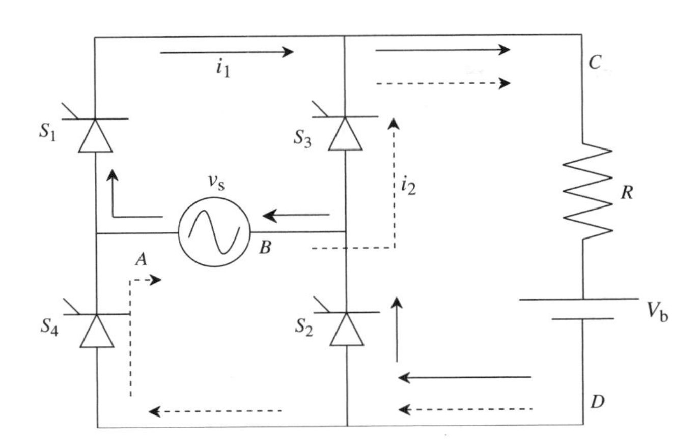

FIGURE 10.26
Full-wave charger circuit.

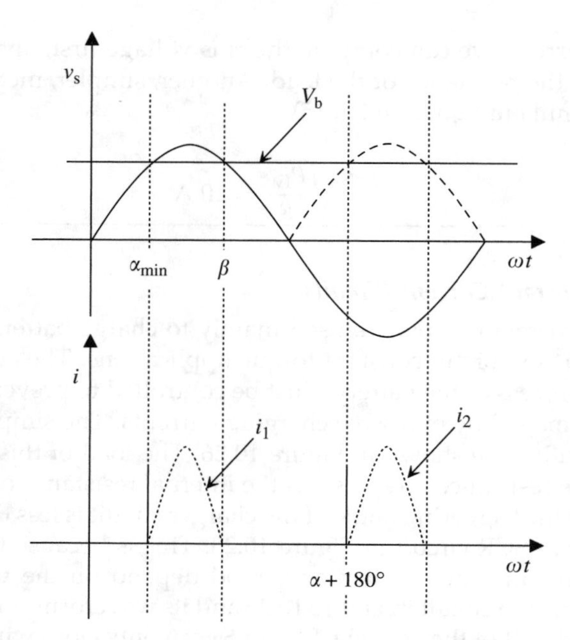

FIGURE 10.27
Waveforms of the full-wave charger circuit.

During the charging process, the voltage of the battery changes, so the triggering angle must change accordingly to maintain the current constant. When the voltage of the battery is low, the triggering angle increases so the current is reduced to the level safe for charging the battery. When the voltage of the battery increases, the current is reduced so the triggering angle is advanced to compensate for the current reduction.

EXAMPLE 10.6 A 120 V full-wave battery charger is designed to provide 0.1 A of charging current. A battery set with 1 Ω internal resistance is connected across the charger. At the beginning of the charging process, the voltage of the battery is 45 V. After 3 h of charging, the voltage of the battery is 50 V. Compute the triggering angle of the converter at these two times. Also compute the conduction periods.

Solution

The instantaneous charging current can be expressed as

$$i = \frac{v_s - V_b}{R}$$

Hence, the average charging current is

$$I_{\text{ave}} = \frac{1}{2\pi}\int_0^{2\pi} i \, d\omega t = \frac{1}{\pi}\int_0^{\pi} i \, d\omega t$$

$$I_{\text{ave}} = \frac{1}{R\pi}\int_{\alpha}^{\beta} (V_{\max}\sin\omega t - V_b) = \frac{1}{R\pi}[V_{\max}(\cos\alpha - \cos\beta) - V_b(\beta - \alpha)]$$

At the beginning of the charging process
We can compute β by using Equations (10.23) and (10.24).

$$\beta = 180° - \alpha_{\min} = 180° - \sin^{-1}\left(\frac{V_b}{V_{\max}}\right)$$
$$= 180° - \sin^{-1}\left(\frac{45}{\sqrt{2}\times 120}\right) = 164.62°$$

Hence,

$$I_{\text{ave}} = \frac{1}{R\pi}[V_{\max}(\cos\alpha - \cos\beta) - V_b(\beta - \alpha)]$$
$$0.1 = \frac{1}{\pi}\left[\sqrt{2}\times 120(\cos\alpha - \cos 164.62) - 45\left(164.62\frac{\pi}{180} - \alpha\right)\right]$$
$$\alpha \approx 161°$$

The conduction period γ is

$$\gamma = \beta - \alpha = 164.62 - 161 = 3.62°$$

After 3 h

$$\beta = 180° - \alpha_{\min} = 180° - \sin^{-1}\left(\frac{V_b}{V_{\max}}\right)$$
$$= 180° - \sin^{-1}\left(\frac{50}{\sqrt{2}\times 120}\right) = 162.86°$$

Hence,

$$I_{\text{ave}} = \frac{1}{R\pi}[V_{\max}(\cos\alpha - \cos\beta) - V_b(\beta - \alpha)]$$
$$0.1 = \frac{1}{\pi}\left[\sqrt{2}\times 120(\cos\alpha - \cos 162.86) - 50\left(162.86\frac{\pi}{180} - \alpha\right)\right]$$
$$\alpha \approx 159.2°$$

The conduction period γ is

$$\gamma = \beta - \alpha = 162.86 - 159.2 = 3.66°$$

The charger control circuit must always track the change in the charging current and compensate for the change by adjusting the triggering angle α.

10.2.2 dc/dc Converters

The dc/dc converters are normally designed to provide adjustable dc voltage waveforms at the output. There are three basic types of dc/dc converters:

- *Buck converter.* This is a step-down converter in which the output voltage is less than the input voltage.
- *Boost converter.* This is a step-up converter in which the output voltage is higher than the input voltage.
- *Buck–Boost converter.* This is a step-down/step-up converter in which the output voltage can be made either lower or higher than the input voltage.

10.2.2.1 Buck Converter

Figure 10.28 shows a simple Buck converter (also known as chopper). The circuit has a transistor as a switching device and the load is connected between the source and the collector of the transistor. The top waveform in the figure is

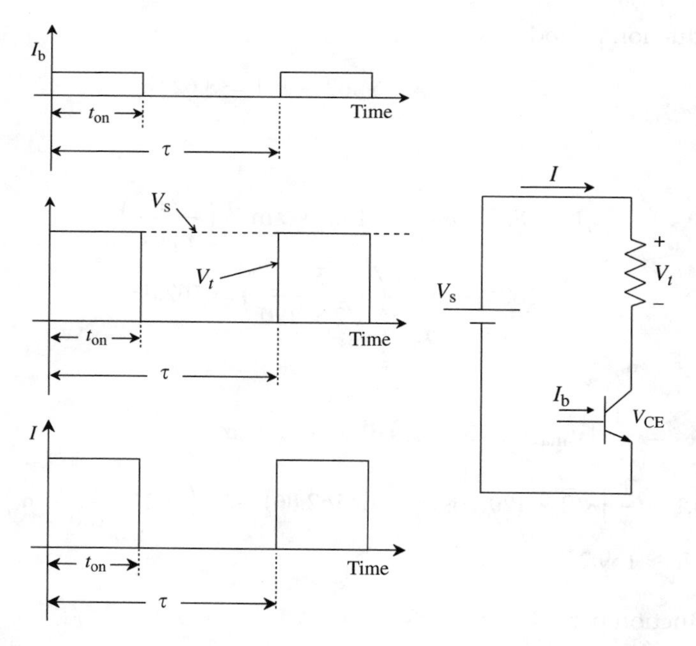

FIGURE 10.28
Simple chopper circuit.

for the base current of the transistor where the transistor is closed throughout the time segment t_{on} of the period τ of one cycle. Only when the transistor is closed, the voltage across the load V_t is equal to the source voltage V_s; otherwise $V_t = 0$. Similarly, the current in the circuit flows only when the transistor is closed.

The average voltage across the load V_{ave} is

$$V_{ave} = \frac{1}{\tau}\int_0^{t_{on}} V_s \, dt = \left(\frac{t_{on}}{\tau}\right) V_s = KV_s \quad (10.25)$$

where $K = t_{on}/\tau$ is called the duty ratio. The maximum value of K is 1 when $t_{on} = \tau$. In this case, the transistor is always closed and the load voltage is equal to the source voltage. For any other value when $t_{on} < \tau$, the load voltage is less than the source voltage. The output voltage of the converter can then be controlled by fixing the period τ and adjusting the on-time t_{on}.

EXAMPLE 10.7 The switching frequency of a chopper is 10 KHz, and the source voltage is 40 V. For an average load voltage of 20 V and a load resistance is 10 Ω, compute the following:

1. The duty ratio.
2. The on-time and the switching period.
3. The average voltage across the transistor.
4. The average current of the load.
5. The rms voltage across the load.
6. The load power.

Solution

1. As given in Equation 10.25, the duty ratio is

$$K = \frac{V_{ave}}{V_s} = \frac{20}{40} = 0.5$$

2. Before we compute the on-time, we need to compute the period.

$$\tau = \frac{1}{f} = \frac{1}{10} = 0.1 \text{ ms}$$

$$t_{on} = K\tau = 0.5 \times 0.1 = 0.05 \text{ ms}$$

3. The average voltage across the transistor V_{ave-tr} is

$$V_{ave-tr} = V_s - V_{ave} = 40 - 20 = 20 \text{ V}$$

4. The average load current is

$$I_{ave} = \frac{V_{ave}}{R} = \frac{20}{10} = 2\text{ A}$$

5. The rms voltage of the load can be computed by using the formula for the rms quantity.

$$V = \sqrt{\frac{1}{\tau}\int_0^{t_{on}} V_s^2\, dt} = \sqrt{\frac{V_s^2}{\tau} t_{on}} = V_s\sqrt{\frac{t_{on}}{\tau}} = 28.28\text{ V}$$

6. $P = V^2/R = 28.28^2/10 = 80$ W

10.2.2.2 Boost converter

The Boost converter can increase the voltage across the load to values higher than the source voltage. A simple circuit for the Boost converter is shown at the top of Figure 10.29. The load of this circuit is the resistance R. The transistor is closed for a time interval t_{on} and is open during a time interval t_{off}. When the transistor is closed, the current in the inductance i_{on} increases and energy is stored in the inductor as depicted in the lower left side of the

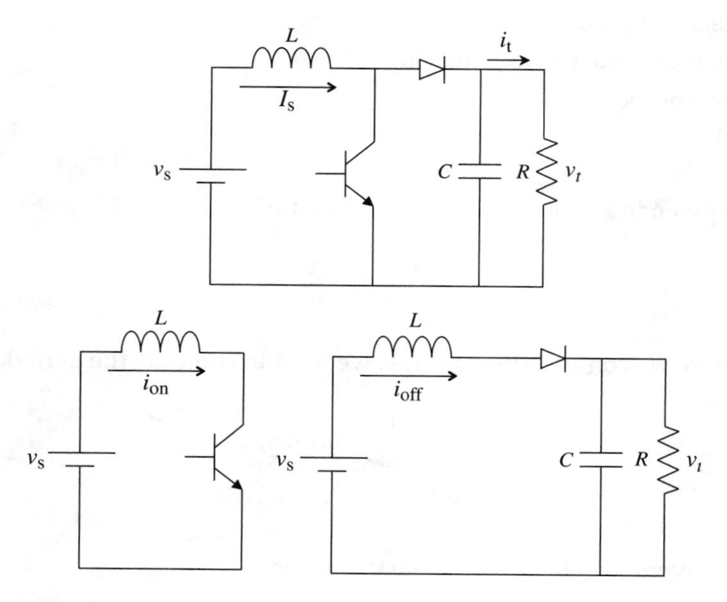

FIGURE 10.29
A simple Boost converter.

figure. In this process, the voltage across the inductance v_L is

$$v_L = L\frac{di_{on}}{dt} \approx L\frac{\Delta i_{on}}{t_{on}} \tag{10.26}$$

where Δi_{on} is the change in the current during the period t_{on}. Since the voltage across the inductance is equal to the source voltage when the transistor is closed, Equation (10.26) can be rewritten as

$$V_s = L\frac{\Delta i_{on}}{t_{on}} \tag{10.27}$$

When the transistor is opened, the stored energy of the inductance is transferred to the load via the diode. The inductor current in this case is i_{off} as depicted in the circuit on the lower right side of Figure 10.29. In this case, the voltage loop can be written as

$$v_t = V_s - v_L = V_s + L\frac{\Delta i_{off}}{t_{off}} \tag{10.28}$$

Because the inductor is producing energy during t_{off}, and because the current through the inductor does not change its direction, the voltage polarity across the inductance must be reversed. This is the reason for the positive sign in front of L in Equations (10.27) and (10.28). If the timing is designed so that $\Delta i_{on} = \Delta i_{off}$ as shown in the current waveforms in Figure 10.30, Equations (10.28) can be combined as

$$v_t = V_s + L\frac{\Delta i_{off}}{t_{off}} = V_s\left(1 + \frac{t_{on}}{t_{off}}\right) \tag{10.29}$$

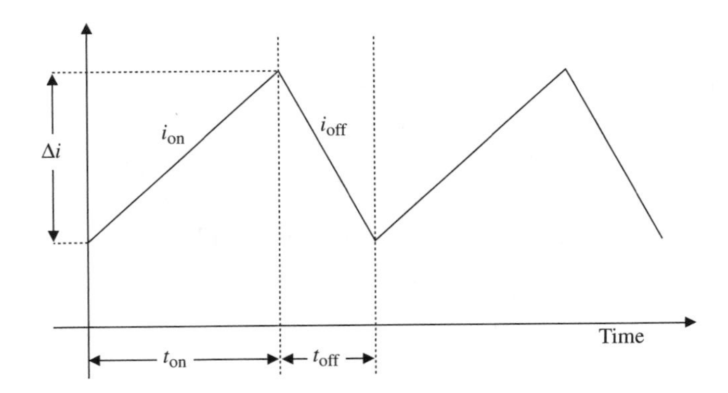

FIGURE 10.30
Waveform of Boost converter.

The above equation shows that the voltage across the load can be adjusted by adjusting t_{on} and t_{off}. Furthermore, the load voltage is always higher than the source voltage when $t_{on}/t_{off} > 0$.

The capacitor in the circuit is used as a filter to reduce the voltage ripples across the load, and the diode is used to block the capacitor from discharging through the transistor when it is closed. If the capacitance is large enough, v_t can be assumed ripple free. Hence, we can rewrite Equation (10.29) as

$$V_t = V_s + L\frac{\Delta i_{off}}{t_{off}} = V_s\left(1 + \frac{t_{on}}{t_{off}}\right) \tag{10.30}$$

where V_t is the average voltage across the load. If we assume that the average currents and voltages are much larger than their ripples, the rms and the average values can be considered equal. Hence, we can compute the input and output powers by using the average values.

$$\begin{aligned} P_{in} &= V_s I_s \\ P_{out} &= V_t I_t \end{aligned} \tag{10.31}$$

where P_{in} is the input power to the converter and P_{out} is the output power consumed by the load. I_s and I_t are the average currents of the source and load, respectively. If we further assume that the components of the circuit are ideal, the input power to the converter is equal to the output power, hence

$$V_t I_t = V_s I_s \tag{10.32}$$

EXAMPLE 10.8 A Boost converter is used to step up 20 V into 50 V. The switching frequency of the transistor is 5 kHz, and the load resistance is 10 Ω. Compute the following:

1. The value of the inductance that would limit the current ripple at the source side to 100 mA.
2. The average current of the load.
3. The power delivered by the source.
4. The average current of the source.

Solution

1. We can use Equation (10.27) to compute the inductance. But first, we need to compute t_{on} using Equation (10.30).

$$\begin{aligned} V_t &= V_s\left(1 + \frac{t_{on}}{t_{off}}\right) \\ 50 &= 20\left(1 + \frac{t_{on}}{t_{off}}\right) \\ t_{on} &= 1.5 t_{off} \end{aligned}$$

Since the switching frequency is 5 kHz,

$$t_{on} + t_{off} = \tfrac{1}{5} = 0.2 \text{ ms}$$

Then,

$$t_{on} = 1.5t_{off} = 1.5(0.2 - t_{on})$$
$$t_{on} = 0.12 \text{ ms}$$

Now to compute the value of the inductance, we can use Equation (10.27).

$$V_s = L\frac{\Delta i_{on}}{t_{on}}$$
$$20 = L\frac{100}{0.12}$$
$$L = 24 \text{ mH}$$

2. $I_t = V_t/R = \frac{50}{10} = 5$ A
3. The power delivered by the source is the same power consumed by the load, assuming that the system's components are all ideal.

$$P = V_t I_t = 50 \times 5 = 250 \text{ W}$$

4. The average current of the source can be computed using the power equation in the previous step.

$$I_s = \frac{P}{V_s} = \frac{250}{20} = 12.5 \text{ A}$$

10.2.2.3 Buck–Boost converter

The Buck–Boost converter has the same components as the Boost converter, but is structured differently as shown in the top circuit in Figure 10.31. When the transistor is closed, as shown in the bottom left side of Figure 10.31, the current i_{on} flows through the inductor, and energy is acquired by the inductor. When the transistor is opened, the inductor delivers its stored energy to the load as shown on the bottom right side of Figure 10.31. A large capacitor is used as a filter to minimize the voltage ripples across the load, and the diode is used to prevent the capacitor from discharging its energy through the inductor.

When the transistor is closed, the voltage across the inductance is

$$v_L = L\frac{di_{on}}{dt} \approx L\frac{\Delta i_{on}}{t_{on}} \tag{10.33}$$

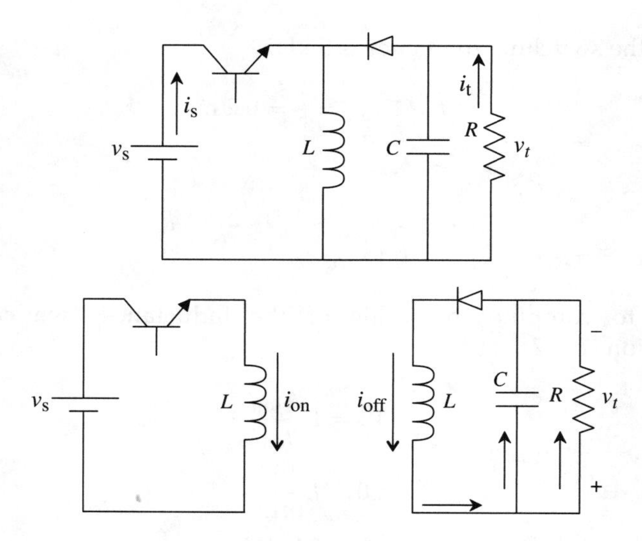

FIGURE 10.31
A simple Buck–Boost converter.

Note that when the transistor is closed, the voltage across the inductance is equal to the source voltage. Hence,

$$V_s = L\frac{\Delta i_{on}}{t_{on}} \tag{10.34}$$

When the transistor is open, the voltage across the inductor reverses because the inductor becomes a source of energy and the voltage across the inductance is

$$v_L = -L\frac{\Delta i_{off}}{t_{off}} \tag{10.35}$$

The voltage across the inductance when the transistor is open is also equal to the load voltage, hence

$$v_t = -L\frac{\Delta i_{off}}{t_{off}} \tag{10.36}$$

Assume that the capacitor is large enough so that the voltage v_t is equal to the average voltage across the load V_t. If the timing is controlled so that $\Delta i_{on} = \Delta i_{off}$, Equations (10.34) and (10.36) can be combined as

$$V_t = -V_s\frac{t_{on}}{t_{off}} \tag{10.37}$$

As seen in Equation (10.37), the average value of the load voltage can be controlled by adjusting the ratio t_{on}/t_{off}. If the on-time is zero, the load voltage is zero. If $0 < t_{on} < t_{off}$, the load voltage is lower than the source voltage. If $t_{on} = t_{off}$, the load voltage is equal to the source voltage. If $t_{on} > t_{off}$, the load voltage is higher than the source voltage. Keep in mind that Equation (10.37) is not valid when $t_{off} = 0$; that is, the transistor is never open. The system in this case is unstable as the current i_{on} will reach very high values because the inductor is a short circuit in steady state dc conditions.

If we assume that the circuit's components are ideal, the input power to the converter is equal to the output power, hence

$$P_{in} = P_{out} = V_t I_t = V_s I_s \tag{10.38}$$

EXAMPLE 10.9 A Buck–Boost converter with an input voltage of 20 V is used to regulate the voltage across a 10 Ω load. The switching frequency of the transistor is 5 kHz. Compute the following:

1. The on-time of the transistor to maintain the output voltage at 10 V.
2. The output power.
3. The on-time of the transistor to maintain the output voltage at 40 V.
4. The output power.

Solution

1. The period can be computed from the switching frequency f.

$$\tau = \frac{1}{f} = \frac{1}{5} = 0.2 \text{ ms}$$

Use Equation (10.37) to compute the time ratio. We only need to use the magnitudes of the voltages.

$$\frac{t_{on}}{t_{off}} = \frac{V_t}{V_s} = \frac{10}{20} = 0.5$$

Hence,

$$t_{on} = 0.0667 \text{ ms}$$

2. The output power is

$$P = \frac{V_t^2}{R} = 10 \text{ W}$$

3. $t_{on}/t_{off} = V_t/V_s = \frac{40}{20} = 2$

Hence,

$$t_{on} = 0.1333 \text{ ms}$$

4. The output power is

$$P = \frac{V_t^2}{R} = \frac{40^2}{10} = 160 \text{ W}$$

10.2.3 dc/ac Converters

The dc/ac converter is also known as an inverter. The inverter is used in applications such as uninterruptible power supplies, variable speed drives, and dc transmission lines. It is also common to use this converter to convert the dc 12 V of automobiles into household alternating voltage to power small equipment such as personal computers and small televisions.

There are two types of dc/ac converters: single phase and multi-phase. The former is used in low to medium power applications and the latter is used for heavy loads.

10.2.3.1 Single-Phase dc/ac Converter

Figure 10.32 shows a simple dc/ac converter in H-bridge configuration. It consists of a dc source, four transistors, and a load. The desired frequency of the output waveform determines the switching period of the transistors. During the first half of the period, Q_1 and Q_2 are closed, and during the second half, Q_3 and Q_4 are closed. The voltage waveform of the circuit is also shown in the figure. During the first half of the period, the current i_1 (in solid arrows) flows through the load. Similarly, during the second half, the current i_2 (in dashed arrows) flows through the load in the opposite direction to i_1. Thus, the load current is alternating. Note that the waveform of Figure 10.32 is not sinusoidal, but still considered an ac waveform.

EXAMPLE 10.10 The source voltage of a single-phase dc/ac converter is 100 V. The switching frequency of the transistors is 1 kHz.

1. Compute the rms voltage across the load.
2. Compute the average voltage across the load.

Solution

1. As given in Chapter 7, the rms voltage of an arbitrary waveform v is

$$V = \sqrt{\frac{1}{\tau}\int_0^{\tau} v^2 \, dt}$$

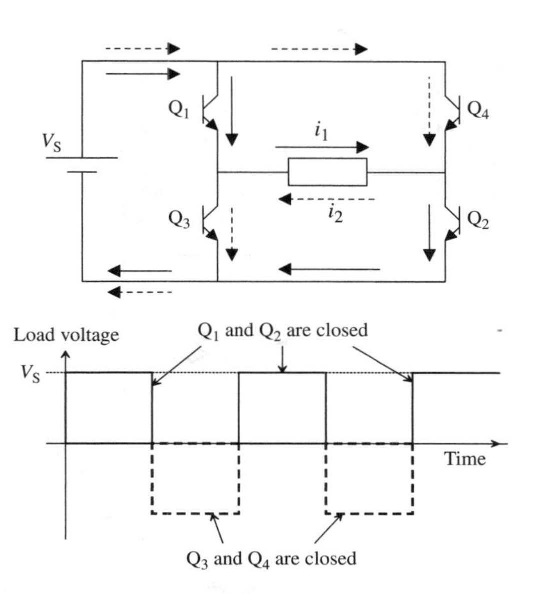

FIGURE 10.32
Single-phase dc/ac converter.

where $\tau = 1/f = 1$ ms.

$$V = \sqrt{\frac{1}{\tau}\int_0^{\tau} v^2\, dt} = \sqrt{\frac{2}{\tau}\int_0^{\tau/2} V_s^2\, dt}$$

where V_s is the source voltage.

$$V = \sqrt{\frac{2}{\tau}10^4 \int_0^{\tau/2} dt} = \sqrt{\frac{2}{\tau}10^4 \frac{\tau}{2}} = 100 \text{ V}$$

The rms voltage across the load is equal to the dc voltage of the source. This is because the ac waveform is rectangular and symmetrical.

2. The average voltage across the load is zero since the waveform is symmetrical around the time axis.

10.2.3.2 Three-Phase dc/ac Converter

Three-phase waveforms can be obtained by using the dc/ac converter shown in Figure 10.33. The converter is composed of six transistors, a dc source, and a three-phase load. This circuit is also known as the six-pulse converter. The transistors are arranged in three legs, each leg having two transistors, and the midpoint of each leg is connected to one of the terminals of the three-phase load. The switching sequence of the transistors is shown at the top part of Figure 10.34. The cycle is divided into six intervals, so each interval

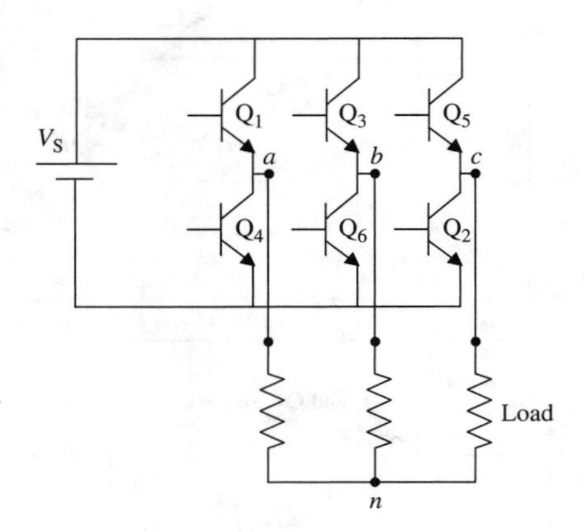

FIGURE 10.33
Three-phase dc/ac inverter.

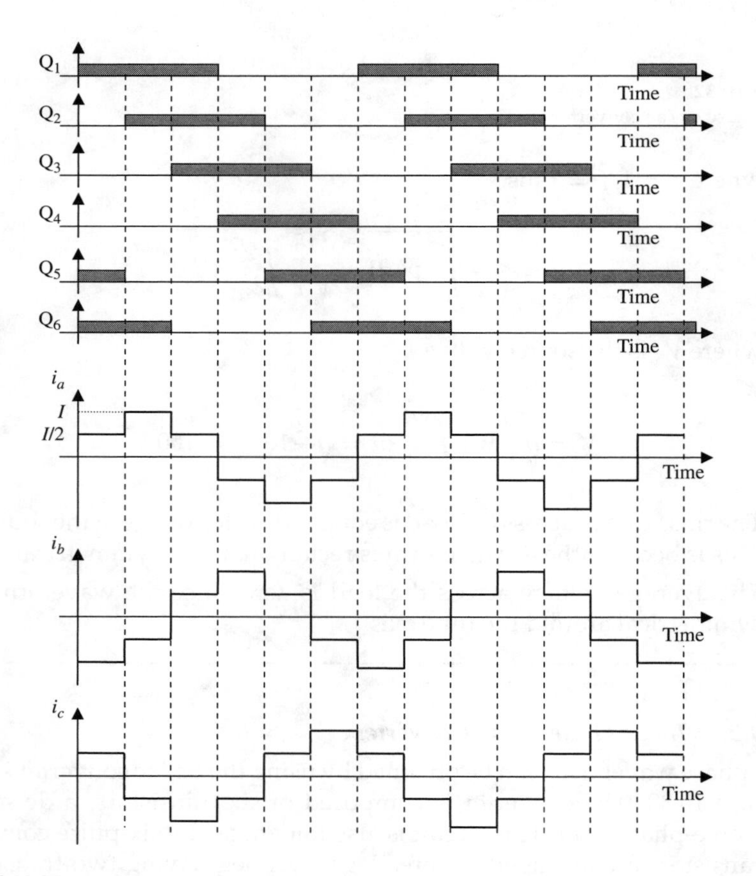

FIGURE 10.34
Timing of transistors and the phase currents.

is 60°. Each transistor is closed for three intervals (180°) and opened for three intervals. The switching of the transistors is based on their ascending order. For example, transistor Q_1 is closed first and then Q_2 after one time interval, Q_3 after additional time interval, and so on. During any given time interval, one transistor per leg is closed. No two transistors on the same leg are closed at the same time as this creates a short circuit across the supply voltage that will damage the transistors.

Now let us study the effect of the switching pattern during the first time interval when transistors Q_1, Q_5, and Q_6 are closed. This interval is depicted on the left side of Figure 10.35, which shows only the transistors that are closed. During this interval, the current from the source I is divided into two equal components; one passes through Q_1 and the other through Q_5. At the neutral node n, the two currents are summed up and returned to the source through Q_6. The load currents during this interval are

$$
\begin{aligned}
i_a &= i_c = \frac{I}{2} \\
i_b &= -I
\end{aligned}
\tag{10.39}
$$

These load currents during the first interval are shown at the bottom part of Figure 10.34. Now let us study the second interval when Q_1, Q_2, and Q_6 are closed. The flow of the current during this interval is shown on the right side of Figure 10.35. The current from the source passes through Q_1 and at the neutral point branches into two equal components; one goes through Q_2 and the other through Q_6. The load currents during this interval are

$$
\begin{aligned}
i_a &= I \\
i_b &= i_c = -\frac{I}{2}
\end{aligned}
\tag{10.40}
$$

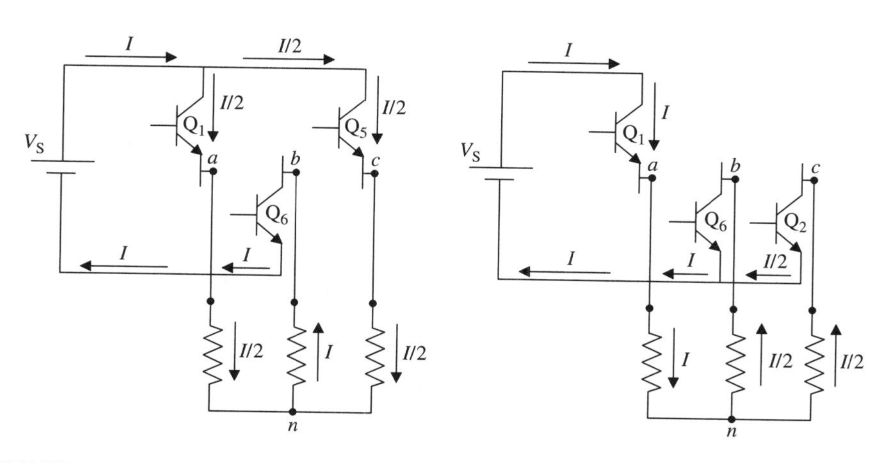

FIGURE 10.35
Active transistors and current flow during the first two intervals.

If you follow the same procedure for the other intervals, you will get the current waveforms in Figure 10.34. Note the following:

- The current waveforms are symmetrical around the time axis.
- All phase currents have the same maximum value.
- i_b lags i_a by two time intervals, that is, 120°.
- i_c lags i_b by two time intervals, that is, 120°.

Hence, the outputs of the converter are balanced three-phase waveforms.

The frequency of the load current is dependent on the time of the interval t_i. Since the cycle is composed of six segments, the frequency of the ac waveform is

$$f = \frac{1}{6t_\mathrm{i}} \tag{10.41}$$

To change the frequency of the ac waveform, the time of the interval must change.

EXAMPLE 10.11 A three-phase dc/ac converter is used to power a three-phase, Y-connected, resistive load of 10 Ω (per phase). The dc voltage is 300 V. Compute the following:

1. The time interval if the desired frequency of the ac waveform is 200 Hz.
2. The current of the dc source.
3. The rms current of the load.
4. The rms voltage across each phase of the load.

Solution

1. Use Equation (10.41) to compute the interval t_i.

$$t_\mathrm{i} = \frac{1}{6f} = \frac{1}{6 \times 200} = 833\ \mu\mathrm{s}$$

2. During any given interval, the load has two of its phases in parallel, and the combination is in series with the third phase. In the circuit on the left side of Figure 10.35, the load resistance of phases a and c are in parallel, and this combination is in series with the resistance of phase b. This is true during any time interval. Hence,

$$R_\mathrm{total} = 10 + \frac{10 \times 10}{10 + 10} = 15\ \Omega$$

Then the source current is

$$I = \frac{V_s}{R_{\text{total}}} = \frac{300}{15} = 20 \text{ A}$$

3. The rms current of phase a is

$$I_a = \sqrt{\frac{1}{6t_i} \int_0^{6t_i} i_a^2 \, dt}$$

If you examine the waveform of i_a for one period, you will discover that the magnitude of the current is equal to I during two time intervals, and is equal to $I/2$ during four intervals. Hence, the rms current of phase a can be written as

$$I_a = \sqrt{\frac{1}{6t_i} \left(\int_0^{2t_i} I^2 dt + \int_0^{4t_i} \left(\frac{I}{2}\right)^2 dt \right)} = \sqrt{\frac{I^2}{6t_i}(3t_i)} = \frac{I}{\sqrt{2}} = 14.14 \text{ A}$$

This rms current is the same as that computed for a purely sinusoidal waveform where I is the maximum current of the instantaneous waveform.

4. The rms phase voltage of the load is

$$V_{an} = I_a R = 14.14 \times 10 = 141.1 \text{ V}$$

10.2.4 ac/ac Converters

The input waveform of the ac/ac converter is typically of fixed frequency and voltage. The output is adjustable frequency and voltage. The conversion is often done indirectly by attaching two converters as shown in Figure 10.36. In the first converter, the input ac waveform is converted into dc. The dc is then converted back to ac waveform through the dc/ac converter. The connection between the two converters is called a dc link.

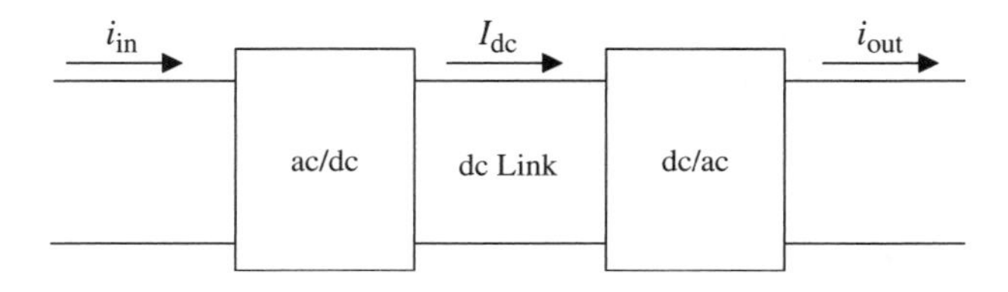

FIGURE 10.36
ac/ac converter with dc link.

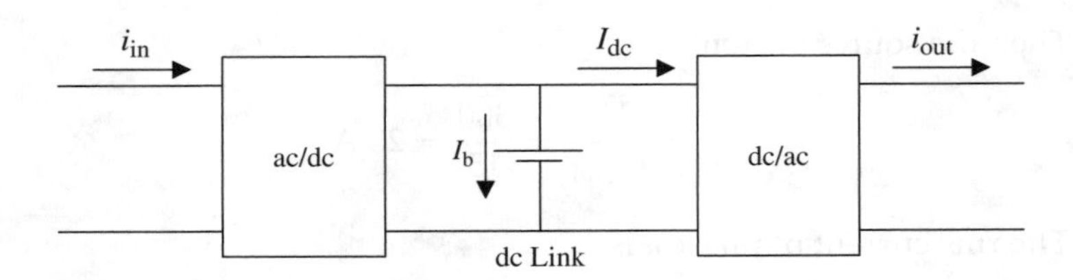

FIGURE 10.37
Normal operation of UPS system.

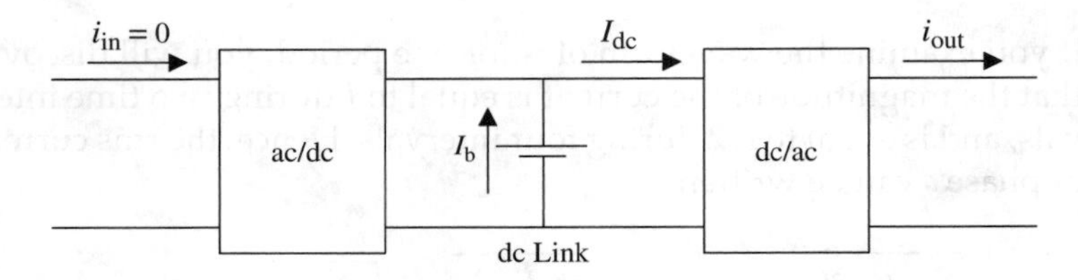

FIGURE 10.38
UPS operation during power outage.

This type of converter is commonly used as an uninterruptible power supply (UPS). A UPS system is shown in Figure 10.37. The system is similar to that shown in Figure 10.36, but a rechargeable battery is added in the dc link. In normal operation, the input ac current i_{in} is converted into dc where part of the current I_b charges the battery and the other part I_{dc} is converted into ac current i_{out}. If the input power is lost due to an outage, the energy stored in the battery is used to feed the load through the dc/ac converter as shown in Figure 10.38. Hence, the load power is not interrupted. Of course, this system can provide temporary power until the battery is discharged.

Exercise

10.1. A bipolar transistor is connected to a resistive load as shown in Figure 10.7. The source voltage $V_{CC} = 40$ V and $R_L = 10\ \Omega$. In the saturation region, the collector–emitter voltage $V_{CE} = 0.1$ V and $\beta = 5$. While the transistor is in the saturation region, calculate the following:

a. Load current.

b. Load power.

c. Losses in the collector circuit.

d. Losses in the base circuit.

e. Efficiency of the circuit.

10.2. For the transistor in the previous problem, compute the load power and the efficiency of the circuit when the transistor is in the cutoff region. Assume that the collector current is 10 mA in the cutoff region.

10.3. A bipolar junction transistor operating in the saturation region has a base current of 10 A and a collector current of 50 A. Compute the following:

a. The current gain of the transistor in the saturation region.

b. The losses of the transistor.

10.4. Compute the rms voltage of the following waveform.

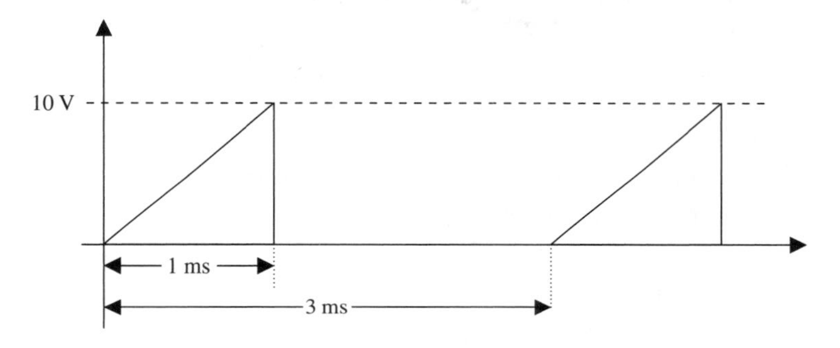

10.5. A half-wave rectifier circuit converts a 120 V (rms) into dc. The load of the circuit is 5 Ω resistance. Compute the following:

a. The average voltage across the load.

b. The average voltage of the source.

c. The rms voltage of the load.

d. The rms current of the load.

e. The power consumed by the load.

10.6. A half-wave SCR converter circuit is used to regulate the power across a 10 Ω resistance. When the triggering angle is 30°, the power consumed by the load is 500 W. Compute the rms voltage of the source and the average and rms currents of the load.

10.7. An ac/dc half wave SCR circuit is used to energize a resistive load. At a triggering angle of 30°, the average voltage across the load is 45 V.

a. Compute the source voltage in rms.

b. If a full-wave circuit is used while the triggering angle is maintained at 30°, compute the average voltage across the load.

10.8. A 120 V full-wave SCR battery charger is designed to provide 1.0 A of charging current. The battery has 1 Ω internal resistance. At the beginning of the charging process the voltage of the battery set is 60 V. Compute the triggering angle of the converter.

10.9. A full-wave, ac/dc converter is connected to a resistive load of 5 Ω. The voltage of the ac source is 120 V (rms). If the triggering angle of the converter is 90°, compute the rms voltage across the load and the power consumed by the load.

10.10. A Boost converter is used to step up 25 V into 40 V. The switching frequency of the transistor is 1 kHz, and the load resistance is 100 Ω. Compute the following:

a. The current ripple when the inductor is 30 mH.
b. The average current of the load.
c. The power delivered by the source.

10.11. A Buck–Boost converter with an input voltage of 40 V is used to regulate the load voltage from 10 to 80 V. The on-time of the transistor is always fixed at 0.1 ms and the switching frequency is adjusted to regulate the load voltage. Compute the range of the switching frequency.

10.12. A three-phase dc/ac converter is used to power a three-phase, Y-connected, resistive load of 50 Ω (per phase). The dc voltage is 150 V. Compute the following:

a. The frequency of the ac waveform if the time interval is 100 μs.
b. The current of the dc source.
c. The rms current of the load.
d. The rms of the line-to-line voltage across the load.

10.13. A full-wave, ac/dc SCR converter circuit is used to power a resistive load of 10 Ω. The ac voltage is 120 V (rms), and the triggering angle of the SCR is adjusted to 60°. Calculate the following:

a. The conduction period.
b. The average voltage across the load.
c. The average voltage across the SCRs.
d. The rms voltage across the load.
e. The average current.
f. The load power.

10.14. A dc/dc, Buck converter consists of a 100 V dc source in series with 10 Ω load resistance and a bipolar transistor. For each cycle, the transistor is turned on for 200 μs and turned off for 800 μs. Calculate the following:

a. The switching frequency of the converter.
b. The average voltage across the load.
c. The average load current.
d. The rms voltage across the load.
e. The rms current of the load.
f. The rms power consumed by the load.

10.15. A dc/dc Buck converter has an input voltage of 100 V and a duty ratio of 0.2. Compute the following:

a. The load voltage.
b. The switching frequency of the converter if the on-time is 0.1 ms.

10.16. The load of the full-wave SCR circuit in Figure 10.23 consumes 130 W. The rms voltage across the load is 80 V, and its average voltage is 50 V.

a. Compute the average voltage across any SCR.

b. The triggering circuit if one SCR failed and that SCR is not conducting anymore. If the triggering angle of the rest of the SCRs is unchanged, compute the average voltage of the load and the load power.

10.17. A resistive load of 5 Ω is connected to an ac source of 120 V (rms) through a back-to-back SCR circuit (two parallel SCRs with the anode of each one is connected to the cathode of the other). The triggering angle of the forward SCR (the one that conducts at the positive half of the cycle) is 30°. The triggering angle of the other SCR is (180° + 30°). Calculate the following:

a. The average voltage across the load.

b. The power consumed by the load.

11

Transformers

The Italian scientist Antonio Pacinotti who invented the transformer in 1860 granted the power industry with one of its most important devices, without which the power grid would not exist. The transformer is used to step up (increase) or step down (decrease) the voltage. This capability allows us to transmit and distribute massive amounts of power in today's extensive power grid. Because the power is the product of current and voltage, large power at a low voltage leads to high current. Since the current determines the cross-section of the transmission line wires, large currents require unrealistic, large cross-section wires. These impractical wires are heavy and expensive to manufacture, and require enormous towers to carry them. The transformer provides us with the solution; it allows us to increase the voltage of the transmission lines so that the current is reduced. Thus, the cross-section of the wire is reasonably small and the cost of the transmission line is dramatically reduced.

Figure 11.1 shows a schematic of a power system with its main transformers. The system has four types of transformers: transmission, distribution, service, and circuit transformers.

1. *Transmission transformers* [Figure 11.2(a)]: These transformers are connected to both ends of the transmission line. At the generating

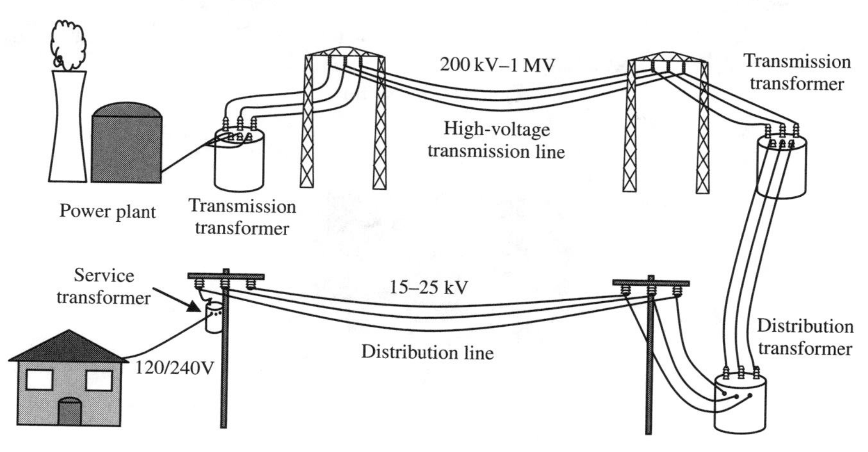

FIGURE 11.1
Various transformers in a power system.

power plants the transformer steps up the voltage of the generators to very high levels (220 kV to 1 MV) so the current can be reduced substantially. At the other end of the transmission line, the transformer steps down the voltage to a lower level, suitable for distribution to load centers. Since the transmission transformers carry the bulk power of the system, they are very large in size.

(a) Transmission transformer

(b) Distribution transformer

(c) Service transformer

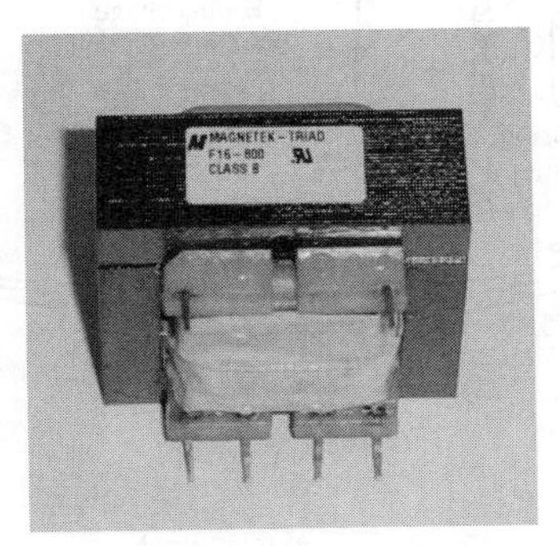

(d) Circuit transformer

FIGURE 11.2
(see color insert following Page 208) Various types of transformers.

The internal losses of the transformer, which are due to the high currents in the windings and core losses, can cause excessive heat inside the container of the transformer. Therefore, the transformer is immersed in a cooling medium such as oil. The transformer container is also equipped with a system of pipes and fans to extract heat from the cooling medium much like the radiator in an automobile. The medium also increases the dielectric strength of the windings so that the transformer can withstand higher voltages.

2. *Distribution transformers* [Figure 11.2(b)]: These transformers are installed in distribution substations near the load centers. They are designed to reduce the output voltage of the transmission transformers to a lower level of 5 to 220 kV as shown in Figure 11.1.
3. *Service transformers* [Figure 11.2(c)]: These transformers are located near the customers' loads. They are designed to lower the distribution voltage to the household level (120/240 V in the United States). They are normally mounted on power poles, placed inside vaults, or installed inside large buildings.
4. *Circuit transformers* [Figure 11.2(d)]: These are small transformers extensively used in power supplies where the household voltage is stepped down to the circuit voltage level of a few volts. Other applications include impedance matching, filters, and electric isolation.

11.1 Theory of Operation

The basic components of a transformer are shown in Figure 11.3. The transformer consists of at least two windings wrapped around a laminated iron alloy core. These windings are electrically isolated from the core by a varnish insulation material capable of withstanding the voltage of the windings. The core provides a low reluctance path for the magnetic flux. An ac voltage source is connected across one of the windings and the current produces flux that passes through the core and links the other windings. Since the current is alternating, the flux in the core changes its direction 120 times per second for the 60 Hz system. The rapid change in the flux induces a voltage in the core itself, thus creating a current that moves through the core in a circular path. This current, which is called *eddy current*, creates heat in the core and results in wasted energy. To reduce the eddy current, the electrical resistance of the core must be increased. This is done by laminating the core as shown in Figure 11.3. Each laminated plate is insulated by oxidized material or varnish. The cross-section of a laminated sheet is much smaller than the total cross-section of the core, thus the circular eddy current is substantially reduced.

Assume that an ac voltage source is connected across one of the windings, and a load is connected across the other winding as shown in Figure 11.4.

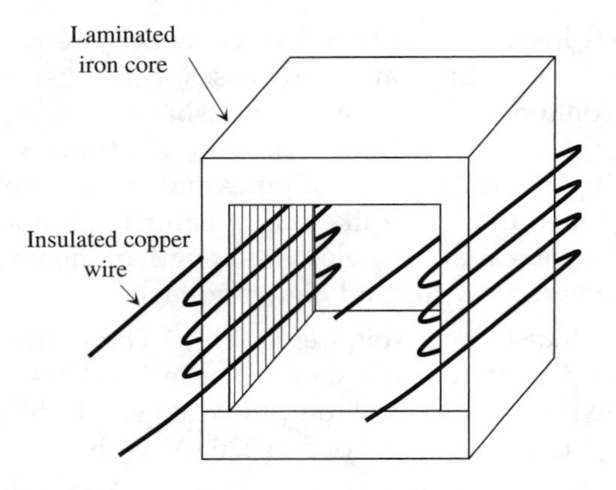

FIGURE 11.3
Main components of a transformer.

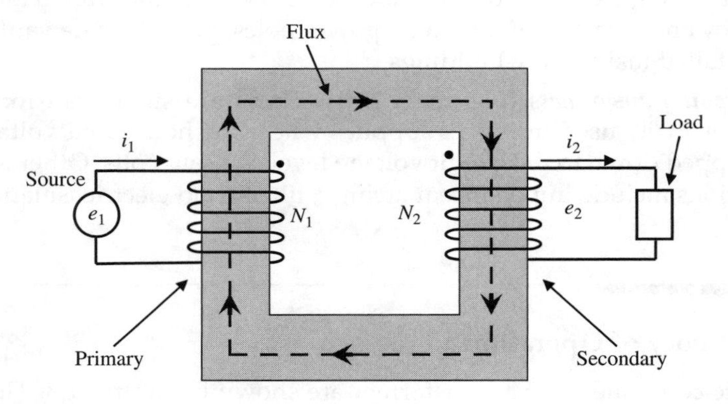

FIGURE 11.4
Flux linkage of the transformer.

The winding connected to the source is called the *primary winding*, and the load winding is called the *secondary winding*. The number of turns of the primary winding is N_1 and of the secondary winding is N_2. The analysis of this transformer is given in the following sections.

11.1.1 Voltage Ratio

According to Faraday's law, the instantaneous voltage e_1 across the primary winding produces a flux ϕ, where

$$
\begin{aligned}
e_1 &= +N_1 \frac{\mathrm{d}\phi}{\mathrm{d}t} \\
\phi &= \frac{1}{N_1} \int e_1 \, \mathrm{d}t
\end{aligned}
\tag{11.1}
$$

Assume that the core captures all the flux produced by e_1, and the flux links the secondary winding. According to Faraday's law, the flux will induce a voltage e_2 across the secondary winding, where

$$e_2 = -N_2 \frac{d\phi}{dt} \tag{11.2}$$

The positive sign in Equation (11.1) indicates that e_1 induces the flux, and the negative sign in Equation (11.2) indicates that the flux induces e_2.

The dot at one end of each winding indicates that the potentials at the dotted terminals are in phase, that is, e_1 and e_2 are in phase. Moreover, the current i_1 flowing into the dotted terminal of the primary winding is in phase with the current i_2 leaving the dotted terminal of the secondary windings.

If we are interested in the magnitudes of e_1 and e_2, Equations (11.1) and (11.2) can be rewritten as

$$\frac{e_1}{e_2} = \frac{N_1}{N_2} \tag{11.3}$$

Equation (11.3) shows that the voltage ratio of the transformer is equal to the ratio of the turns. In rms quantities, the voltage ratio is

$$\frac{E_1}{E_2} = \frac{N_1}{N_2} \tag{11.4}$$

where E is the rms value of e. If $N_2 < N_1$, the secondary voltage E_2 is less than the primary voltage E_1, and the transformer is called a *step-down* transformer. When $N_2 > N_1$, the secondary voltage E_2 is higher than the primary voltage E_1, and the transformer is called a *step-up* transformer.

EXAMPLE 11.1 A transformer has 4000 turns in the primary winding and 100 turns in the secondary winding. If 120 V is applied to the primary winding, compute the voltage across the secondary winding.

Solution

The voltage ratio of the transformer is

$$\frac{E_1}{E_2} = \frac{N_1}{N_2}$$

Hence,

$$E_2 = E_1 \frac{N_2}{N_1} = 120 \frac{100}{4000} = 3 \text{ V}$$

Equation (11.4) can also be rewritten as

$$\frac{E_1}{N_1} = \frac{E_2}{N_2} \tag{11.5}$$

The equation shows that the voltage per turn is the same for the primary or secondary winding.

EXAMPLE 11.2 Compute the voltage per turn for the transformer in Example 11.1.

Solution

The voltage per turn is

$$\frac{E_1}{N_1} = \frac{E_2}{N_2} = \frac{120}{4000} = 30 \text{ mV/turn}$$

11.1.2 Current Ratio

If we assume the transformer is ideal without losses, the input power to the transformer is equal to the output power consumed by the load. In complex number form, the apparent power of the primary winding S_1 is equal to the apparent power of the secondary winding S_2.

$$\begin{aligned} \overline{S}_1 &= \overline{S}_2 \\ \overline{S}_1 &= \overline{E}_1 \overline{I}_1^* \\ \overline{S}_2 &= \overline{E}_2 \overline{I}_2^* \end{aligned} \tag{11.6}$$

$\overline{I}^*$ is the conjugate of the current $\overline{I}$. Based on Equations (11.4) and (11.6), we can write the current relationship as

$$\frac{I_1}{I_2} = \frac{E_2}{E_1} = \frac{N_2}{N_1} \tag{11.7}$$

Note that the current ratio is the inverse of the voltage ratio. Hence, the winding with higher voltage carries lower current. If you open a transformer, you will find two sets of windings: one has a large number of turns made of small cross-section wire, and the other has a smaller number of turns of thicker wire. The number of turns is determined by the voltage–turn ratio of the transformer, and the cross-section of the wire is determined by the amount of current passing through the windings. Can you tell which winding has a high voltage and which one has a high current?

11.1.3 Reflected Load Impedance

Assume that the load impedance in Figure 11.4 is Z_{load}, where

$$Z_{load} = \frac{E_2}{I_2} \tag{11.8}$$

This load impedance when seen from the primary winding is Z'_{load}, where

$$Z'_{load} = \frac{E_1}{I_1} \tag{11.9}$$

Hence,

$$\begin{aligned} \frac{Z'_{load}}{Z_{load}} &= \frac{E_1}{E_2}\frac{I_2}{I_1} = \left(\frac{N_1}{N_2}\right)^2 \\ Z'_{load} &= Z_{load}\left(\frac{N_1}{N_2}\right)^2 \end{aligned} \tag{11.10}$$

Z'_{load} is known as the *reflected impedance* of the load.

EXAMPLE 11.3 If the load of the transformer in Example 11.1 is 3 VA, compute the reflected impedance of the load as seen from the primary winding.

Solution

The magnitude of the load current is

$$I_2 = \frac{S}{E_2} = \frac{3}{3} = 1 \text{ A}$$

The magnitude of the load impedance is

$$Z_{load} = \frac{E_2}{I_2} = \frac{3}{1} = 3\ \Omega$$

The magnitude of the reflected impedance of the load is

$$Z'_{load} = Z_{load}\left(\frac{N_1}{N_2}\right)^2 = 3\left(\frac{4000}{100}\right)^2 = 4.8\ \text{k}\Omega$$

EXAMPLE 11.4 A transformer with turns ratio $N_1/N_2 = 2$ is connected to a load impedance of $3 + j4\ \Omega$. Compute the reflected impedance of the load.

Solution

$$\overline{Z}'_{\text{load}} = \overline{Z}_{\text{load}} \left(\frac{N_1}{N_2}\right)^2 = (3 + j4)\,4 = 12 + j16\ \Omega$$

11.1.4 Transformer Ratings

The transformer has electrical and mechanical ratings. The electrical ratings include voltages, currents, and powers. The mechanical ratings include thermal limits, container dimensions, weight, volume, etc. The rated voltages and currents are the values at which the transformer can operate for a long time without any damage to its components. Exceeding these values can cause instant or gradual damage to the transformer. Excessive voltage can cause the insulation of the windings to fail very rapidly leading to internal short circuits. Excessive current causes heat to build up inside the transformer leading to an eventual melting down of the insulation varnish of the winding, leading to short circuits.

Two important electrical ratings are of particular importance to the analysis of the transformer and are always included in the nameplate data: the voltage ratio and the apparent power. For example, the nameplate data may read "10 kVA, 8 kV/240 V." With this data, you can extract the following information:

1. The rated apparent power of the transformer is 10 kVA.
2. The voltage ratio is $E_1/E_2 = 8{,}000/240 = 33.33$. Keep in mind that the voltage ratio is the same as the turns ratio N_1/N_2. However, we cannot assume that $N_1 = 8{,}000$ or $N_2 = 240$, but we can assume that the ratio $N_1/N_2 = 33.33$.
3. The rated current of the primary winding is $I_1 = S/E_1 = 10{,}000/8{,}000 = 1.25$ A.
4. The rated current of the secondary winding is $I_2 = S/E_2 = 10{,}000/240 = 41.67$ A.
5. At full load, meaning the current and voltage are at their rated values, the magnitude of the load impedance is $Z_{\text{load}} = E_2/I_2 = 240/41.67 = 5.76\ \Omega$.
6. The magnitude of the reflected impedance of the load is $Z'_{\text{load}} = Z_{\text{load}}(N_1/N_2)^2 = 5.76(33.33)^2 = 6.4\ \text{k}\Omega$.

11.2 Multi-Winding Transformer

The transformer can have multiple secondary windings to provide different voltage levels to different loads. An example of a transformer with two secondary windings is shown in Figure 11.5. The number of turns of the two secondary windings are N_2 and N_3.

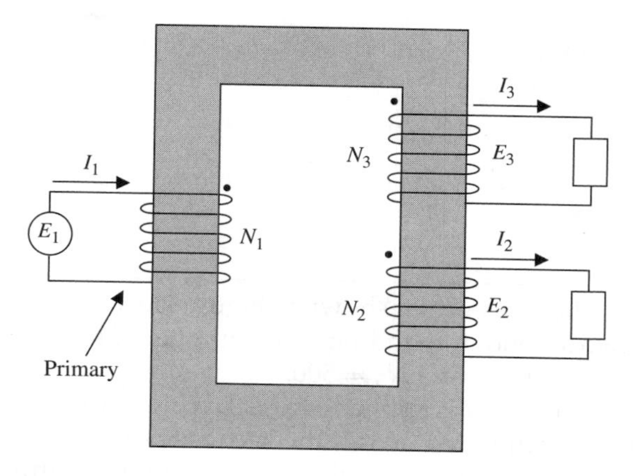

FIGURE 11.5
Transformer with multiple windings.

The voltage ratios of the transformer are

$$\frac{E_1}{E_2} = \frac{N_1}{N_2}$$
$$\frac{E_1}{E_3} = \frac{N_1}{N_3} \tag{11.11}$$

The computations of the current ratios require more attention. If we use the superposition theory and assume that the load of the winding N_3 is zero, then the current of the primary winding I_{12} due to I_2 alone is

$$I_{12} = I_2 \frac{N_2}{N_1} \tag{11.12}$$

Now let us assume that the load of the winding N_2 is removed and that of N_3 is connected. The primary current I_{13} due to I_3 is

$$I_{13} = I_3 \frac{N_3}{N_1} \tag{11.13}$$

Hence, the total primary current I_1 is the phasor sum of I_{12} and I_{13}.

$$\bar{I}_1 = \bar{I}_{12} + \bar{I}_{13} = \bar{I}_2 \frac{N_2}{N_1} + \bar{I}_3 \frac{N_3}{N_1}$$
$$\bar{I}_1 N_1 = \bar{I}_2 N_2 + \bar{I}_3 N_3 \tag{11.14}$$

Equation (11.14) is known as the *Ampere-turns rule*, according to which the current is shared between the windings according to their respective number of turns. Moreover, the sum of the apparent powers of all loads in the secondary windings is equal to the apparent power delivered by the source in the primary winding.

$$\overline{S}_1 = \overline{S}_3 + \overline{S}_3$$
$$\overline{E}_1\overline{I}_1^* = \overline{E}_2\overline{I}_2^* + \overline{E}_3\overline{I}_3^* \tag{11.15}$$

EXAMPLE 11.5 The transformer shown in Figure 11.6 consists of one primary winding and two secondary windings. The number of turns of the windings are $N_1 = 4000, N_2 = 1000$, and $N_3 = 500$.

A voltage source of 120 V is applied across the primary winding, and purely resistive loads are connected across the secondary windings. A wattmeter placed in the primary circuit measures 300 W. Another wattmeter placed in the secondary winding N_2 measures 90 W. Compute the following:

1. The voltages of the secondary windings.
2. The current in N_3.
3. The power of the load connected across N_3.

Solution

1. Using the ratios in Equation (11.11), we can compute the voltages of the secondary windings.

$$E_2 = E_1\frac{N_2}{N_1} = 120\frac{1000}{4000} = 30 \text{ V}$$
$$E_3 = E_1\frac{N_3}{N_1} = 120\frac{500}{4000} = 15 \text{ V}$$

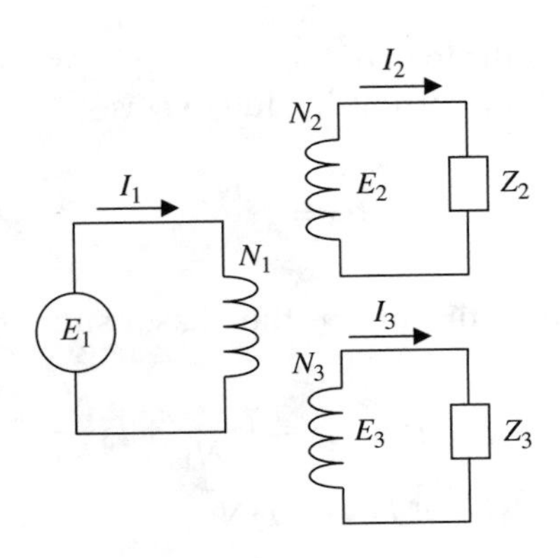

FIGURE 11.6
Three-winding transformer.

2. Before we compute the currents of the secondary winding N_3, we need to compute I_1 and I_2. Since the loads are purely resistive, the power factor everywhere is unity.

$$I_1 = \frac{P_1}{E_1 \cos \theta_1} = \frac{300}{120} = 2.5 \text{ A}$$
$$I_2 = \frac{P_2}{E_2 \cos \theta_2} = \frac{90}{30} = 3 \text{ A}$$

Since all loads are resistive, we do not have to include the phase angles in Equation (11.14).

$$I_1 N_1 = I_2 N_2 + I_3 N_3$$
$$2.5 \times 4000 = 1000 \times 3 + 500 I_3$$
$$I_3 = 14 \text{ A}$$

We can arrive at the same result by using Equation (11.15).

$$\overline{S}_1 = \overline{S}_2 + \overline{S}_3$$

Since the loads are resistive, the angles of all currents are zero.

$$\overline{E}_1 \overline{I}_1^* = \overline{E}_2 \overline{I}_2^* + \overline{E}_3 \overline{I}_3^*$$
$$E_1 I_1 = E_2 I_2 + E_3 I_3$$
$$120 \times 2.5 = 30 \times 3 + 15 I_3$$
$$I_3 = 14 \text{ A}$$

3. Since the power factor is unity and the transformer is lossless,

$$P_1 = P_2 + P_3$$
$$300 = 90 + P_3$$
$$P_3 = 210 \text{ W}$$

EXAMPLE 11.6 For the transformer in Example (11.5), assume the load impedances connected across the secondary windings are $\overline{Z}_2 = 3 + j4\ \Omega$ and $\overline{Z}_3 = 5 + j0\ \Omega$.

A voltage source of 120 V is applied across the primary winding. Compute the following:

1. The currents in all windings.
2. The real and reactive powers in the primary windings.

Solution

1. Assume that the voltage of the primary winding is the reference voltage, thus all voltages have zero phase angles. Hence, the currents in the secondary windings are

$$\bar{I}_2 = \frac{\bar{E}_2}{\bar{Z}_2} = \frac{30\angle 0^\circ}{3 + j4} = 6\angle -53.1^\circ \text{ A}$$

$$\bar{I}_3 = \frac{\bar{E}_3}{\bar{Z}_3} = \frac{15\angle 0^\circ}{5 + j0} = 3\angle 0^\circ \text{ A}$$

Using Equation (11.14),

$$\bar{I}_1 N_1 = \bar{I}_2 N_2 + \bar{I}_3 N_3$$

$$\bar{I}_1 4000 = (6\angle -53.1^\circ) \times 1000 + 3 \times 500$$

$$\bar{I}_1 = 1.276 - j1.2 = 1.75\angle -43.24^\circ \text{ A}$$

2. The complex power of the primary winding is

$$\bar{S}_1 = \bar{E}_1 \bar{I}_1^* = (120\angle 0^\circ)(1.75\angle 43.24) = 153.12 + j144 \text{ VA}$$

Hence,

$$P_1 = 153.12 \text{ W}$$

$$Q_1 = 144 \text{ VAr}$$

11.3 Autotransformer

The autotransformer has its primary and secondary windings connected in series as shown in Figure 11.7. The primary terminal A_2 is connected to the secondary terminal B_1, and the source voltage V_1 is connected between A_1 and B_2. The load is connected between B_1 and B_2. This way, almost any transformer can be connected as an autotransformer.

The wiring diagram of the autotransformer is shown in Figure 11.8. Let us assume that the rated voltage of the primary winding is E_1, and that for the secondary winding is E_2. Also assume that the rated current of the primary winding is I_1 and that for the secondary winding is I_2.

As stated in Equation (11.5), the volt/turn ratio is constant for any winding. Hence,

$$\frac{E_1}{N_1} = \frac{E_2}{N_2} = \frac{V_1}{(N_1 + N_2)} = \frac{V_2}{N_2} \tag{11.16}$$

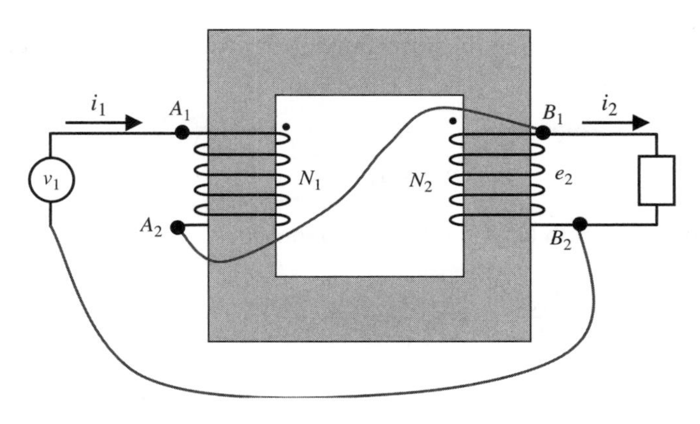

FIGURE 11.7
Connecting a regular transformer as an autotransformer.

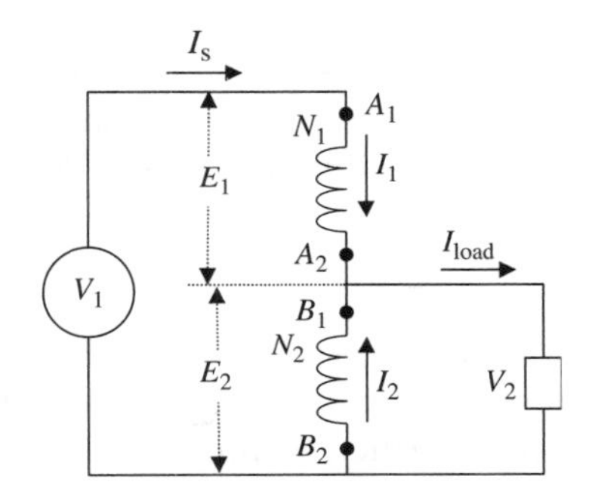

FIGURE 11.8
Wiring connection of the autotransformer in Figure 11.7.

The voltage ratio (input voltage versus output voltage) is

$$\frac{V_1}{V_2} = \frac{N_1 + N_2}{N_2} \tag{11.17}$$

Using the Ampere-turns law, we can write the current equation as

$$N_1 I_1 = N_2 I_2 \tag{11.18}$$

As shown in Figure 11.8, the load and source currents are

$$\begin{aligned} I_{load} &= I_1 + I_2 \\ I_s &= I_1 \end{aligned} \tag{11.19}$$

Hence, the current ratio of the autotransformer (output current versus input current) is

$$\frac{I_{\text{load}}}{I_s} = \frac{I_1 + I_2}{I_1} = \left(1 + \frac{I_2}{I_1}\right) \tag{11.20}$$

Substituting Equation (11.18) into (11.20) yields

$$\frac{I_{\text{load}}}{I_s} = 1 + \frac{I_2}{I_1} = 1 + \frac{N_1}{N_2} = \frac{N_1 + N_2}{N_2} \tag{11.21}$$

The apparent power can be computed for the input or output circuits of the autotransformer.

$$\overline{S} = \overline{V}_1 \overline{I}_s^* = (\overline{E}_1 + \overline{E}_2)\overline{I}_1^* = \overline{E}_1 \overline{I}_1^* + \overline{E}_2 \overline{I}_1^* \tag{11.22}$$

Note that $E_1 I_1$ is the apparent power of the regular transformer as given in Equation (11.6). Hence, the autotransformer can handle more power than the regular transformer without exceeding the rated voltage or the rated current of any winding.

EXAMPLE 11.7 A 1 kVA transformer has a voltage ratio of 120/240 V. It is desired to change the voltage ratio of the transformer to 360/240 V by connecting it as an autotransformer. Compute the new power rating of the autotransformer.

Solution

If you assume that the voltage of winding $A_1 - A_2$ is 120 V in Figure 11.7 and $B_1 - B_2$ is 240 V, the voltage ratio is

$$\frac{V_1}{V_2} = \frac{E_1 + E_2}{E_2} = \frac{120 + 240}{240} = \frac{360}{240}$$

As a regular transformer, the rated currents of the windings are

$$I_1 = \frac{S}{E_1} = \frac{1000}{120} = 8.33 \text{ A}$$

$$I_2 = \frac{S}{E_2} = \frac{1000}{240} = 4.167 \text{ A}$$

The power of the autotransformer is

$$S = E_1 I_1 + E_2 I_1 = 1000 + 240 \times 8.33 = 3.0 \text{ kVA}$$

This new power rating of the autotransformer is three times the power rating of the regular transformer.

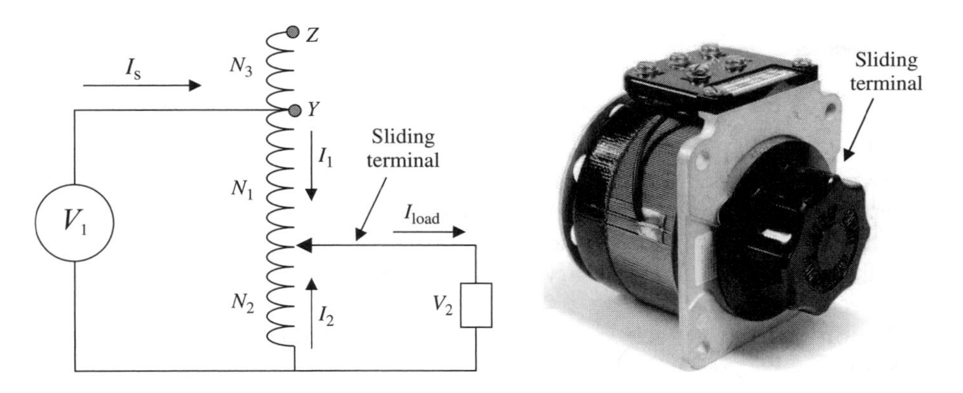

FIGURE 11.9
(see color insert following Page 208) Autotransformer with adjustable voltage.

The autotransformer can be built to provide adjustable output voltage. This type of autotransformer, shown in Figure 11.9, consists of one winding where the primary voltage is applied across a part of the winding. The number of turns of the secondary winding can be changed by sliding one of the secondary terminals along the winding. This requires a mechanism by which the winding and the sliding terminal are always connected electrically. In the position shown in the figure, the secondary winding is N_2. Hence, the secondary voltage is

$$V_2 = V_1 \frac{N_2}{N_1 + N_2} \tag{11.23}$$

In Equation (11.23), $V_2 < V_1$. When the slider is moved to position Y, the voltage across the load is

$$V_2 = V_1 \frac{N_1 + N_2}{N_1 + N_2} \tag{11.24}$$

where $V_2 = V_1$. If the slider moves to point Z, the load voltage is

$$V_2 = V_1 \frac{N_1 + N_2 + N_3}{N_1 + N_2} \tag{11.25}$$

where $V_2 > V_1$.

This autotransformer, which is also known as *Variac*, can adjust the load voltage from zero to greater than the supply voltage. The variac is very useful in laboratories and test set-ups where the load voltage is easily adjusted to any desired value. It can also be used in speed control and start-up for electric motors.

11.4 Three-Phase Transformer

The three-phase transformers can be constructed out of a three-legged core as shown in Figure 10.10. The primary and secondary windings of each phase are placed on the same leg of the core. The primary windings are labeled *aa′*, *bb′*, and *cc′*. The secondary windings are labeled *AA′*, *BB′*, and *CC′*. N_1 is the number of turns of any primary winding, and N_2 is the number of turns of any secondary winding.

A more convenient schematic representation of the three-phase windings is shown in Figure 11.11.

11.4.1 Three-Phase Transformer Ratings

The ratings of three-phase transformers are given in three-phase quantities. The nameplate data can read, for example, "60 kVA, 8 kV(Δ)/416 V(Y)." This nameplate data can provide us with the following information:

1. The rated apparent power of the three phases is 60 kVA.
2. The rated apparent power of each phase is $\frac{60}{3} = 20$ kVA.

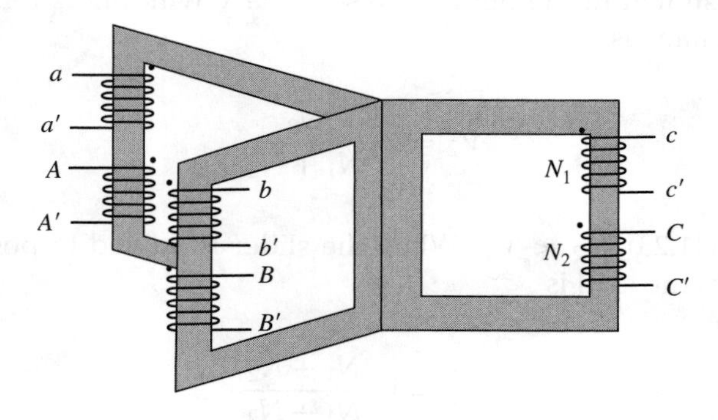

FIGURE 11.10
Configuration for three-phase transformer.

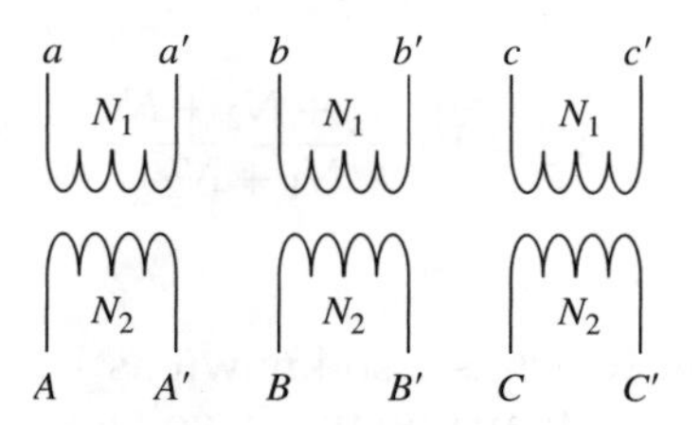

FIGURE 11.11
Schematic of three-phase transformer.

3. The primary windings are connected in delta.
4. The rated line-to-line voltage of the primary winding is 8 kV.
5. The secondary windings are connected in wye.
6. The rated line-to-line voltage of the secondary winding is 416 V.
7. The line-to-line voltage ratio of the transformer is $8000/416 = 19.23$.
8. The voltage per turn is constant in any winding of the transformer. Hence, the turns ratio N_1/N_2 is equal to the ratio of the phase voltages $V_{\text{phase of primary}}/V_{\text{phase of secondary}}$. The turns ratio is not necessarily equal to the ratio of the line-to-line voltages.

11.4.1.1 Y–Y Transformer

A three-phase transformer connected in Y–Y is shown in Figure 11.12. The top figure shows the wiring of the transformer where the terminals a', b', and c' of the primary windings are connected to a common point called neutral n. The secondary windings are similarly connected. A more convenient schematic is the one at the bottom of Figure 11.12.

The turns ratio of the transformer is N_1/N_2, while the voltage ratio of the transformer is the ratio of the line-to-line voltages V_{ab}/V_{AB}. These two ratios are equal only if the connection of the primary and secondary windings are

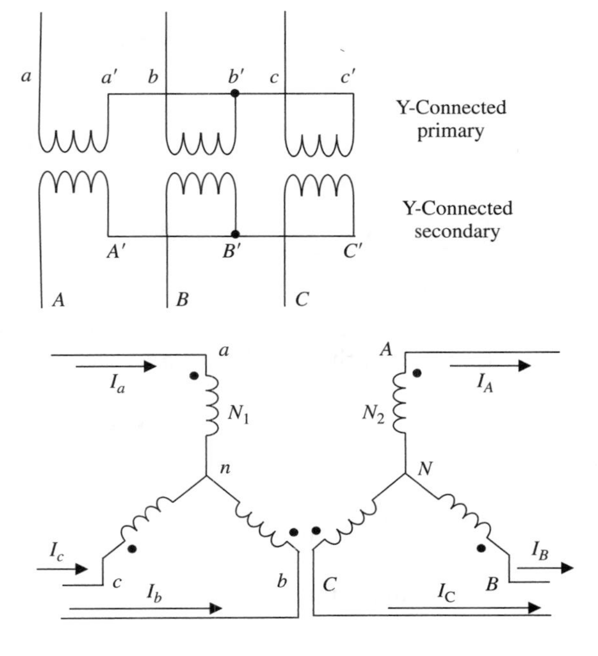

FIGURE 11.12
Y–Y transformer.

the same; that is, both windings are connected in wye or delta. The best way to compute the turns ratio is to use the voltage per turn constant,

$$\frac{V_{an}}{N_1} = \frac{V_{AN}}{N_2}$$
$$\frac{N_1}{N_2} = \frac{V_{\text{phase of primary}}}{V_{\text{phase of secondary}}} = \frac{V_{an}}{V_{AN}} = \frac{V_{ab}}{V_{AB}} \tag{11.26}$$

The current ratio of the phase voltages is the inverse of the turns ratio. We can arrive at this conclusion by using the Ampere-turns equation where any winding current multiplied by its number of turns is constant for any winding.

$$I_a N_1 = I_A N_2$$
$$\frac{N_1}{N_2} = \frac{I_A}{I_a} \tag{11.27}$$

EXAMPLE 11.8 A 12 kVA transformer has a voltage ratio of 13.8 kV(Y)/416 V(Y).

1. Compute the ratio of the phase voltages.
2. Compute the turns ratio.
3. Compute the rated current of the primary and secondary winding.

Solution

1. The transformer is connected in Y–Y. Consider Figure 11.12. The phase voltage ratio is V_{an}/V_{AN}. Hence,

$$\frac{V_{an}}{V_{AN}} = \frac{V_{ab}/\sqrt{3}}{V_{AB}/\sqrt{3}} = \frac{V_{ab}}{V_{AB}} = \frac{13{,}800}{416} = 33.17$$

   Note that the line-to-line voltage ratio is the same as the phase voltage ratio. This is only true if both the primary and secondary windings have the same connection.

2. The turns ratio is N_1/N_2, which is the same as the phase voltage ratio V_{an}/V_{AN}.

$$\frac{N_1}{N_2} = \frac{V_{an}}{V_{AN}} = 33.17$$

3. The current of the primary winding is

$$I_a = \frac{S}{\sqrt{3}V_{ab}} = \frac{12}{\sqrt{3} \times 13.8} = 0.502 \text{ A}$$

The current of the secondary winding is

$$I_A = \frac{S}{\sqrt{3} V_{AB}} = \frac{12{,}000}{\sqrt{3} \times 416} = 16.65 \text{ A}$$

11.4.1.2 Δ–Δ Transformer

The Δ–Δ connection of the three-phase transformer is shown in Figure 11.13. Each phase of the delta winding is connected between two lines of the three-phase system. Hence, the phase voltage of any winding is equal to the line-to-line voltage. The currents in the transformer windings are known as phase currents. The currents of the lines feeding the transformers are the line currents.

The turns ratio of the transformer is N_1/N_2. The voltage ratio of the transformer is the ratio of the line-to-line voltages V_{ab}/V_{AB}. These two ratios are equal since the primary and secondary windings have the same connection. Keep in mind that the voltage per turn is constant in any winding; hence,

$$\frac{V_{ab}}{N_1} = \frac{V_{AB}}{N_2}$$
$$\frac{N_1}{N_2} = \frac{V_{ab}}{V_{AB}} \tag{11.28}$$

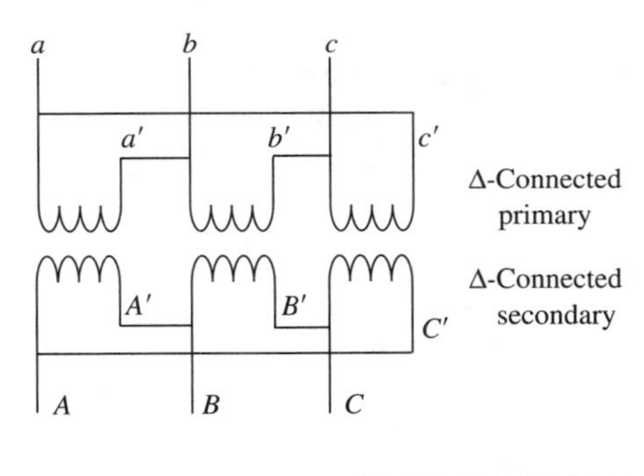

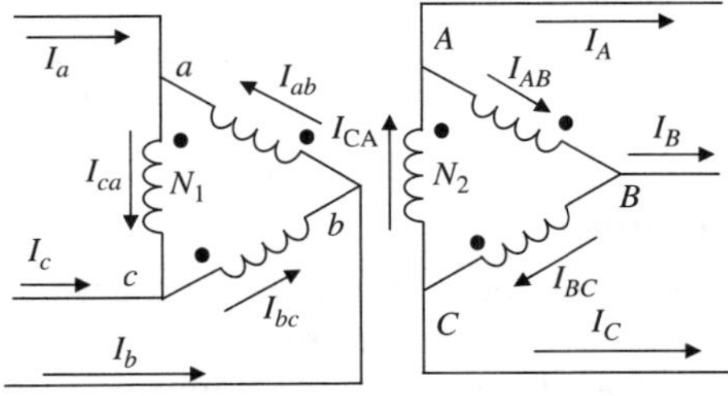

FIGURE 11.13
Transformer connected in Δ–Δ.

Using the Ampere-turns equation, we can compute the current ratio.

$$I_{ab}N_1 = I_{AB}N_2$$
$$\frac{N_1}{N_2} = \frac{I_{AB}}{I_{ab}} \tag{11.29}$$

EXAMPLE 11.9 A 15 kVA transformer has a voltage ratio of 25 kV(Δ)/5 kV(Δ).

1. Compute the turns ratio.
2. Compute the line current in the primary circuit.
3. Compute the currents of the primary and secondary windings.

Solution

1. The transformer is connected in Δ–Δ. As shown in Figure 11.13, the turns ratio is the ratio of phase voltages which are equal to the line-to-line voltages.

$$\frac{N_1}{N_2} = \frac{V_{ab}}{V_{AB}} = \frac{25}{5} = 5$$

2. The line current of the source is

$$I_a = \frac{S}{\sqrt{3}V_{ab}} = \frac{15}{\sqrt{3} \times 25} = 0.346 \text{ A}$$

3. The current of the primary winding is

$$I_{ab} = \frac{I_a}{\sqrt{3}} = \frac{0.346}{\sqrt{3}} = 0.2 \text{ A}$$

The current of the secondary winding is

$$\frac{N_1}{N_2} = \frac{I_{AB}}{I_{ab}}$$
$$I_{AB} = \frac{N_1}{N_2} I_{ab} = 5 \times 0.2 = 1 \text{ A}$$

11.4.1.3 Y–Δ Transformer

The Y–Δ transformer has one set of windings (primary or secondary) connected in wye and the other windings connected in delta as shown in Figure 11.14. The analysis of the Y–Δ transformer requires some attention when computing the turns ratio. In Figure 11.14, the voltage across N_1 is the phase voltage of the primary circuit V_{an}, and the voltage across N_2 is the line-to-line voltage

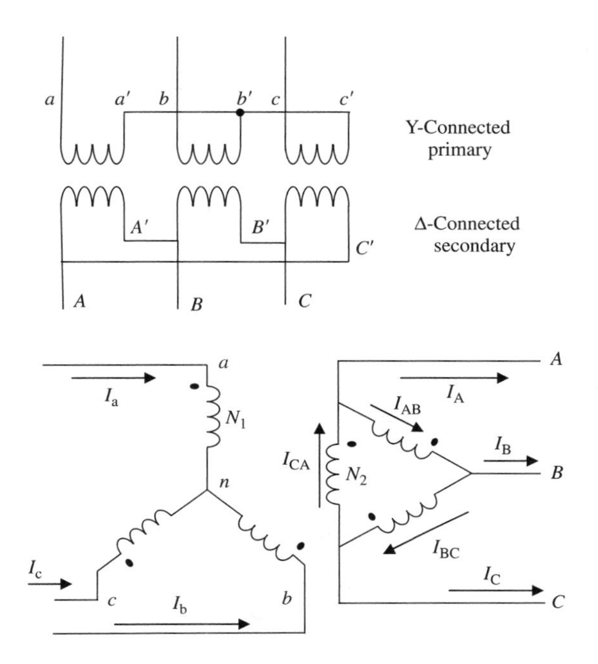

FIGURE 11.14
Y–Δ transformer.

of the secondary circuit V_{AB}. Since the voltage per turn is constant in any winding, we can write the turns ratio as

$$\frac{V_{an}}{N_1} = \frac{V_{AB}}{N_2}$$
$$\frac{N_1}{N_2} = \frac{V_{an}}{V_{AB}} \tag{11.30}$$

Similarly, the current through N_1 is the line current of the primary circuit I_a, and the current through N_2 is the phase current of the secondary winding I_{AB}. Hence, we can compute the current ratio using the Ampere-turns equation.

$$I_a N_1 = I_{AB} N_2$$
$$\frac{N_1}{N_2} = \frac{I_{AB}}{I_a} \tag{11.31}$$

EXAMPLE 11.10 A 25 kVA transformer has a voltage ratio of 20 kV(Y)/ 10 kV(Δ).

1. Compute the phase voltages.
2. Compute the turns ratio.

3. Compute the line current of the primary and secondary circuits.
4. Compute the phase currents of the transformer.

Solution

1. The primary windings of the transformer are connected in wye. Hence, its phase voltage is

$$V_{an} = \frac{V_{ab}}{\sqrt{3}} = \frac{20}{\sqrt{3}} = 11.55 \text{ kV}$$

The secondary windings are connected in delta. Hence, its phase voltage is equal to its line-to-line voltage.

$$V_{AB} = 10 \text{ kV}$$

2. The turns ratio is the ratio of the phase voltages across the windings.

$$\frac{N_1}{N_2} = \frac{11.55}{10} = 1.155$$

3. The line current of the primary circuit is

$$I_a = \frac{S}{\sqrt{3}\,V_{ab}} = \frac{25}{\sqrt{3} \times 20} = 0.7217 \text{ A}$$

The line current of the secondary circuit is

$$I_A = \frac{S}{\sqrt{3}\,V_{AB}} = \frac{25}{\sqrt{3} \times 10} = 1.4434 \text{ A}$$

4. Since the primary windings are connected in wye, the phase current of the primary winding is the same as the line current of the primary circuit.

The secondary windings are connected in delta, hence the phase current of the secondary winding is

$$I_{AB} = \frac{I_A}{\sqrt{3}} = \frac{1.4434}{\sqrt{3}} = 0.8333 \text{ A}$$

Note that the turns ratio of this transformer is

$$\frac{N_1}{N_2} = \frac{V_{an}}{V_{AB}} = \frac{I_{AB}}{I_a}$$

$$\frac{N_1}{N_2} = \frac{11.55}{10} = \frac{0.8333}{0.7217}$$

11.4.2 Transformer Bank

A transformer bank is two or three single-phase transformers connected as a three-phase transformer. These transformer banks are often used in distribution systems. Take for example the cases shown in Figure 11.15. The first photograph, on the left, shows a power pole with one single-phase transformer. The load served by this transformer is small enough that one transformer is adequate. If more loads are added, an additional single-phase transformer may be needed as shown in the middle photograph. Further increase in loads may demand a third single-phase transformer as shown in the photograph on the right.

The three single-phase transformers can be connected in several configurations. One of these is the Δ–Y connection shown in Figure 11.16. The analysis of the transformer bank is the same as that for the three-phase transformers.

EXAMPLE 11.11 To provide electric power to a residential area, three single-phase transformers are connected in the Δ–Y configuration shown in Figure 11.16. Each single-phase transformer is rated at 10 kVA, 15 kV/240 V. Compute the ratings of the transformer bank and the turns ratio.

Solution

The ratings of the transformer bank are the same as the ratings of the three-phase transformer; that is, the power rating is for the three phases and the voltage ratings are for the line-to-line quantities.

The power rating of the transformer bank is

$$S = 3 \times 10 = 30 \text{ kVA}$$

The primary windings are connected in delta. Hence, the voltage across N_1 is the same as the line-to-line voltage. Hence

$$V_{ab} = 15 \text{ kV}$$

The secondary windings are connected in wye. The voltage across N_2 is the phase voltage, hence

$$V_{AB} = \sqrt{3} V_{AN} = \sqrt{3} \times 240 = 415.7 \text{ V}$$

The ratings of the transformer bank are 30 kVA, 15 kV(Δ)/415.7 V(Y)

The turns ratio of the transformer bank is the ratio of its phase voltages.

$$\frac{N_1}{N_2} = \frac{V_{ab}}{V_{AN}} = \frac{15{,}000}{240} = 62.5$$

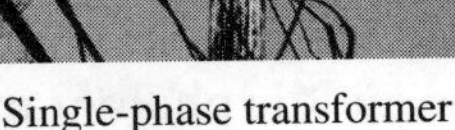
Single-phase transformer

Two transformer bank

Three transformer bank

FIGURE 11.15
(see color insert following Page 208) Transformer banks.

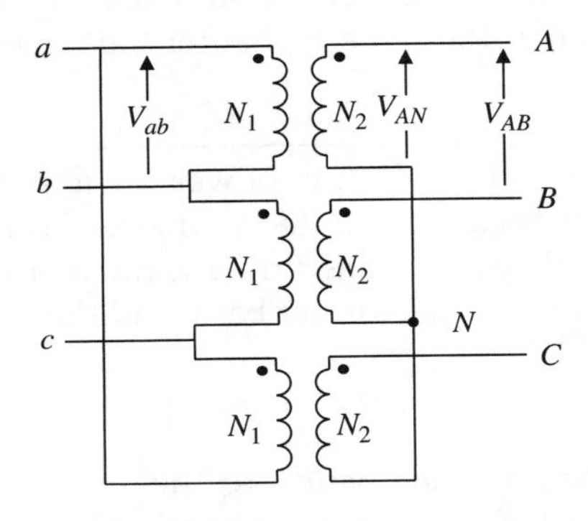

FIGURE 11.16
Δ–Y connection of a transformer bank.

11.5 Actual Transformer

The transformers we have discussed so far are ideal without any internal losses. Actual transformers have parasitic parameters such as the resistances and inductances of the windings and core. The resistance and inductance of the windings can be modeled as shown in Figure 11.17. The dashed box shows the ideal transformer we have discussed so far, and R_1, X_1 and R_2, X_2 are the resistances and inductive reactances of the primary and secondary windings, respectively. The input voltage of the transformer is V_1 and the load voltage is V_2. V_1 and V_2 are known as the *terminal voltages*. One major difference between the ideal transformer and actual transformer is that the turns ratio of the actual transformer is not equal to the ratio of its terminal voltages. Let us explain this by looking at the turns ratio of the transformer, which is

$$\frac{N_1}{N_2} = \frac{E_1}{E_2} \tag{11.32}$$

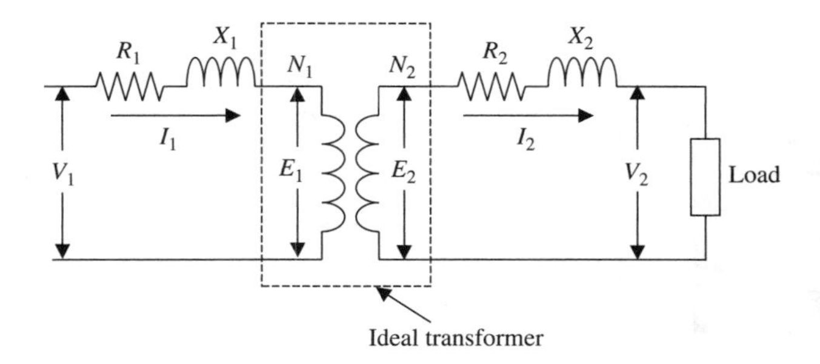

FIGURE 11.17
Winding impedance of the transformer.

Since, due to the presence of the resistances and inductive reactances, E_1 and V_1 are not equal, and neither are E_2 and V_2, then

$$\frac{N_1}{N_2} \neq \frac{V_1}{V_2} \tag{11.33}$$

Modeling the core of the transformer is a more involved process because of the nonlinear behavior of the flux in the core. Consider the case of a coil wrapped around a core shown on the left side of Figure 11.18. Let us assume that the current of the coil is sinusoidal as shown in the middle figure. This current produces a sinusoidal flux ϕ in the core that has a flux density B, where

$$B = \frac{\phi}{A} \tag{11.34}$$

A is the cross-section area of the core. The magnetic field intensity in the core H is

$$H = \mu_0 \mu_r B \tag{11.35}$$

where $\mu_0 = 4\pi 10^{-7}$ H/m. It is a constant value known as the absolute permeability. μ_r is the relative permeability; its magnitude depends on the material of the core. For air, $\mu_r = 1$ and for unsaturated iron, $\mu_r \approx 5000$.

The value of μ_r varies with the flux density. Hence, the relationship between the flux density B versus the magnetic field intensity H is nonlinear as shown on the right side of Figure 11.18. The curve is known as the hysteresis loop. Let us assume that the core was never used before. If the current starts from zero (point 0), the flux starts to flow in the core as shown by the dashed line in the hysteresis loop. Unlike air, the core cannot carry an unlimited amount of flux. When the current exceeds a certain limit (e.g., above or below the dashed lines in the middle figure), no appreciable increase occurs in the flux, and the core is said to be "saturated." This is shown in the area around point 1 in the figure. When the current starts to fall toward point 2, the flux starts to

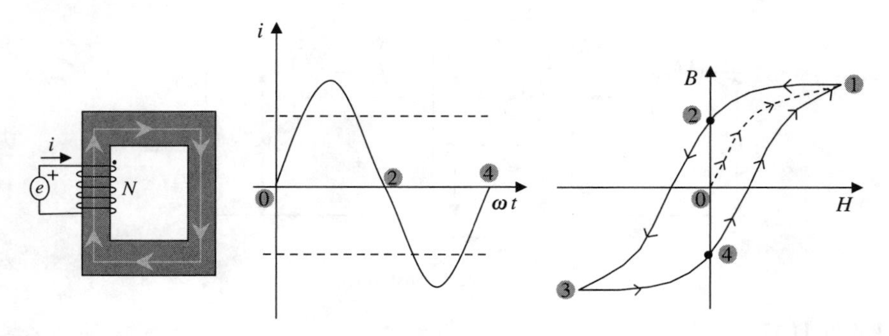

FIGURE 11.18
B–H relationship and core hysteresis.

decrease. At point 2, the current is zero, but the flux density is nonzero. This is known as the residual flux. The core at this point acts as a magnet. When the current is further reduced, the flux is also reduced to zero, and then reverses its direction. This hysteresis loop is repeated every current cycle.

To model the hysteresis, we can use an approximate process with linear elements: resistance and inductance. Using Faraday's law, we can write the expression of the flux as

$$e = \frac{d\phi}{dt} \tag{11.36}$$

Hence, the flux density is

$$B = \frac{\phi}{A} \sim \int e\, dt \tag{11.37}$$

Equation (11.37) shows that the flux density is directly proportional to the integral of the voltage across the winding. Keep in mind that the magnetic field intensity is directly proportional to the current. Hence, the *B* versus *H* characteristics in Figure 11.18 can be approximated by $\int e\, dt$ versus i relationship. This relationship can be obtained by the two circuits in Figure 11.19 and Figure 11.20. In Figure 11.19 the resistive circuit on the left side is connected across a sinusoidal voltage source.

$$e = E_{max} \sin \omega t \tag{11.38}$$

The voltage integral and currents of this circuit are

$$\int e\, dt = -\frac{E_{max}}{\omega} \cos \omega t \tag{11.39}$$

$$i_R = \frac{e}{R} = \frac{E_{max}}{R} \sin \omega t \tag{11.40}$$

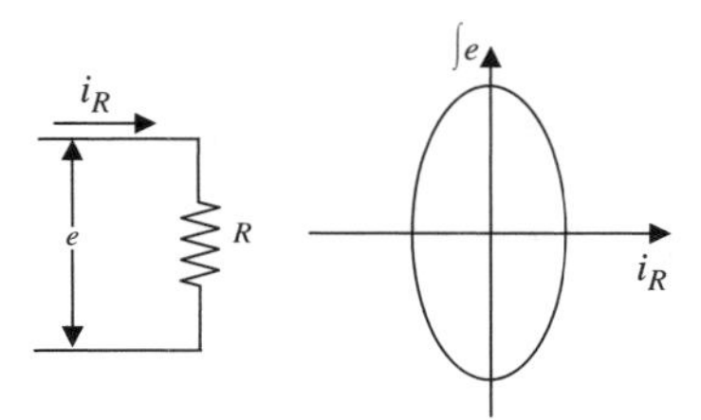

FIGURE 11.19
Voltage integral versus current of resistive element.

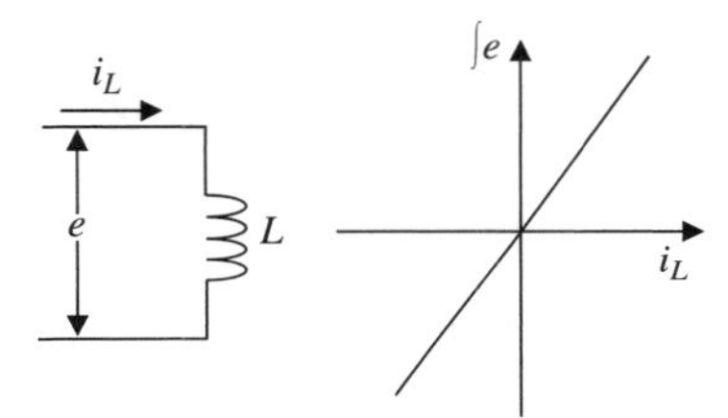

FIGURE 11.20
Voltage integral versus current of inductive element.

Note that $\int e\,dt$ and i_R are phase-shifted by 90°. The relationship of $\int e\,dt$ versus i_R is shown on the right side of Figure 11.19. Because of the phase shift, the relationship has an elliptical shape with two radii that are functions of the resistance and the angular frequency.

For the inductive element in Figure 11.20, $\int e\,dt$ and i_L are computed as follows:

$$\int e\,dt = -\frac{E_{max}}{\omega} \cos \omega t \tag{11.41}$$

$$e = L\frac{di_L}{dt}$$

$$i_L = \frac{1}{L}\int e\,dt = -\frac{E_{max}}{\omega L} \cos \omega t \tag{11.42}$$

where L is the inductance. $\int e\,dt$ and i_L are in-phase, forming a straight line relationship as shown in Figure 11.20.

Now let us add the two elements in parallel as shown in Figure 11.21. The total current in this case is

$$i = i_R + i_L \tag{11.43}$$

Combining the two characteristics in Figure 11.19 and Figure 11.20 gives the elliptically bent shape shown at the lower right side of Figure 11.21. This

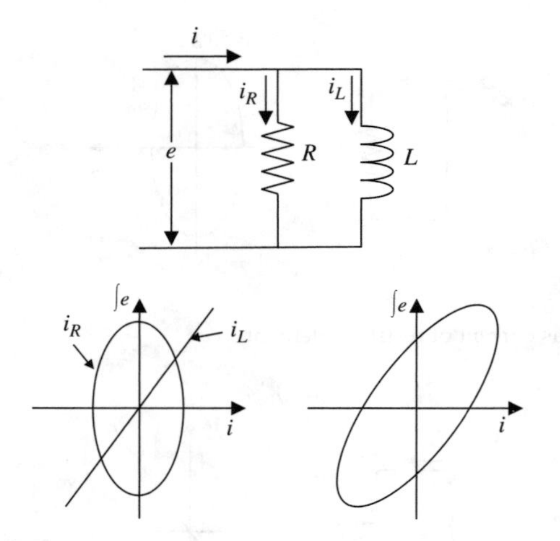

FIGURE 11.21
Approximate representation of hysteresis.

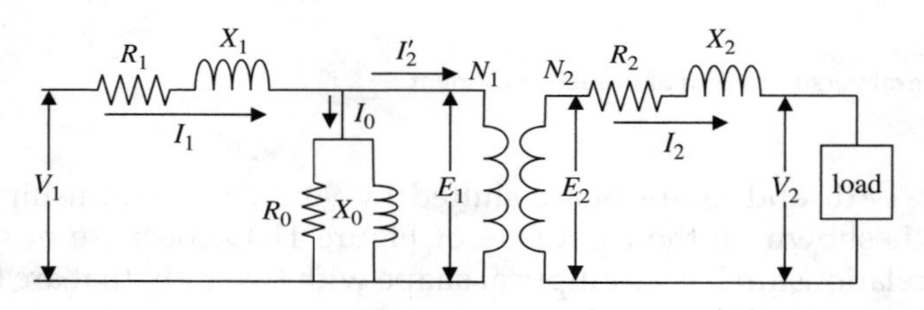

FIGURE 11.22
Equivalent circuit of a transformer.

shape is similar enough to the hysteresis loop in Figure 11.18, and is used as an approximate representation of the core. The parameters of the core model are often computed from the primary side of the transformer. Hence, the core is often included in the primary circuit as shown in Figure 11.22. I_0 is known as the excitation current representing the magnetic field intensity of the hysteresis loop. R_0 and X_0 are the equivalent core resistance and core inductive reactance of the hysteresis model.

11.5.1 Analysis of Actual Transformer

The transformer model in Figure 11.22 can be used to compute the relationships between all currents and voltages. For example, the primary current of the transformer I_1 is

$$\bar{I}_1 = \bar{I}_0 + \bar{I}'_2 \tag{11.44}$$

The current I_2' is equal to the load current as seen from the primary side. This is also known as the *reflected load current*. The relationship between I_2 and I_2' is the turns ratio of the transformer as given by the Ampere-turns equation.

$$I_2' N_1 = I_2 N_2$$
$$\frac{I_2'}{I_2} = \frac{N_2}{N_1} \tag{11.45}$$

The voltage equations of the primary and secondary circuits are

$$\overline{V}_1 = \overline{E}_1 + \overline{I}_1(R_1 + jX_1) \tag{11.46}$$
$$\overline{E}_2 = \overline{V}_2 + \overline{I}_2(R_2 + jX_2) \tag{11.47}$$

We can rewrite Equation (11.46) by substituting Equation (11.47) into Equation (11.46).

$$\overline{V}_1 = \overline{E}_2 \frac{N_1}{N_2} + \overline{I}_1(R_1 + jX_1)$$
$$\overline{V}_1 = \left[\overline{V}_2 + \overline{I}_2(R_2 + jX_2)\right] \frac{N_1}{N_2} + \overline{I}_1(R_1 + jX_1)$$
$$\overline{V}_1 = \left[\overline{V}_2 + \overline{I}_2' \frac{N_1}{N_2}(R_2 + jX_2)\right] \frac{N_1}{N_2} + \overline{I}_1(R_1 + jX_1) \tag{11.48}$$
$$\overline{V}_1 = \left[\overline{V}_2 \frac{N_1}{N_2} + \overline{I}_2' \left(\frac{N_1}{N_2}\right)^2 (R_2 + jX_2)\right] + \overline{I}_1(R_1 + jX_1)$$

Define the following variables:

$$V_2' = V_2 \frac{N_1}{N_2}$$
$$R_2' = R_2 \left(\frac{N_1}{N_2}\right)^2 \tag{11.49}$$
$$X_2' = X_2 \left(\frac{N_1}{N_2}\right)^2$$

where V_2' is known as the *reflected voltage of the secondary winding* (or *secondary voltage referred to primary*), R_2' is the *reflected resistance of the secondary* winding (or *secondary resistance referred to the primary*), and X_2' is the *reflected inductive reactance of the secondary* winding (or *secondary inductive reactance referred to the primary*).

Substituting Equation (11.49) into Equation (11.48) yields

$$\overline{V}_1 = \overline{V}_2' + \overline{I}_2' \left(R_2' + jX_2'\right) + \overline{I}_1(R_1 + jX_1) \tag{11.50}$$

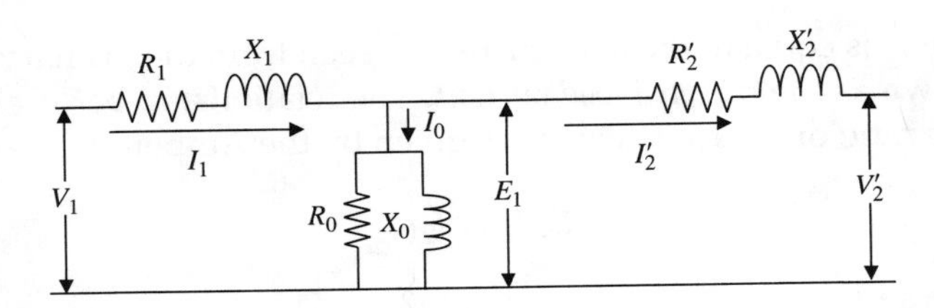

FIGURE 11.23
Modified equivalent circuit of the transformer.

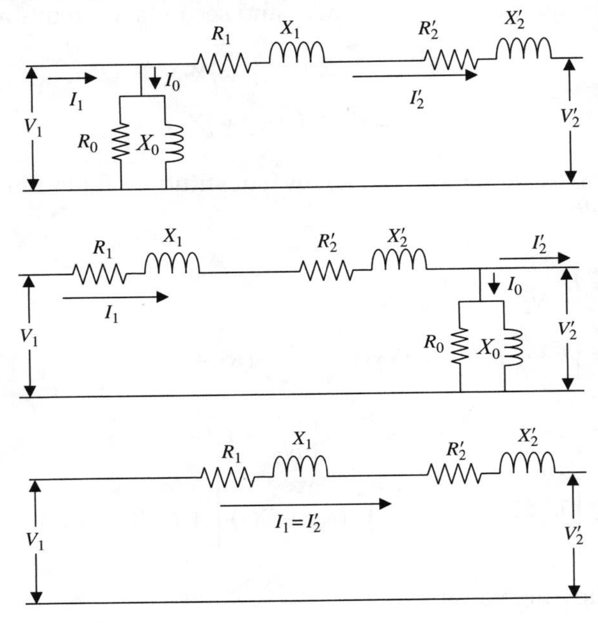

FIGURE 11.24
Acceptable equivalent circuits for the transformer.

The transformer model representing Equation (11.50) is shown in Figure 11.23. This circuit is easier to use than the one shown in Figure 11.22. Further approximation of the transformer model can be made since the excitation current I_0 is often less than 5% of the rated primary current. In this case, moving the core branch anywhere in the model, or even eliminate it altogether, as shown in Figure 11.24, should not introduce meaningful errors.

EXAMPLE 11.12 A single-phase transformer has the following parameters:

$$\frac{N_1}{N_2} = 10; \quad R_{eq} = R_1 + R_2' = 1\ \Omega; \quad X_{eq} = X_1 + X_2' = 10\ \Omega;$$

$$R_0 = 1000\ \Omega; \quad X_0 = 5000\ \Omega$$

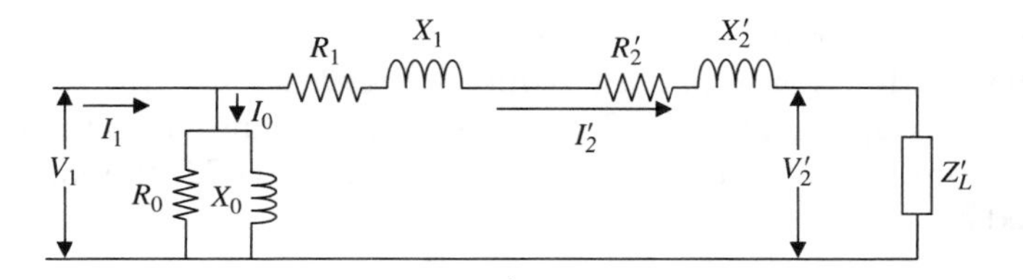

FIGURE 11.25
Equivalent circuit of the transformer with load.

The rated voltage of the primary winding is 1000 V. A $0.5\angle 30°\,\Omega$ load is connected across the secondary terminals. Compute the load voltage.

Solution

Let us use the equivalent circuit in Figure 11.25. The load impedance referred to the primary side is

$$Z_L' = Z_L \left(\frac{N_1}{N_2}\right)^2 = 50\ \Omega$$

The current I_2' is

$$\bar{I}_2' = \frac{\bar{V}_1}{(R_{\text{eq}} + jX_{\text{eq}}) + Z_L'} = \frac{1000\angle 0°}{(1 + j10) + 50\angle 30°} = 17.7\angle{-38.31°}\ \text{A}$$

The load voltage referred to the primary side V_2' is

$$\bar{V}_2' = \bar{I}_2' Z_L' = (17.7\angle{-38.31°})(50\angle 30°) = 885\angle{-8.31°}\ \text{V}$$

Use the turns ratio in Equation (11.49) to compute the actual voltage across the load.

$$V_2 = V_2' \frac{N_2}{N_1} = 885 \frac{1}{10} = 88.5\ \text{V}$$

11.5.2 Transformer Efficiency

The efficiency η of any equipment is defined by

$$\begin{aligned}\eta &= \frac{\text{output power}}{\text{input Power}} = \frac{\text{output energy}}{\text{input energy}} = \frac{\text{output power}}{\text{output power} + \text{losses}} \\ &= \frac{\text{input power} - \text{losses}}{\text{input power}}\end{aligned} \tag{11.51}$$

The losses of the transformer are due to R_1, R_2, and R_0. The losses in R_1 and R_2 are the winding losses and are named *copper losses* P_{cu} (the windings are made out of copper material). The losses in R_0 are known as the *iron losses* P_{iron} (the core is made out of iron). Using Figure 11.22, the losses can be computed as

$$P_{cu} = I_1^2 R_1 + I_2^2 R_2 \tag{11.52}$$

$$P_{iron} = \frac{E_1^2}{R_0} \tag{11.53}$$

These losses can also be computed using the approximate equivalent circuits in Figure 11.24. A good approximation for the losses is given below:

$$P_{cu} \approx I_1^2 \left(R_1 + R_2'\right) \approx \left(I_2'\right)^2 \left(R_1 + R_2'\right) \tag{11.54}$$

$$P_{iron} \approx \frac{V_1^2}{R_0} \approx \frac{\left(V_2'\right)^2}{R_0} \tag{11.55}$$

Using Equation (11.51), the efficiency of the transformer can be computed as

$$\eta = \frac{P_{out}}{P_{input}} = \frac{V_2' I_2' \cos\theta_2}{V_2' I_2' \cos\theta_2 + P_{cu} + P_{iron}} = \frac{V_1 I_1 \cos\theta_1 - P_{cu} - P_{iron}}{V_1 I_1 \cos\theta_1} \tag{11.56}$$

where θ_1 is the power factor angle of I_1, and θ_2 is the power factor angle of I_2. θ_2 is also the angle of the load impedance.

EXAMPLE 11.13 Compute the efficiency of the transformer in Example 11.12.

Solution

$$P_{cu} = \left(I_2'\right)^2 R_{eq} = 17.7^2 \times 1 = 313.29 \text{ W}$$

$$P_{iron} = \frac{V_1^2}{R_0} = \frac{1000^2}{1000} = 1 \text{ kW}$$

The output power is

$$P_{out} = V_2' I_2' \cos\theta_2 = 885 \times 17.7 \times \cos 30 = 13.57 \text{ kW}$$

The efficiency of the transformer is

$$\eta = \frac{P_{out}}{P_{input}} = \frac{V_2' I_2' \cos\theta_2}{V_2' I_2' \cos\theta_2 + P_{cu} + P_{iron}} = \frac{13{,}570}{13{,}570 + 313.29 + 1000} = 91.2\%$$

11.5.3 Voltage Regulation

The voltage regulation VR of the transformer is defined as

$$VR = \frac{|V_{\text{no load}}| - |V_{\text{full load}}|}{|V_{\text{full load}}|} \tag{11.57}$$

where $|V_{\text{no load}}|$ is the magnitude of the open circuit voltage measured at the load terminals, and $|V_{\text{full load}}|$ is the magnitude of the voltage at the load terminals when the rated current is delivered to the load. The VR represents the change in the load voltage from no load to full load. It indicates the voltage reduction due to the various parameters of the transformer.

Consider the equivalent circuit in Figure 11.25. At no load (open circuit), $I_2' = 0$. Hence, the voltage measured at the load terminals is

$$V_{\text{no load}} = V_1 \tag{11.58}$$

At full load,

$$V_{\text{full load}} = V_2' \tag{11.59}$$

Hence, the voltage regulation of the transformer is

$$VR = \frac{V_1 - V_2'}{V_2'} \tag{11.60}$$

EXAMPLE 11.14 Compute the voltage regulation of the transformer in Example 11.12. Assume that the secondary winding of the transformer carries the full load current when $0.5\angle 30^\circ\ \Omega$ is connected across the secondary terminals.

Solution

$$VR = \frac{V_1 - V_2'}{V_2'} = \frac{1000 - 885}{885} = 13\%$$

Exercise

11.1. A single-phase transformer has a turns ratio of 10,000/5,000. A direct current voltage of 30 V is applied to the primary winding. Compute the voltage of the secondary winding.

11.2. A single-phase transformer is rated at 2 kVA, 240/120 V. The transformer is fully loaded by an inductive load of 0.8 power factor lagging.

a. Compute the real power delivered to the load.
b. Compute the load current.

11.3. A single-phase transformer is rated at 10 KVA, 220/110 V.

a. Compute the rated current of each winding.
b. If a 2 Ω load resistance is connected across the 110 V winding, what are the currents in the high voltage and low voltage windings?
c. What is the equivalent load resistance referred to the 220 V side?

11.4. Three single-phase transformers are connected in Y–Δ configuration. Each single-phase transformer has an identical number of turns in the primary and secondary windings. If a line-to-line voltage of 480 V is applied on the wye windings, compute the line-to-line voltage on the delta windings.

11.5. Three single-phase transformers, Each is rated at 10 kVA, 400/300 V are connected in Y–Δ configuration.

a. Compute the rated power of the transformer bank.
b. Compute the line-to-line voltage ratio of the transformer bank.

11.6. A single-phase transformer has a voltage regulation of 5%. The input voltage of the transformer is 120 V, and the turns ratio N_1/N_2 is 2. Compute the voltage across the load V_2.

11.7. Three single-phase transformers are connected as a transformer bank rated at 18 MVA, 13.8 kV(Δ)/120 kV(Y). One side of the transformer bank is connected to a 120 kV transmission line, and the other side is connected to a three-phase load of 12 MVA at 0.8 lagging power factor.

a. Determine the turns ratio of the transformer bank.
b. What is the line current at the 120 kV side?

11.8. A single-phase 10 kVA, 2300/230 V, two-winding transformer is connected as an autotransformer to step up the voltage from 2300 to 2530 V.

a. Draw the schematic diagram of the autotransformer showing the winding connections and all voltages and currents at full-load.
b. Find the kilovolts amperes rating of the autotransformer. Do not allow the currents of the windings to exceed their rated values.

11.9. A single phase, 240/120 V transformer has the following parameters:

$$R_1 = 1\ \Omega;\quad R_2 = 0.5\ \Omega;\quad X_1 = 6\ \Omega;\quad X_2 = 2\ \Omega$$
$$R_0 = 500\ \Omega;\quad X_o = 1.5\ k\Omega$$

A load of 10 Ω at 0.8 power factor lagging is connected across the low-voltage terminals of the transformer. The voltage measured across the load side is 110 V. Compute the following:

a. The load voltage referred to the primary side.
b. The currents of the primary and secondary windings.
c. The source voltage.

d. The voltage regulation.

e. The load power.

f. The efficiency of the transformer.

11.10. Three identical single-phase transformers, each rated at 100 kVA, 7 kV/3.5 kV, are connected as a three-phase transformer bank. The high-voltage side of the transformer is connected in delta and the low voltage is connected in Y. A 200 kVA, Y-connected load is attached across the secondary winding.

a. Compute the ratio of the line-to-line voltages.

b. Compute the line current on both sides of the transformer.

11.11. The transformer shown in Figure 11.6 consists of one primary winding and two secondary windings. The number of turns of the windings are: $N_1 = 10{,}000; N_2 = 5{,}000$; and $N_3 = 1{,}000$. A voltage source of 120 V is applied to the primary winding. The load of winding N_2 consumes 600 W and 300 VAr inductive power. The load of winding N_3 consumes 24 W and 36 VAr capacitive power. Compute the following:

a. The voltages of the secondary windings.

b. The currents of all windings.

11.12. A single-phase transformer has three windings. The primary winding N_1 carries a current of $I_1 = 10$ A. One of the secondary windings has 3000 turns and carries 2 A; the other secondary winding has 6000 turns and carries 1 A. All currents are in phase. Compute the number of turns in the primary winding.

11.13. Three single-phase transformers are used to form a three-phase transformer bank rated 13.8 kV(Δ)/120 kV(Y). One side of the transformer bank is connected to a 120 kV transmission line, and the other side is connected to a three-phase load of 12 MVA at 0.8 power factor lagging. Determine the turns ratio of each transformer, and the line current of the bank at the 120 kV side.

12

Electric Machines

Most electric machines are dual-action electromechanical converters. The machines that convert electric energy into mechanical energy are called *motors,* and the machines that convert mechanical energy into electric energy are called *generators.* Electric motors are probably the most used among power devices anywhere. You can find them in almost every device with a mechanical movement — from children's toys to spacecrafts. They play incredible roles in our daily lives by performing numerous vital tasks in various applications. Refrigerators, washers, dryers, stoves, air conditioners, hair dryers, computers, printers, clocks, electric toothbrushes, electric shavers, and fans are some of the household devices that use motors. In the industrial and commercial sectors, motors are used in numerous applications such as transportation, elevators, forklifts, blowers, robots, actuators, electric and hybrid cars, machine tooling, paper mills, cooking machines, medical tools, assembly lines, and conveyor belts. The computer hard disk, for example, has at least two motors and the printer has at least four motors. NASA's Mars exploration rover built by the Jet Propulsion Laboratory (JPL) has about 200 motors; most are used for actuation, sampling, and control. Many of them are for one-time use only for the cruising phase and the *Entry, Descent, and Landing* (EDL) phase.

In total, about 65% of the electric energy in the United States is consumed by electric motors, and over 99% of the energy produced by utilities worldwide is produced by electrical generators. In stand-alone systems such as aircraft, ships, and automobiles, the generators are the main source of electric power for these mobile systems.

Electrical machines come in various sizes and power levels. Their weights range from micrograms for motors installed inside silicon chips to 7700 tonnes for the generator built by the consortium from Hewitt, Siemens, and General Electric for China's Yangtze River Three Gorges Hydraulic Power Plant. The power capacity of the machines is also of a wide range, from the microwatts to more than 2 GW.

12.1 Rotating Magnetic Field

A multi-phase alternating current source produces a rotating magnetic field. Consider the conceptual diagram shown in Figure 12.1. The figure shows three windings mounted symmetrically inside a hollow metal tube called *stator* and which are shifted by 120° from each other. Each winding goes along the length of the stator on one side then returns back from the other side. If we apply three-phase balanced voltage across the terminals of the windings, the currents of the windings create balanced three-phase magnetic fields inside the tube as shown in Figure 12.2.

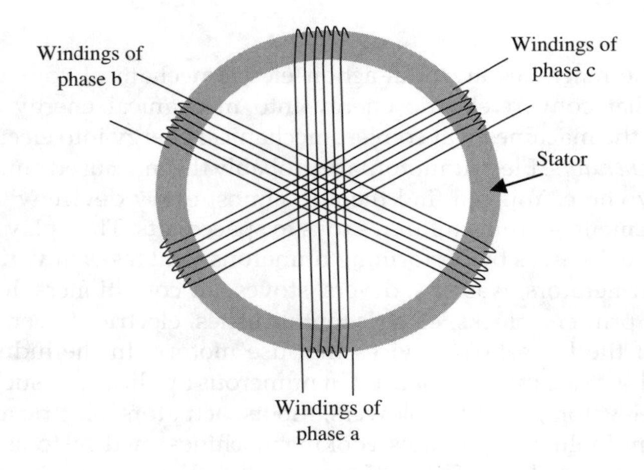

FIGURE 12.1
Three-phase windings mounted on a stator.

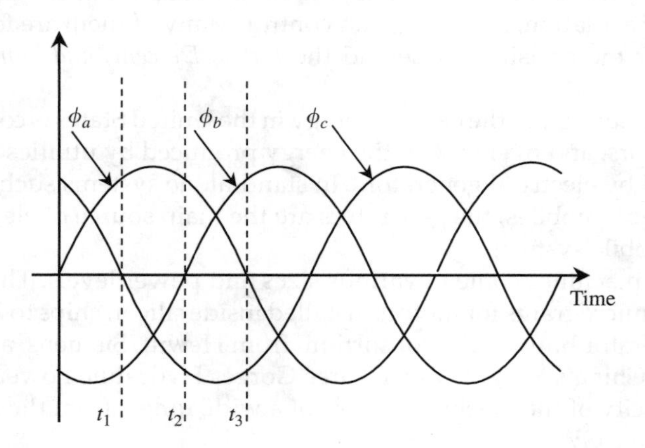

FIGURE 12.2
Airgap flux of the three phases.

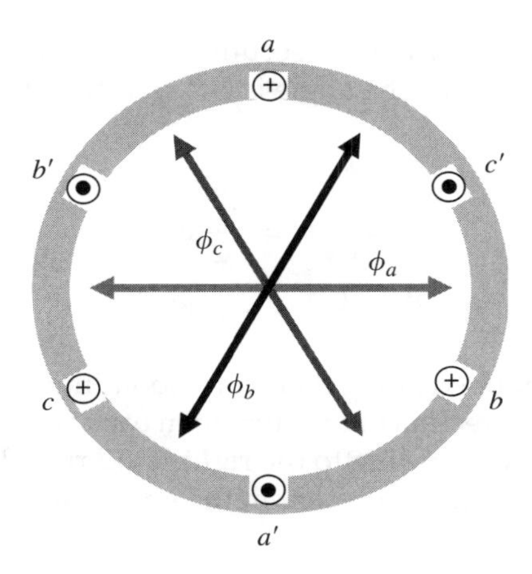

FIGURE 12.3
Loci of airgap fluxes.

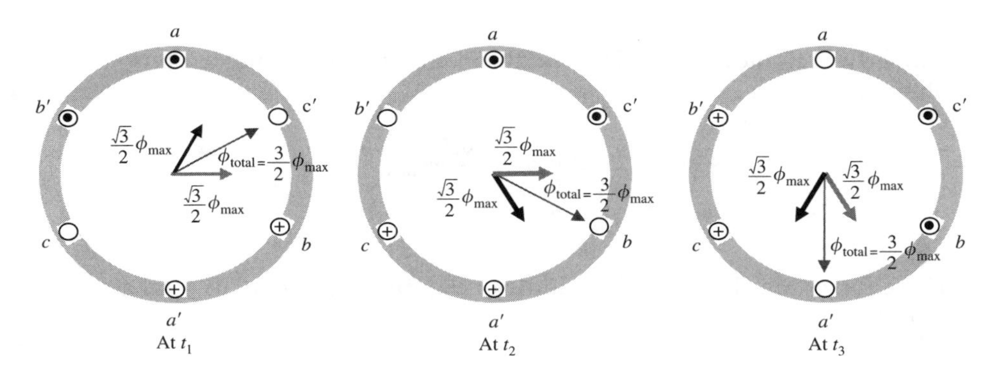

FIGURE 12.4
Rotating airgap flux.

Figure 12.3 shows a more convenient representation of the stator windings. The circles embedded inside the stator represent the winding *a*–*a*′ represents the winding of phase *a*, *b*–*b*′ represents the winding of phase *b*, and *c*–*c*′ represents the winding of phase *c*. The crosses and dots inside the circles represent current entering and leaving the windings, respectively.

Because of the mechanical arrangement of the stator windings, the flux of each phase, according to the right-hand rule, travels along its axis as shown in Figure 12.3. This flux is inside the tube, thus called an *airgap flux*.

Now, let us consider any three consecutive time instances (e.g., t_1, t_2, and t_3 shown in Figure 12.2). At t_1, the angle is 60°, and the magnitudes of the flux of the three phases are

$$\phi_a = -\phi_b = \frac{\sqrt{3}}{2}\phi_{\max}$$
$$\phi_c = 0 \tag{12.1}$$

The fluxes at t_1 are mapped along their corresponding axes on the left side in Figure 12.4. Note the direction of the current in each winding as it determines the direction of its flux according to the right-hand rule. The total airgap flux is the phasor sum of all fluxes present in the airgap at t_1.

$$\overline{\phi}_{\text{total}}(t_1) = \overline{\phi}_a(t_1) + \overline{\phi}_b(t_1) + \overline{\phi}_c(t_1) = \frac{\sqrt{3}}{2}\phi_{\max}\angle 0^\circ + \frac{\sqrt{3}}{2}\phi_{\max}\angle 60^\circ + 0$$
$$\overline{\phi}_{\text{total}}(t_1) = \frac{3}{2}\phi_{\max}\angle 30^\circ \tag{12.2}$$

Similarly, at t_2, the angle is 120°, and the magnitudes of the fluxes of the three phases are

$$\phi_a = -\phi_c = \frac{\sqrt{3}}{2}\phi_{\max}$$
$$\phi_b = 0 \tag{12.3}$$

These fluxes are shown in the middle of Figure 12.4, where the total airgap flux at t_2 is

$$\overline{\phi}_{\text{total}}(t_1) = \overline{\phi}_a(t_1) + \overline{\phi}_b(t_1) + \overline{\phi}_c(t_1) = \frac{\sqrt{3}}{2}\phi_{\max}\angle 0^\circ + 0 + \frac{\sqrt{3}}{2}\phi_{\max}\angle{-60^\circ}$$
$$\overline{\phi}_{\text{total}}(t_1) = \frac{3}{2}\phi_{\max}\angle{-30^\circ} \tag{12.4}$$

Finally, at t_3, the angle of the waveform is 180°, and the magnitudes of the fluxes of the three phases are

$$\phi_a = 0$$
$$\phi_b = -\phi_c = \frac{\sqrt{3}}{2}\phi_{\max} \tag{12.5}$$

These fluxes are shown on the right side of Figure 12.4. In this case, the total airgap flux at t_3 is

$$\overline{\phi}_{\text{total}}(t_1) = \overline{\phi}_a(t_1) + \overline{\phi}_b(t_1) + \overline{\phi}_c(t_1) = 0 + \frac{\sqrt{3}}{2}\phi_{\max}\angle{-120^\circ} + \frac{\sqrt{3}}{2}\phi_{\max}\angle{-60^0}$$

$$\overline{\phi}_{\text{total}}(t_1) = \frac{3}{2}\phi_{\max}\angle{-90^\circ} \tag{12.6}$$

By examining Equations (12.2), (12.4), and (12.6) we can conclude the following:

- The magnitude of the total flux in the airgap is always equal to 1.5 $\phi_{\max}$.
- The angle of the total airgap flux is changing with time. The flux in the above case is rotating in the clockwise direction. This rotating flux is one of the main reasons for the development of the three-phase systems.
- The total flux in the airgap completes one revolution every ac cycle. Hence, the mechanical speed of the total flux in the airgap n_s is one revolution per ac cycle.

$$n_s = f \text{ rev/s} \tag{12.7}$$

where f is the frequency of the ac supply in hertz. The mechanical speed is often measured in revolution/minute or rpm. Hence,

$$n_s = 60 f \text{ rpm} \tag{12.8}$$

The speed n_s is known as the *synchronous speed* because its magnitude is determined by the frequency of the supply voltage; that is, the speed is synchronized with the supply frequency.

The machine in Figure 12.3 is considered a two-pole one because every phase has one winding that creates one North and one South pole. If each phase is composed of two windings (i.e., a_1–a_1' and a_2–a_2') arranged symmetrically along the inner circumference of the stator as shown in Figure 12.5, the machine is considered a 4-pole one. In this arrangement, the mechanical angle between the phases is 60° instead of the 120° for the 2-pole machines. If you repeat the analyses in Equations (12.1) to (12.6), you will find that the flux moves by a mechanical angle of 180° for each complete ac cycle. Hence, we can write a general expression for the mechanical speed of the airgap flux as

$$n_s = \frac{60 f}{pp} = 120\frac{f}{p} \text{ rpm} \tag{12.9}$$

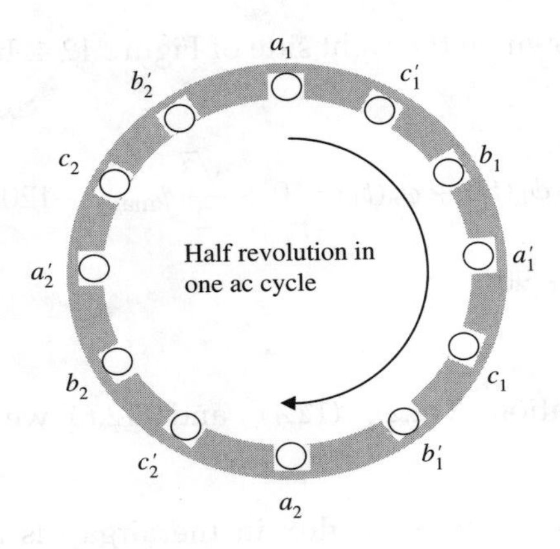

FIGURE 12.5
Four-pole arrangement.

where pp is the number of pole-pairs, and p is the number of poles ($p = 2\ pp$).

EXAMPLE 12.1 Compute the synchronous speed for 2, 4, 6, 8, and 10-pole machines operating at 50 and 60 Hz.

Solution

With the direct substitution in Equation (12.9), you can get the data tabulated below.

Number of Poles	Synchronous Speed (rpm)	
	50 Hz	60 Hz
2	3000	3600
4	1500	1800
6	1000	1200
8	750	900
10	600	720

Note that the synchronous speed for the 50 Hz system is slower than that for the 60 Hz systems.

12.2 Rotating Induction Motor

About 65% of the electric energy in the United States is consumed by electric motors. In the industrial sector alone, about 75% of the total energy

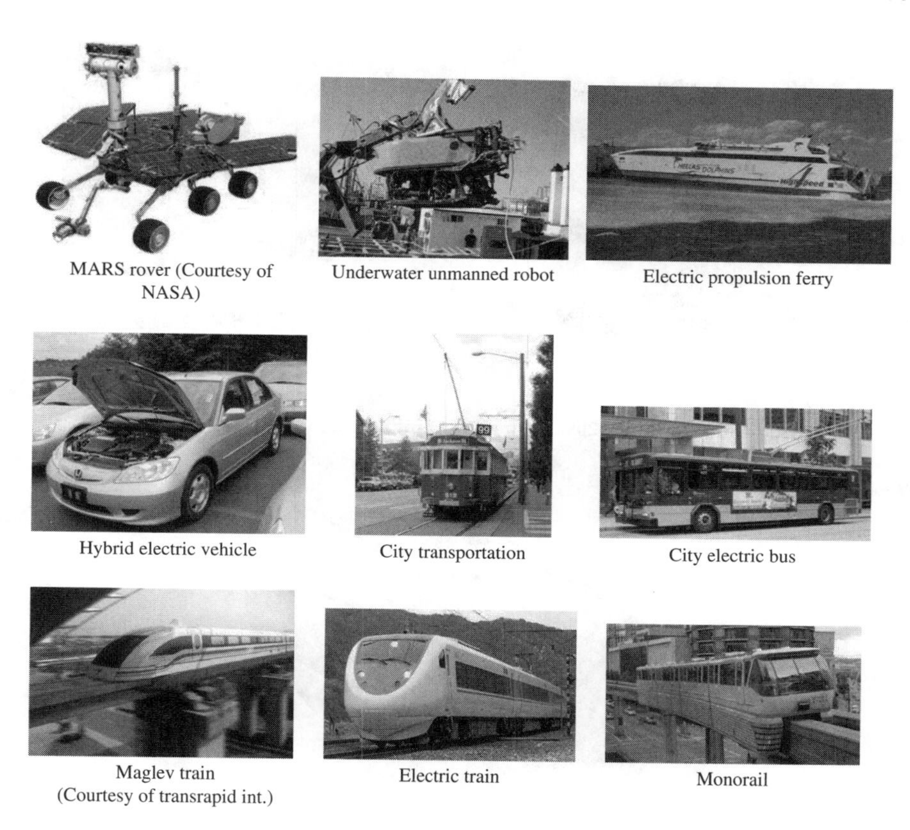

FIGURE 12.6
(see color insert following Page 208) Induction machines are the prime movers of these vehicles.

is consumed by motors. Some of the uses of electric motors are shown in Figure 12.6. Over 90% of the energy consumed by electric motors is consumed by induction motors because they are rugged, reliable, easy to maintain, and relatively inexpensive.

The induction motor is composed of one stator and one rotor; a small induction motor is shown in Figure 12.7. The stator has three-phase windings that are excited by a three-phase supply. The rotor, as the name implies, is the rotating part of the motor. It is mounted inside the airgap of the motor through two sets of ball bearings, one on each end of the rotor. The mechanical load of the motor is attached to the shaft of the rotor. The rotor circuit consists of windings that are shorted either permanently as part of the rotor structure, or externally through a system of slip rings and brushes. The rotor with an external short, called a *slip-ring* rotor, has three rings mounted on the rotor shaft inside the motor structure as shown in Figure 12.7. These slip rings are electrically isolated from one another, but each is connected to one terminal of the three phase windings of the rotor; most rotor windings are connected in wye. Carbon brushes mounted on the stator structure are placed to continuously touch

FIGURE 12.7
(see color insert following Page 208) Main parts of induction machine.

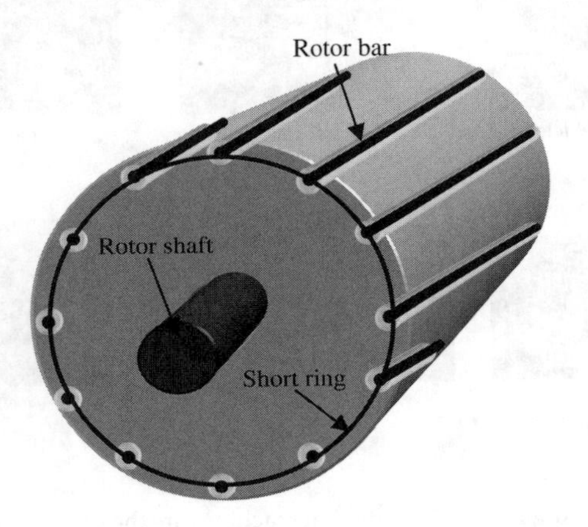

FIGURE 12.8
Squirrel cage rotor.

the rotating slip rings, thus achieving connectivity between the rotor windings and any external equipment. The rotor circuit with an internal short, called a *squirrel cage* rotor, consists of slanted wire bars shorted on both ends of the rotor as shown in Figure 12.8.

12.2.1 Rotation of Induction Motor

The rotation of the induction motor can be explained using Faraday's law and the Lorentz equation. When a conductor carries a current in a uniform magnetic field, the electromechanical relationships can be represented by the following two equations:

$$\begin{aligned} e &= Bl\Delta v \\ \text{force} &= Bli \end{aligned} \tag{12.10}$$

where e is the induced voltage across the conductor, B is the flux density, l is the length of the conductor, Δv is the relative speed between the conductor and the field, i is the current, and *force* is the mechanical force of the conductor. If we generalize these equations for rotating fields, we can rewrite them in the following form:

$$e = f(\phi, \Delta n) \tag{12.11}$$

$$T = f(\phi, i) \tag{12.12}$$

where T is the torque of the conductors, and Δn is the relative angular speed between the conductor and the airgap flux.

To understand the mechanism by which the rotor spins, let us assume that the rotor is at a standstill. When a three-phase voltage is applied across the stator windings, a rotating flux at a synchronous speed n_s is produced in the airgap. The relative speed Δn is the difference between the rotor speed n and the speed of the airgap flux n_s.

$$\Delta n = n_s - n \tag{12.13}$$

Since the rotor is stationary, Δn is equal to the synchronous speed n_s. Hence, the rotating field induces a voltage e in the rotor windings as given in Equation (12.11). Since the rotor windings are shorted, the induced voltage causes a current i to flow in the rotor windings producing the Lorentz torque T in Equation (12.12). This torque spins the rotor.

As you may conclude from Equations (12.11) and (12.12), the induction motor cannot spin at the synchronous speed. This is because Δn in this case is zero resulting in no induced voltage across the rotor windings, so no current flow in the windings and therefore no torque develops to spin the motor. So what is the steady state speed of the induction motor? The answer depends on the load torque as explained in the following paragraphs.

The third law of motion developed by Isaac Newton in 1686 stated that "whenever one body exerts a force on another, the second exerts a force on the first that is equal in magnitude, opposite in direction and has the same line of action." For the induction motor, this theory means that the motor develops a torque equal in magnitude and opposite in direction to the load torque. Hence, the steady state operation is achieved when the motor, on a continuous basis, provides the torque needed by the load. Assuming that the flux is unchanged, this developed torque needs a certain magnitude of rotor current i as shown in Equation (12.12). The magnitude of this rotor current requires a certain value of induced voltage e in the rotor windings that is equal to the rotor current i multiplied by the rotor impedance. This voltage in turn requires a certain relative speed Δn as given in Equation (12.11). Hence, the steady state speed of the rotor must always be less than the synchronous speed to maintain the balance between the load torque and the motor's developed torque. Δn is large for heavy load torques and small for light load torques.

The per unit value of the relative speed is known as the slip S.

$$S = \frac{\Delta n}{n_s} = \frac{n_s - n}{n_s} \tag{12.14}$$

Keep in mind that the unit of n is rpm. If we use the angular speed ω (in radian per second or rad/s) instead of n, the slip of the motor is

$$S = \frac{\Delta \omega}{\omega_s} = \frac{\omega_s - \omega}{\omega_s} \tag{12.15}$$

where

$$\omega = \frac{2\pi}{60} n \tag{12.16}$$

The slip while starting, when the rotor speed is zero, is equal to one. At no load, when the motor speed is very close to the synchronous speed, the slip is close to zero. During normal steady state operation, the slip is small and often less than 0.1.

EXAMPLE 12.2 A 2-pole, 60 Hz induction motor operates at a slip of 0.02. Compute its rotor speed.

Solution

First, let us compute the synchronous speed.

$$n_s = 120\frac{f}{p} = 120\frac{60}{2} = 3600 \text{ rpm}$$

Use Equation (12.14) to compute the speed of the rotor.

$$S = \frac{n_s - n}{n_s}; \qquad n = n_s(1 - S) = 3600(1 - 0.02) = 3528 \text{ rpm}$$

EXAMPLE 12.3 The steady state speed of a 60 Hz induction motor is 1150 rpm. Compute the number of poles of the induction motor, and the slip.

Solution

As stated earlier, during steady state operation, the slip of the motor is just below the synchronous speed. Using the table in Example 12.1, you will find that the synchronous speed that is just above 1150 is 1200. Hence the motor is a 6-pole machine.

$$S = \frac{n_s - n}{n_s} = \frac{1200 - 1150}{1200} = 0.0417 \text{ or } 4.17\%$$

12.2.2 Equivalent Circuit of Induction Motor

Since the induction motor is considered a balanced three-phase load, a single-phase equivalent circuit is adequate to model the machine. The stator of the machine consists of a set of copper windings mounted on an iron core. This is almost the same as the primary circuit of the transformer. The stator winding can be represented by a resistance R_1 and inductive reactance X_1. Since the stator copper windings are imbedded in the stator's iron core, the core can be represented by a parallel combination of a core resistance R_c and a core inductive reactance X_c. By using these parameters, the model for the stator circuit shown in Figure 12.9 is very similar to the model for the primary circuit of the transformer shown in Figure 11.22. The sum of the currents in R_c and X_c is called the *magnetizing* (or *core*) current I_c. E_1 in the figure is equal to the source voltage V minus the drop across the copper impedance.

$$\bar{E}_1 = \bar{V} - \bar{I}_1(R_1 + jX_1) \tag{12.17}$$

The equivalent circuit of the rotor is also similar to the equivalent circuit for the secondary of the transformer, but only at stand still ($n = 0$) and when the secondary winding is shorted (because the rotor winding is shorted). The rotor can be represented by its winding resistance R_2 and inductive reactance X_2 as shown in Figure 12.10. E_2 is the induced voltage across the rotor winding

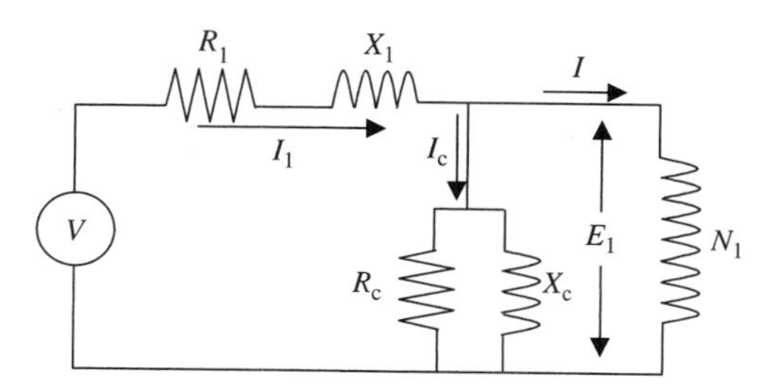

FIGURE 12.9
Equivalent circuit of the stator.

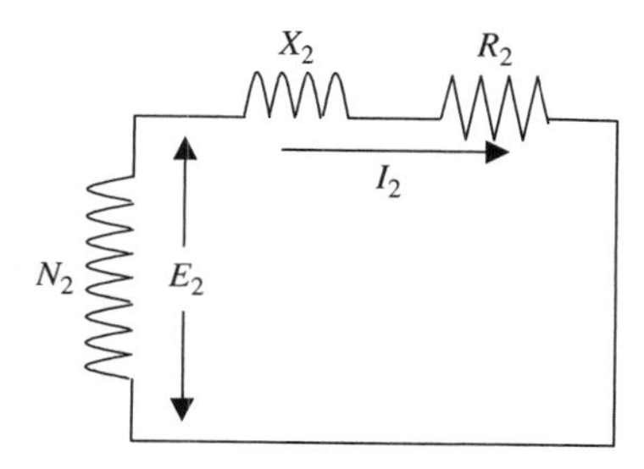

FIGURE 12.10
Equivalent circuit of the rotor at standstill.

at standstill. If you assume that the number of turns of the stator winding is N_1 and that for the rotor winding is N_2, then

$$\frac{E_2}{E_1} = \frac{N_2}{N_1} \tag{12.18}$$

The inductive reactance X_2 of the rotor winding is

$$X_2 = 2\pi f L_2 \tag{12.19}$$

where f is the frequency of the supply voltage and L_2 is the inductance of the rotor winding.

When the rotor spins, two variables change:

1. The induced voltage across the rotor winding.
2. The frequency of the rotor current (or voltage).

As shown in Equation (12.11), the induced voltage across the rotor winding is proportional to the relative speed Δn. Hence, the induced voltage E_2 at standstill when $n = 0$ is

$$E_2 \sim \Delta n = n_s \tag{12.20}$$

At any other speed, the induced voltage across the rotor winding E_r is

$$E_r \sim \Delta n = n_s - n \tag{12.21}$$

Dividing Equations (12.21) by (12.20), we get the relationship between the induced voltage at any speed and the induced voltage at standstill.

$$\frac{E_r}{E_2} = \frac{n_s - n}{n_s} = S \tag{12.22}$$

Hence, the voltage across the rotor winding at any speed E_r is equal to the rotor voltage at standstill E_2 multiplied by the slip S. In other words, the induced voltage across the rotor windings is always equal to

$$E_r = SE_2 = \frac{N_2}{N_1} SE_1 \tag{12.23}$$

The frequency of the rotor current is directly proportional to the rate by which the magnetic field cuts the rotor winding. At standstill, the airgap flux cuts the rotor windings at a rate proportional to the synchronous speed n_s. Hence, the frequency of the rotor current at standstill is f_s and is proportional to the synchronous speed. This makes f_s equal to the frequency of the stator voltage f. If the rotor is spinning at a speed n, the flux cuts the rotor windings at a rate proportional to Δn. Hence, the frequency of the rotor current at any speed f_r is proportional to Δn.

$$\begin{aligned} f &= f_s \sim n_s \\ f_r &\sim \Delta n \end{aligned} \tag{12.24}$$

Hence, the frequency of the rotor current at any speed is

$$\begin{aligned} \frac{f_r}{f} &= \frac{\Delta n}{n_s} = S \\ f_r &= Sf_s = Sf \end{aligned} \tag{12.25}$$

The equivalent circuit for the rotor at any speed is shown in Figure 12.11. Because the rotor voltage and rotor frequency are changing with speed, it is hard to use the equivalent circuit in Figure 12.11 and we need to modify the circuit to make it easier to analyze at any speed. The rotor inductive reactance X_r is different from X_2 in Equation (12.19) because the rotor frequency at any speed is f_r, hence

$$X_r = 2\pi f_r L_2 \tag{12.26}$$

Substituting the value of f_r in Equation (12.25) into Equation (12.26) yields

$$X_r = 2\pi f_r L_2 = 2\pi (Sf) L_2 \tag{12.27}$$

Using the value of X_2 in Equation (12.19) yields

$$X_r = 2\pi (Sf) L_2 = SX_2 \tag{12.28}$$

The rotor current I_r for the circuit in Figure 12.11 is

$$\bar{I}_r = \frac{\bar{E}_r}{R_2 + jX_r} = \frac{S\bar{E}_2}{R_2 + jSX_2} = \frac{\bar{E}_2}{(R_2/S) + jX_2} \tag{12.29}$$

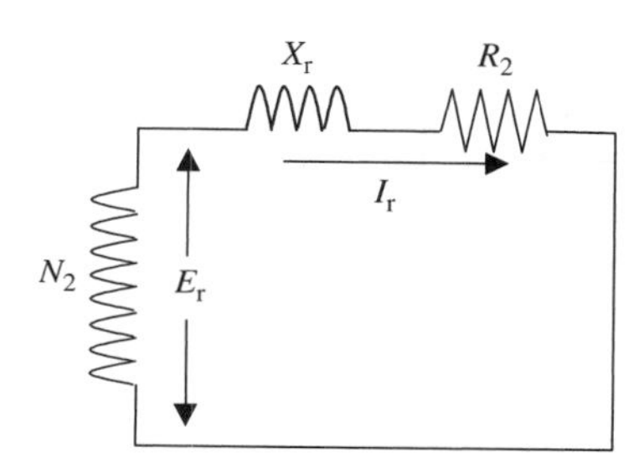

FIGURE 12.11
Equivalent circuit of the rotor at any speed.

Using Equation (12.29) the equivalent circuit of the rotor can be modified as shown in Figure 12.12. Now we can add the stator circuit to complete the model for the motor as shown in Figure 12.13. This circuit can further be modified as shown in Figure 12.14 by referring the rotor to the stator using the turns ratio. This is the same process used to refer the secondary circuit of the transformer to the primary side. Thus, the resistance R_2' and inductive reactance X_2' of the rotor winding referred to the stator circuit are computed

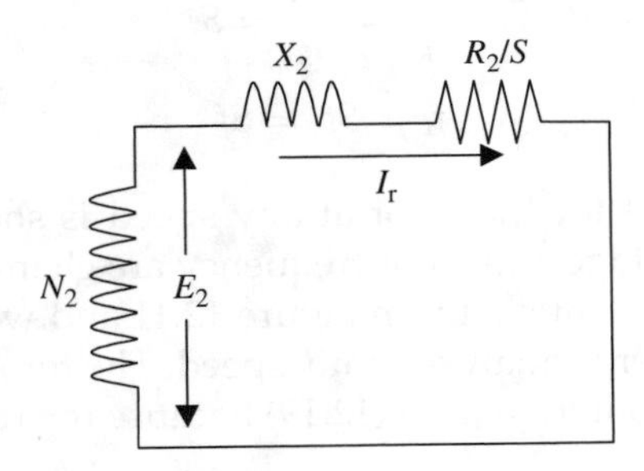

FIGURE 12.12
Modified equivalent circuit of the rotor at any speed.

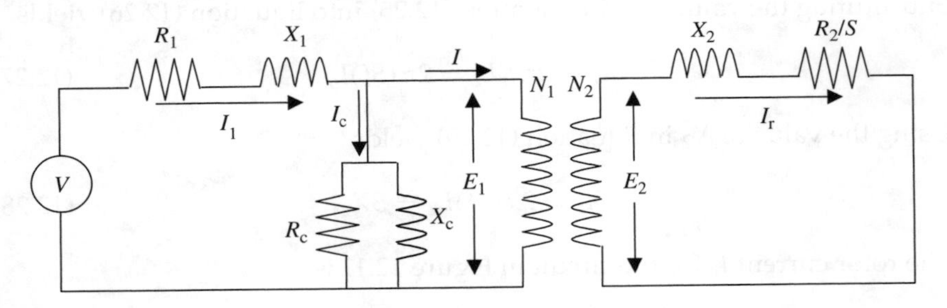

FIGURE 12.13
Equivalent circuit for induction motor.

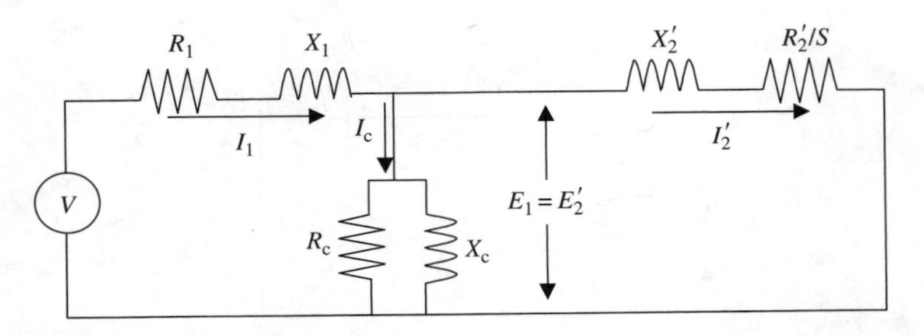

FIGURE 12.14
Equivalent circuit for induction motor referred to the stator.

as follows:

$$R_2' = R_2 \left(\frac{N_1}{N_2}\right)^2 \tag{12.30}$$

$$X_2' = X_2 \left(\frac{N_1}{N_2}\right)^2 \tag{12.31}$$

where N_1 and N_2 are the number of turns of the stator and rotor windings, respectively. Also, the rotor current referred to the stator circuit I_2' can be computed as

$$I_2' = I_r \left(\frac{N_2}{N_1}\right) \tag{12.32}$$

The resistance R_2'/S has two components; one of them is the resistance of the rotor winding R_2'.

$$\frac{R_2'}{S} = R_2' + \frac{R_2'}{S}(1 - S) \tag{12.33}$$

The resistive element $(R_2'/S)(1 - S)$ is an electrical representation of the mechanical load of the motor and is therefore known as the *load resistance*. The parsing of R_2'/S leads to the equivalent circuit in Figure 12.15(a). We can further modify the equivalent circuit by assuming that the core current I_c is much smaller than I_1; that is, $I_1 \approx I_2'$. Hence, we can assume that the impedances of the stator and rotor windings are in series as shown in Figure 12.15(b).

R_{eq} and X_{eq} in Figure 12.15(c) are defined as

$$\begin{aligned} R_{eq} &= R_1 + R_2' \\ X_{eq} &= X_1 + X_2' \end{aligned} \tag{12.34}$$

12.2.3 Power Analysis

Figure 12.16 shows the power flow through the induction motor. The motor receives input power P_{in} from the electric source, where

$$P_{in} = 3V\, I_1 \cos\theta_1 \tag{12.35}$$

P_{in} is the input power of the three phases of the motor, V is the phase voltage of the source, I_1 is the phase current of the stator, and θ_1 is the phase angle of the current (the angle between the phase voltage and phase current). Part of the input power is wasted in the stator circuit in the form of copper losses

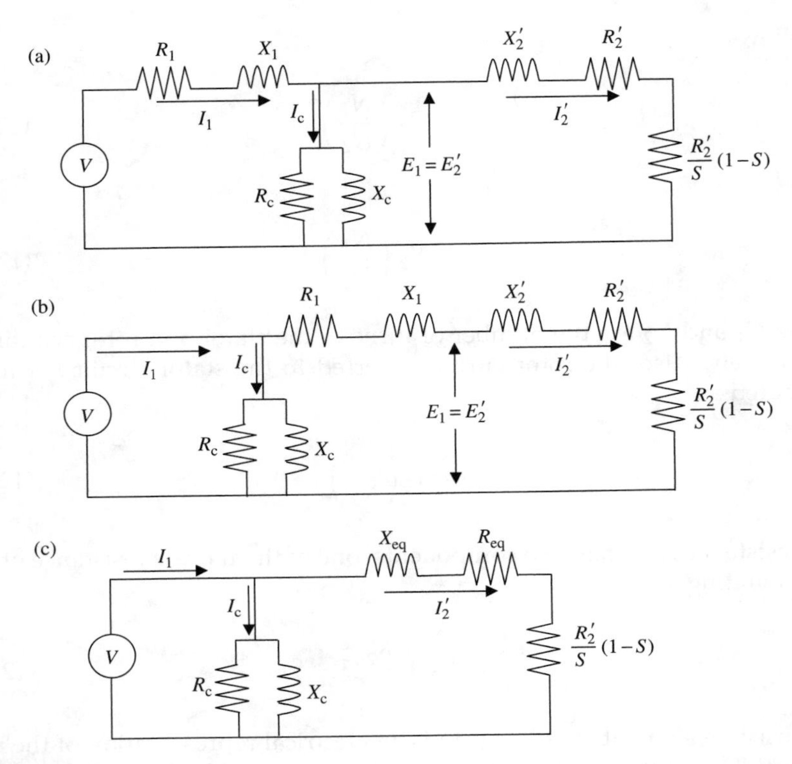

FIGURE 12.15
More equivalent circuits for the induction motor.

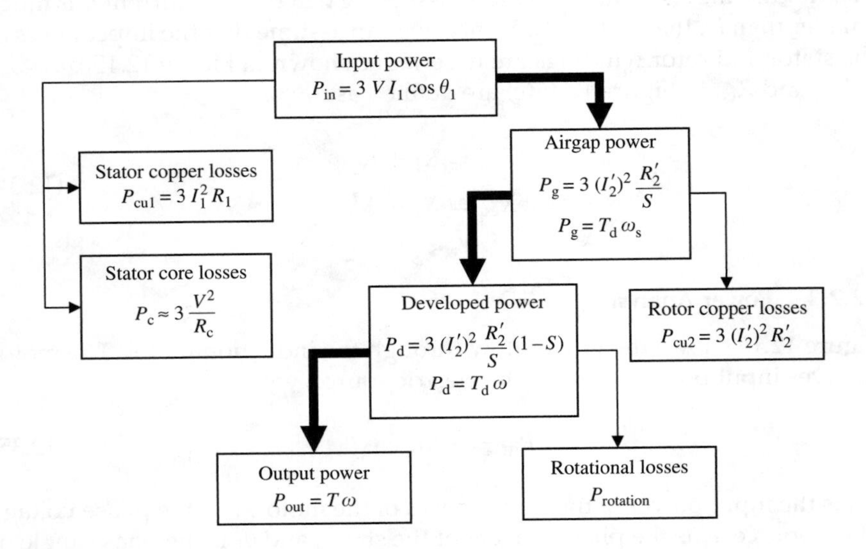

FIGURE 12.16
Power flow of the induction motor.

inside the windings P_{cu1} and core losses P_c. These losses can be computed by using the equivalent circuit in Figure 12.15(a).

$$P_{cu1} = 3I_1^2 R_1$$
$$P_c \approx 3\frac{V^2}{R_c} \tag{12.36}$$

The rest of the power is transmitted to the rotor through the airgap flux, thus it is called the *airgap power* P_g. P_g is consumed in R_2'/S.

$$P_g = 3(I_2')^2 \frac{R_2'}{S} \tag{12.37}$$

The airgap power is also a form of mechanical power since it involves the mechanical rotation of the airgap flux. Therefore, P_g can also be written in the mechanical form

$$P_g = T_d \omega_s \tag{12.38}$$

where T_d is the developed torque of the airgap flux and ω_s is the synchronous speed of the flux. T_d is the torque produced by all three phases. When the airgap power enters the rotor circuit, part of it is wasted in the copper windings of the rotor P_{cu2}.

$$P_{cu2} = 3(I_2')^2 R_2' \tag{12.39}$$

The rest of the power is the *developed power* P_d consumed by $(R_2'/S)(1 - S)$.

$$P_d = 3(I_2')^2 \frac{R_2'}{S}(1 - S) \tag{12.40}$$

This developed power can be represented in the mechanical form

$$P_d = T_d \omega \tag{12.41}$$

where ω is the speed of the rotor (not the synchronous speed). The developed power P_d is not completely converted to useful power (shaft power) as part of it is wasted in the form of rotational losses, such as friction and windage. The output power of the motor is equal to the output torque of the motor shaft T (not the developed torque) multiplied by the rotor speed.

$$P_{out} = T\omega \tag{12.42}$$

Note the relationships between the airgap power, developed power, and rotor copper losses.

$$P_{cu2} = SP_g$$
$$P_d = (1 - S)P_g \quad (12.43)$$

EXAMPLE 12.4 A 60 Hz, three-phase, Y-connected induction motor produces 100 hp at 1150 rpm. The friction and windage losses of the motor are 1.2 kW, the stator copper and core losses are 2.1 kW. Compute the input power and the efficiency of the motor.

Solution

The mechanical unit for the output power is horse power (hp), which can be converted into kilowatts by the conversion factor given in the Appendix.

$$P_{out} = \frac{100}{1.34} = 74.63 \text{ kW}$$

To compute the input power, we can use the power diagram in Figure 12.16. We can work the problem from the output power toward the input power. First, calculate the developed power.

$$P_d = P_{out} + P_{rotation} = 74.63 + 1.2 = 75.83 \text{ kW}$$

To compute the airgap power, we need to know the rotor copper losses, which are not explicitly provided. Another alternative is to compute the slip and use the relationship in Equation (12.43) to compute the airgap power. To calculate the slip, we need the synchronous speed and the actual speed of the motor. The actual speed is given, but not the synchronous speed. However, from the principle of operation, the speed of the induction motor is just below the synchronous speed. Using the table in Example 12.1, we can find that the number of poles of this machine is six, and its synchronous speed is 1200 rpm. Hence,

$$S = \frac{n_s - n}{n_s} = \frac{1200 - 1150}{1200} = 0.0417$$

Using Equation (12.43), we can compute the airgap power.

$$P_g = \frac{P_d}{1 - S} = \frac{75.83}{0.9583} = 79.13 \text{ kW}$$

Now use the diagram in Figure 12.16 to compute the input power.

$$P_{in} = P_g + P_{cu1} + P_c = 79.13 + 2.1 = 81.23 \text{ kW}$$

The motor efficiency η is

$$\eta = \frac{P_{\text{out}}}{P_{\text{in}}} = \frac{74.63}{81.23} = 91.87\%$$

12.2.4 Speed–Torque Relationship

By using the mechanical expression for the developed power in Figure 12.16, we can compute the developed torque of the induction motor.

$$T_{\text{d}} = \frac{P_{\text{d}}}{\omega} = \frac{3(I_2')^2 R_2'(1-S)}{S\omega} \tag{12.44}$$

From the equivalent circuit in Figure 12.15(c), the rotor current is

$$I_2' = \frac{V}{\sqrt{\left(R_1 + (R_2'/S)\right)^2 + X_{\text{eq}}^2}} \tag{12.45}$$

Substituting Equation (12.45) into (12.44) leads to

$$T_{\text{d}} = \frac{3V^2 R_2'(1-S)}{S\omega\left[\left(R_1 + (R_2'/S)\right)^2 + X_{\text{eq}}^2\right]} \tag{12.46}$$

From Equation (12.15), the speed of the motor ω can be written in terms of the synchronous speed and slip.

$$\omega = \omega_{\text{s}}(1-S) \tag{12.47}$$

Substituting Equation (12.47) into (12.46) we can obtain the following equation for the developed torque.

$$T_{\text{d}} = \frac{3V^2 R_2'}{S\omega_{\text{s}}\left[\left(R_1 + (R_2'/S)\right)^2 + X_{\text{eq}}^2\right]} \tag{12.48}$$

Keep in mind that V is the phase voltage, and the torque is developed by all three phases. Equation (12.48) gives the torque as a function of the slip, which can be modified to represent the torque as a function of speed.

$$T_{\text{d}} = \frac{3V^2 R_2'}{(\omega_{\text{s}} - \omega)\left[\left(R_1 + \omega_{\text{s}} R_2'/(\omega_{\text{s}} - \omega)\right)^2 + X_{\text{eq}}^2\right]} \tag{12.49}$$

The speed–torque, or slip–torque, characteristic of the induction motor based on Equations (12.48) and (12.49) is shown in Figure 12.17. The characteristic

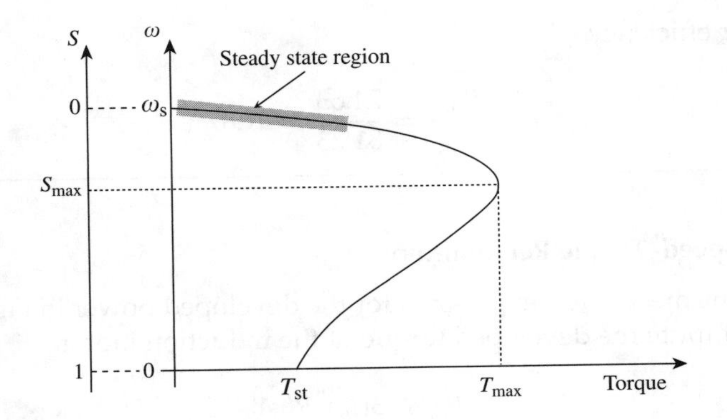

FIGURE 12.17
Speed–torque characteristic of the induction motor.

is nonlinear and has several key operating points. When the motor starts (the speed is zero), the motor develops a starting torque T_{st} that spins the motor. When the motor accelerates while starting, the developed torque increases until it reaches its maximum value T_{max} at the slip S_{max}. The speed of the motor continues to increase until it reaches the steady state operating region in the shaded area where the speed of the motor is constant and the developed torque of the motor is equal to the load torque.

EXAMPLE 12.5 A 480 V, 60 Hz, three-phase, four-pole induction motor has the following parameters:

$$R_1 = 0.3\ \Omega; \quad R_2' = 0.2\ \Omega; \quad X_{eq} = 2.0\ \Omega$$

At full load, the motor speed is 1760 rpm. Calculate the following:

a. The slip of the motor.
b. The developed torque of the motor at full load.
c. The developed power in horsepower.
d. The rotor current.
e. The copper losses of the motor.

Solution

a. The synchronous speed of the motor can be computed by using Equation (12.9).

$$n_s = 120\frac{f}{p} = 120\frac{60}{4} = 1800 \text{ rpm}$$

$$\omega_s = \frac{2\pi}{60}n_s = \frac{2\pi}{60}1800 = 188.5 \text{ rad/s}$$

Using Equation (12.15), we can compute the slip of the motor.

$$S = \frac{n_s - n}{n_s} = \frac{1800 - 1760}{1800} = 0.0222$$

b. Equation (12.48) or (12.49) can be used to compute the load torque.

$$\begin{aligned} T_d &= \frac{3V^2 R_2'}{S\omega_s \left[\left(R_1 + (R_2'/S) \right)^2 + X_{eq}^2 \right]} \\ &= \frac{3(480/\sqrt{3})^2 0.2}{0.0222 \times 188.5 \left[\left(0.3 + (0.2/0.0222) \right)^2 + 4 \right]} = 121.46 \text{ Nm} \end{aligned}$$

c. The angular speed of the motor is

$$\omega = \frac{2\pi}{60} n = \frac{2\pi}{60} 1760 = 184.3 \text{ rad/s}$$

The developed power of the motor is

$$P_d = T_d \omega = 121.46 \times 184.3 = 22.386 \text{ kW}$$

The developed power in horsepower units is

$$P_d = 22.386 \times 1.34 = 30 \text{ hp}$$

d. The rotor current can be computed using Equation (12.45).

$$I_2' = \frac{V}{\sqrt{\left(R_1 + (R_2'/s) \right)^2 + X_{eq}^2}} = \frac{480/\sqrt{3}}{\sqrt{\left(0.3 + (0.2/0.0222) \right)^2 + 4}} = 29.1 \text{ A}$$

e. The copper losses of the motor are in the stator and rotor windings; that is,

$$P_{cu} = P_{cu1} + P_{cu2} = 3 \left(I_2' \right)^2 R_{eq} = 3 \times (29.1)^2 (0.3 + 0.2) = 1.27 \text{ kW}$$

12.2.5 Starting Torque and Current

At starting, the speed of the motor is zero and $S = 1$. Substituting these values in either Equation (12.48) or (12.49) yields the starting torque T_{st}.

$$T_{st} = \frac{3V^2 R_2'}{\omega_s \left[(R_1 + R_2')^2 + X_{eq}^2 \right]} = \frac{3V^2 R_2'}{\omega_s Z_{eq}^2} \tag{12.50}$$

where

$$Z_{eq} = \sqrt{(R_1 + R_2')^2 + X_{eq}^2} = \sqrt{R_{eq}^2 + X_{eq}^2} \tag{12.51}$$

The current of the motor at starting I_{st} can also be computed using Equation (12.45) when $S = 1$.

$$I_{st}' = \frac{V}{\sqrt{(R_1 + R_2')^2 + X_{eq}^2}} = \frac{V}{Z_{eq}} \tag{12.52}$$

EXAMPLE 12.6 For the motor in Example 12.5, compute the following:

a. The starting torque.
b. The starting current.
c. The ratio of the starting torque to the full load torque.
d. The ratio of the starting current to the full load current.

Solution

a. The starting torque can be computed using Equation (12.50).

$$T_{st} = \frac{3V^2 R_2'}{\omega_s Z_{eq}^2} = \frac{480^2 \times 0.2}{188.5(0.5^2 + 4)} = 57.52 \text{ Nm}$$

b. Using Equation (12.52), we can compute the starting current.

$$I_{st}' = \frac{V}{Z_{eq}} = \frac{480/\sqrt{3}}{\sqrt{0.5^2 + 4}} = 134.42 \text{ A}$$

c. The full load torque computed in Example 12.5 is 121.46 Nm. The ratio of the starting torque to the full load torque is

$$\frac{T_{st}}{T_d} = \frac{57.52}{121.46} = 0.4736$$

As seen, the starting torque is less than half the full load torque. Heavily loaded induction motors may not start with low starting torque. In the following sections, we shall discuss some techniques used to increase the starting torque.

d. The full load current computed in Example 12.5 is 29.1 A. The ratio of the starting current to the full load current is

$$\frac{I_{st}'}{I_2'} = \frac{134.42}{29.1} = 4.62$$

The starting current of this motor is more than four times the rated current. This excessive current could damage the motor at starting. In the next sections, we shall discuss methods by which the starting current can be reduced.

12.2.6 Maximum Torque

The maximum torque of the induction motor can be computed by setting the derivative of the torque equation with respect to slip equal to zero. The slip that satisfies Equation (12.53) is the slip at maximum torque S_{max}.

$$\frac{dT_d}{dS} = 0; \qquad S \Rightarrow S_{max}$$
$$(R_1^2 + X_{eq}^2)S_{max} - (R_2')^2 = 0 \tag{12.53}$$

Hence,

$$S_{max} = \frac{R_2'}{\sqrt{R_1^2 + X_{eq}^2}} \tag{12.54}$$

Substituting the value of S_{max} into Equation (12.48) leads to the maximum torque of the motor.

$$T_{max} = \frac{3V^2}{2\omega_s \left[R_1 + \sqrt{R_1^2 + X_{eq}^2}\right]} \tag{12.55}$$

EXAMPLE 12.7 For the motor in Example 12.5, compute the following:

a. The slip at maximum torque.
b. The speed at maximum torque.
c. The maximum torque.
d. The ratio of the maximum torque to the full load torque.
e. The current at the maximum torque.

Solution

a. $$S_{max} = \frac{R_2'}{\sqrt{R_1^2 + X_{eq}^2}} = \frac{0.2}{\sqrt{0.3^2 + 4}} = 0.0989$$

b. Using Equation (12.47), we can compute the speed at maximum torque.

$$n = n_s(1 - S_{max}) = 1800(1 - 0.0989) = 1621.98 \text{ rpm}$$

c. Using Equation (12.55), we can compute the maximum torque.

$$T_{max} = \frac{3V^2}{2\omega_s\left[R_1 + \sqrt{R_1^2 + X_{eq}^2}\right]} = \frac{480^2}{2\times188.5\left[0.3 + \sqrt{0.2^2 + 4}\right]} = 264.57 \text{ Nm}$$

d. The full load torque computed in Example 12.5 is 121.46 Nm. The ratio of the maximum torque to the full load torque is

$$\frac{T_{max}}{T_d} = \frac{264.57}{121.46} = 2.18$$

The maximum torque is larger than the full load torque.

e. The rotor current at S_{max} can be computed using Equation (12.45).

$$I_2' = \frac{V}{\sqrt{\left(R_1 + (R_2'/S_{max})\right)^2 + X_{eq}^2}} = \frac{480/\sqrt{3}}{\sqrt{\left(0.3 + (0.2/0.0989)\right)^2 + 4}} = 90.42 \text{ A}$$

The current at the maximum torque is smaller than the starting current, but is larger than the full load current.

12.2.7 Starting Methods

The induction motor has two problems during starting: the starting current is large and the starting torque is small. The large starting current can damage the rotor windings due to excessive thermal heat inside the windings. The small starting torque may not be enough to start the motor under heavy loading conditions. As shown in Equation (12.52), the starting current can be reduced if we reduce the terminal voltage of the motor or increase its equivalent impedance Z_{eq}. These two methods are explored below.

12.2.7.1 Voltage Reduction

If an ac/ac converter is used to drive the induction motor as shown in Figure 12.18, the voltage across the motor at starting can be reduced, thus reducing the starting current. However, the reduction of the stator voltage can also reduce the starting torque and the maximum torque as shown in Figure 12.19 and explained by Equations (12.50) and (12.55); a 20% reduction in the stator voltage reduces the starting torque and the maximum torque by 36% each.

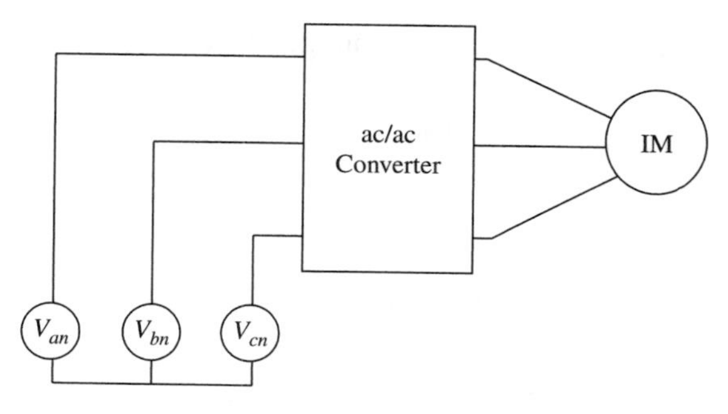

FIGURE 12.18
Voltage control of the induction motor.

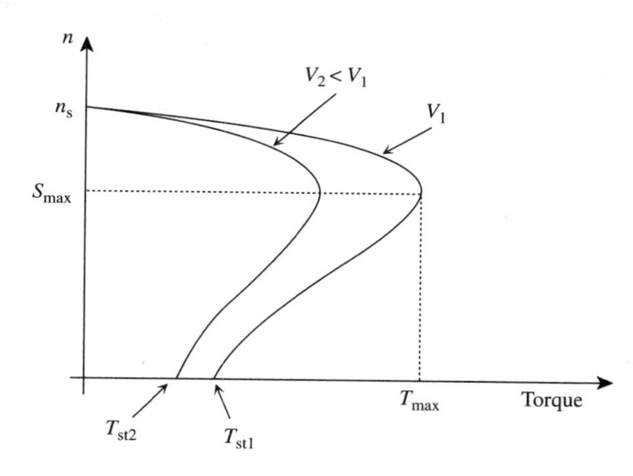

FIGURE 12.19
Speed–torque characteristics at different voltage levels.

EXAMPLE 12.8 For the motor in Example 12.5, compute the following:

a. The terminal voltage that limits the starting current to twice the full load current.
b. The reduction in the starting torque.
c. The reduction in the maximum torque.

Solution

a. The full load current computed in Example 12.5 is 29.1 A. Hence, the starting current is limited to 58.2 A.

$$I'_{st} = \frac{V_{st}}{Z_{eq}} = \frac{V_{st}}{\sqrt{0.5^2 + 4}} = 58.2\ \text{A}$$

where V_{st} is the reduced phase voltage at starting.

$$V_{st} = 119.98 \text{ V}$$

The line-to-line voltage at starting is $\sqrt{3} \times 119.98 = 207.8$ V, which is about 43% of the rated voltage.

b. The starting torque can be computed using Equation (12.50).

$$T_{st} = \frac{3V_{st}^2 R_2'}{\omega_s Z_{eq}^2} = \frac{207.8^2 \times 0.2}{188.5(0.5^2 + 4)} = 10.78 \text{ Nm}$$

This starting torque is about 19% of the starting torque computed in Example 12.6 for the full voltage. Hence, the reduction in the starting torque is about 81%.

c. Using Equation (12.55), we can compute the maximum torque for the new voltage.

$$T_{max} = \frac{V_{st}^2}{2\omega_s\left[R_1 + \sqrt{R_1^2 + X_{eq}^2}\right]} = \frac{207.8^2}{2 \times 188.5\left[0.3 + \sqrt{0.2^2 + 4}\right]} = 49.58 \text{ Nm}$$

This maximum torque is also about 19% of the maximum torque computed in Example 12.7 for the full voltage. The reduction is about 81%.

12.2.7.2 Insertion of Resistance

As seen in Example 12.8, the voltage reduction at starting reduces the starting current, which is desirable. However, the method leads to undesirable reduction in the starting torque. Therefore the motor may produce less torque at starting than what is needed by the load. To address this problem, we should use other methods that can increase the starting torque while reducing the starting current. One of these methods is based on increasing the rotor resistance. For the squirrel cage rotor, the windings can be designed from material that exhibits the skin effect at the frequency of the source, thus increasing its resistance at starting only (keep in mind that the rotor frequency is very low at full speed). For the slip-ring motor, the rotor windings are accessible from the stator through the slip-ring mechanism shown in Figure 12.7. Thus an external resistance can be added to the rotor circuit as shown in Figure 12.20 at starting. In either type of motor, when the rotor resistance increases, the starting current is decreased as shown in Equation (12.52). But how do the starting and maximum torques change? Upon the first examination of Equation (12.50) it is hard to tell whether the increase in the rotor resistance would increase the starting torque. However, because most

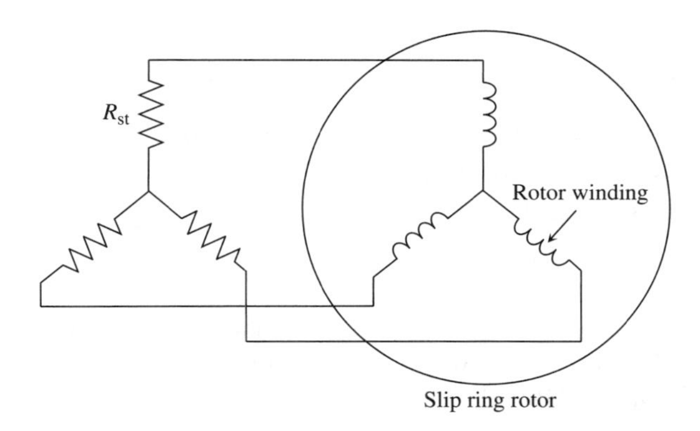

FIGURE 12.20
Starting of the induction motor by inserting a resistance in the rotor circuit.

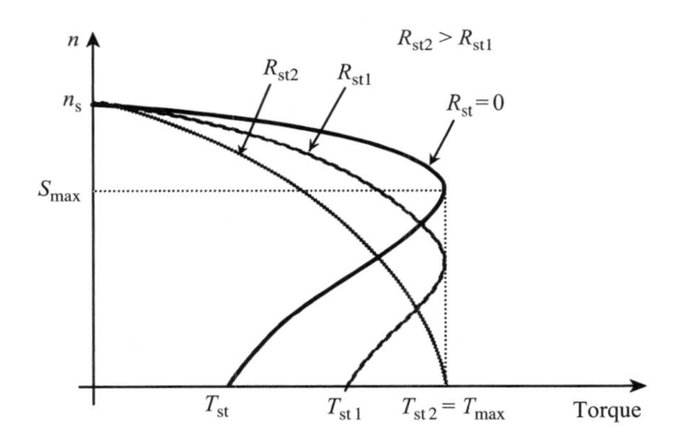

FIGURE 12.21
Speed–torque characteristics when a resistance is added to the rotor circuit.

machines have their $R_{eq}^2 \ll X_{eq}^2$, we can approximate the starting torque equation as

$$T_{st} = \frac{3V^2 R_2'}{\omega_s \left[(R_1 + R_2')^2 + X_{eq}^2 \right]} \approx \frac{3V^2 R_2'}{\omega_s X_{eq}^2} \tag{12.56}$$

Now it is obvious that the increase in the rotor resistance would result in an increase in the starting torque. This is another desirable feature.

The maximum torque, as shown in Equation (12.55), is unaffected by the increase in rotor resistance. However, the slip at maximum torque S_{max} in Equation (12.54) increases when the rotor resistance is increased. This means the maximum torque occurs at lower speeds. In fact, if enough resistance is added to the rotor circuit so that $S_{max} = 1$, the starting torque is equal to the maximum torque of the motor as shown in Figure 12.21.

EXAMPLE 12.9 For the motor in Example 12.5 compute the following:

a. The resistance that should be added to the rotor circuit to achieve the maximum torque at starting.
b. The starting current.

Solution

a. To compute the value of the inserted resistance in the rotor circuit to achieve the maximum torque at starting, we can set the maximum slip in Equation (12.54) to 1.

$$S_{max} = \frac{R_2' + R_{st}'}{\sqrt{R_1^2 + X_{eq}^2}} = 1$$

Hence,

$$R_{st}' = \sqrt{R_1^2 + X_{eq}^2} - R_2' = \sqrt{0.3^2 + 4} - 0.2 = 1.822\ \Omega$$

This is the starting resistance referred to the stator circuit. The actual rotor resistance can be computed if we know the turns ratio of the motor windings.

b. The starting current in Equation (12.52) can be modified to include the starting resistance in the rotor circuit.

$$I_{st}' = \frac{V}{\sqrt{(R_1 + (R_2' + R_{st}'))^2 + X_{eq}^2}} = \frac{(480/\sqrt{3})}{\sqrt{(0.3 + (0.2 + 1.822))^2 + 4}} = 90.43\ \text{A}$$

Comparing this starting current to that computed in Example 12.6, you will find that the starting current is decreased because of the insertion of the starting resistance.

12.3 Linear Induction Motor

The *linear induction motors* (LIM) are very effective drive mechanisms for transportation and actuation systems. The high-power linear motors are used in rapid transportation, baggage handling, conveyers, crane drives, theme park rides, and flexible manufacturing systems. The low-power ones are used in robotics, gate control, guided trajectories (e.g., aluminum can propulsion), and stage and curtain movement. NASA envisions the use of such motors in launching spacecrafts in the future.

Roller coaster

Maglev launcher
(Courtesy of NASA Marshall Space Flight Center)

Maglev rapid transportation

FIGURE 12.22
(see color insert following Page 208) Example of the applications of linear induction motors. (Images courtesy of Transrapid International.)

Linear induction motors come in two general designs: the *wheeled linear induction motor* (WLIM) and the *magnetically levitated linear induction motor* (Maglev). The Maglev technology, which was developed in 1934 by Hermann Kemper, is becoming highly popular for high-speed transportation. Some of the LIM applications are shown in Figure 12.22.

12.3.1 Wheeled Linear Induction Motor

The linear induction motor is similar to the rotating induction motor except that the LIM has a flat structure instead of the cylindrical structure of the rotating motor. Consider the rotating induction motor on the left side of Figure 12.23. If you imagine that we cut the motor along the dashed line and flatten the machine, we will get the wheeled linear motor on the right side of the figure, which can now be used to propel a train. In this case, the rotor is the track of the train and the stator is the train's engine. The stator of the linear motor is called the *primary circuit* and the rotor is called the *secondary circuit*. In rotating motors, the separation between the rotor and stator is

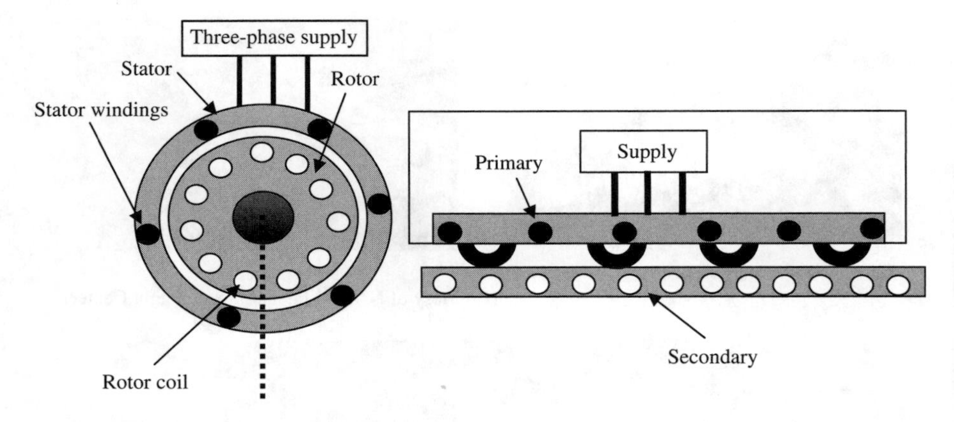

FIGURE 12.23
Wheeled linear induction motor.

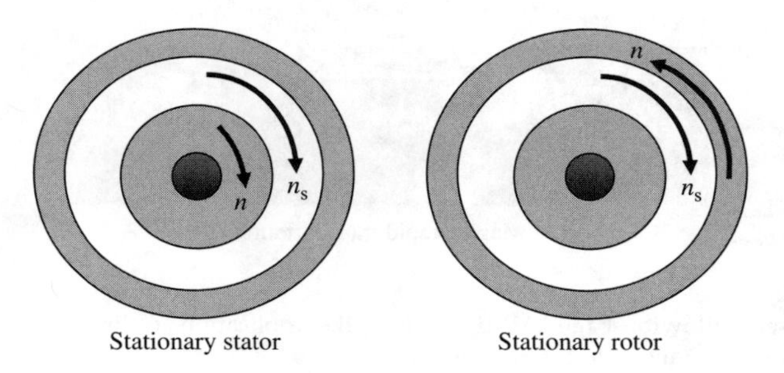

FIGURE 12.24
Relative speed of the rotating induction motor.

maintained by the ball bearings of the rotor's shaft. In wheeled linear motors, the separation is maintained by the wheels of the train.

For the rotating induction motors on the left side of Figure 12.24, when the stator is anchored and the rotor is allowed to freely rotate, the speed of the rotor is in the same direction as the speed of the magnetic field. However, if you hold the rotor stationary and allow the stator to spin, the stator rotates in the opposite direction to the synchronous speed of the field as shown on the right side of Figure 12.24. Similarly, for the LIM the primary circuit moves opposite to the synchronous speed of the field.

The synchronous speed of the rotating field of the induction motor is given in Equation (12.9). The linear speed v_s of this rotating field at any point located along the inner surface of the stator is

$$v_s = \omega_s r = \left(\frac{2\pi}{60} n_s\right) r = \left(\frac{2\pi}{60}\frac{120f}{p}\right) r = 2\left(\frac{2\pi r}{p}\right) f = 2\tau_p f \qquad (12.57)$$

where r is the inner radius of the stator in meters (Figure 12.25), n_s is the synchronous speed in rpm, and ω_s is the radian per second. The term τ_p is called *pole pitch*. It is the circumference of the inner circle of the stator divided by the number of poles. This is the separation between a and a', b and b', or c and c', as shown in Figure 12.25.

Let us consider the train that is powered by the WLIM in Figure 12.26. The primary windings are mounted under the floor of the train's compartment (called *bogie*). The secondary consists of metal alloy bars impeded along the track's guideway and that are perpendicular to the track (called *reaction plates*). The length of the secondary circuit is the length of the track itself. The train is powered through a dc or ac power line alongside the track or above it. The train taps its energy from the power line through brushes that are always in contact with the line. A converter is mounted on the vehicle to convert the waveform of the power line into balanced multi-phase waveforms with variable voltage and frequency.

The three-phase currents of the linear induction motor produce a magnetic field traveling in the direction shown in Figure 12.26. The speed of this

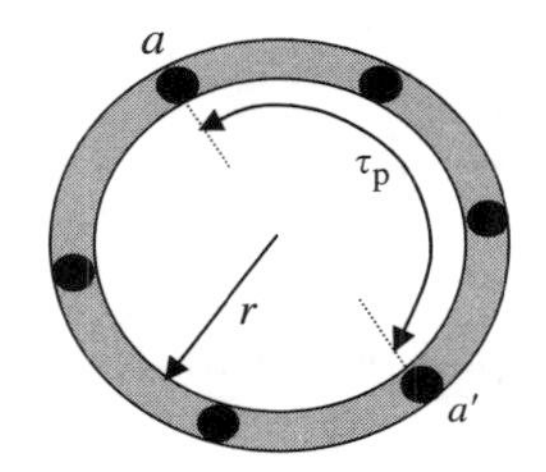

FIGURE 12.25
Pole pitch of a motor.

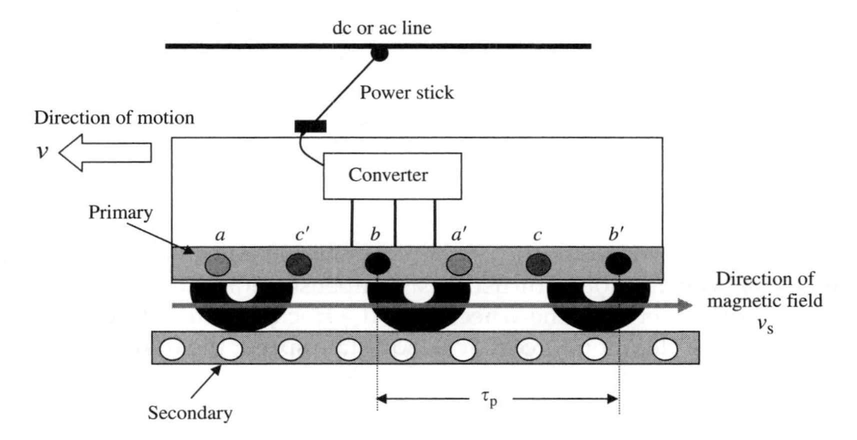

FIGURE 12.26
(see color insert following Page 208) Motion of the linear induction motor.

magnetic field is given in Equation (12.57). The train itself moves opposite to the direction of the magnetic field as explained earlier. The pole pitch of the linear motor is the distance between the opposite poles of any phase, that is, the distance between a and a'.

The equivalent circuit of the linear induction motor is basically the same as that for the rotating induction motor in Figure 12.15. One key difference is that the resistance of the secondary winding of the linear induction motor is larger than the rotor resistance of the rotating motor.

In linear motion, force is used instead of torque. The developed force of the linear induction motor F_d is also known as the thrust of the motor. It can be computed from Equation (12.48), but modified for a linear motion where angular velocities are replaced by linear speeds.

$$P_d = \frac{3V^2 R_2' v}{S v_s \left[\left(R_1 + (R_2'/S) \right)^2 + X_{eq}^2 \right]} \tag{12.58}$$

where v is the actual speed of the train and v_s is the synchronous speed of the magnetic field. The slip S is defined as

$$S = \frac{v_s - v}{v_s} \tag{12.59}$$

In linear motion, the developed power in the mechanical form is the force multiplied by the speed.

$$P_d = F_d v \tag{12.60}$$

Substituting Equation (12.60) into (12.58), we can compute the thrust of the linear induction motor.

$$F_d = \frac{P_d}{v} = \frac{3V^2 R_2'}{S v_s \left[\left(R_1 + (R_2'/s) \right)^2 + X_{eq}^2 \right]} \tag{12.61}$$

The speed–thrust characteristic of the linear induction motor is shown in Figure 12.27. The developed thrust must compensate for three main components: the friction between the wheels and the track, the drag force, and the acceleration/deceleration force. For steady state speed, the third component is zero.

The friction force $F_{friction}$ is a function of the normal force F_{normal} on the track.

$$F_{friction} = \mu F_{normal} \tag{12.62}$$

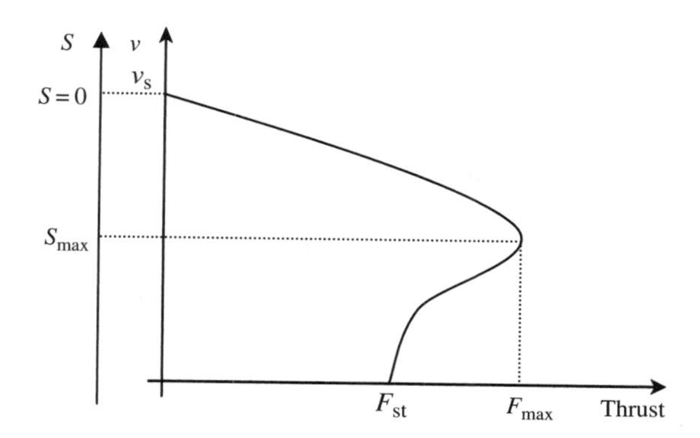

FIGURE 12.27
Speed–thrust of the linear induction motor.

where μ is the coefficient of friction between the wheels and the track. For a leveled track, the normal force is the weight of the vehicle.

$$F_{normal} = mg \tag{12.63}$$

where m is the mass of the moving vehicle and g is the acceleration of gravity. For an inclined track, the normal force is the component of the weight that is perpendicular to the track.

$$F_{normal} = mg \cos \theta \tag{12.64}$$

where θ is the inclination angle of the track. Example 12.12 deals with an inclined track.

The air drag force, which slows down the moving vehicle, is due to many factors such as the frontal area of the vehicle, the shape of the vehicle, the velocity of the vehicle, and the air density. The air drag force F_{air} can be computed by

$$F_{air} = 0.5 \delta v^2 A C_d \tag{12.65}$$

where δ is the air density (about 1 kg/m^3), v is the velocity of the vehicle in meters per second, A is the frontal area of the vehicle that is perpendicular to the air flow in meters, and C_d is the coefficient of drag. C_d depends on the shape of the frontal area of the vehicle; C_d can be about 1 for train, 0.4 for roller coaster, and 0.5 for a typical passenger car.

EXAMPLE 12.10 A train with a wheeled linear induction motor is traveling at 100 km/h. The frontal area of the train is 20 m^2 and its coefficient of drag is 0.8. The friction coefficient between the wheels of the train and the track is 0.05, and the weight of the train is 100,000 kg. Compute the thrust of the motor at steady state. Also compute the developed power of the motor in hp.

Solution

At steady state, the induction motor produces enough thrust to compensate for the friction force plus the drag force. The frictional force is computed based on the normal force of the vehicle and the coefficient of friction μ.

$$F_{\text{normal}} = mg = 10^5 \times 9.8 = 980 \text{ kN}$$

$$F_{\text{friction}} = \mu F_{\text{normal}} = 0.05 \times 980 = 49 \text{ kN}$$

The drag force can be computed using Equation (12.65).

$$F_{\text{air}} = 0.5\delta v^2 A C_{\text{d}} = 0.5 \times 1 \times \left(100 \times \frac{1000}{3600}\right)^2 \times 20 \times 0.8 = 6.174 \text{ kN}$$

The developed force is the sum of the frictional force and the drag force.

$$F_{\text{d}} = F_{\text{friction}} + F_{\text{air}} = 49 + 6.174 = 55.174 \text{ kN}$$

The developed power of the linear induction motor is

$$P_{\text{d}} = F_{\text{d}} v = 55.174 \times 27.78 = 1532.7 \text{ kW}$$

$$P_{\text{d}} = 1532.7 \times 1.34 = 2054 \text{ hp}$$

EXAMPLE 12.11 A vehicle with a wheeled linear induction motor has a mass of 4000 kg. The friction coefficient between the wheels of the vehicle and the track is 0.05. The pole pitch of the vehicle is 3 m. At the steady state, the slip of the motor is 0.1 when the frequency of the primary windings is 10 Hz. Assume that the drag force at steady state is 5000 N. Compute the developed power of the motor.

Solution

At steady state, there is no acceleration, so the induction motor produces enough thrust to compensate for the friction force plus the drag force.

$$F_{\text{normal}} = mg = 4000 \times 9.8 = 39.2 \text{ kN}$$

$$F_{\text{friction}} = \mu F_{\text{normal}} = 0.05 \times 39.2 = 1.96 \text{ kN}$$

To compute the developed power, we need the speed of the vehicle. The synchronous speed of the magnetic field is

$$v_{\text{s}} = 2\tau_{\text{p}} f = 2 \times 3 \times 10 = 60 \text{ m/s}$$

Hence, the speed of the vehicle is

$$v = v_s(1 - S) = 60(1 - 0.1) = 54 \text{ m/s}$$

The developed power is

$$P_d = F_d v = (F_{\text{friction}} + F_{\text{air}})\, v = (1960 + 5000) \times 54 = 375.84 \text{ kW}$$

The equations for the maximum developed force (maximum thrust) and the slip at maximum force for the LIM are the same as Equations (12.55) and (12.54) for the rotating motor.

$$S_{\max} = \frac{R_2'}{\sqrt{R_1^2 + X_{\text{eq}}^2}} \tag{12.66}$$

$$F_{\max} = \frac{3V^2}{2v_s\left[R_1 + \sqrt{R_1^2 + X_{\text{eq}}^2}\right]} \tag{12.67}$$

For the linear induction motor, the rotor resistance is higher than that for the rotating motor. Hence, $S_{\max}$ for the linear motor is large and occurring near the starting region. This way, the linear motor develops enough thrust when starting to speed up the train even at heavy loading conditions. Therefore, a reasonable approximation is to assume that the speed–thrust characteristic of the motor is linear between the operating slip S and $S_{\max}$, that is,

$$S = \frac{F_d}{F_{\max}} S_{\max} \tag{12.68}$$

EXAMPLE 12.12 A theme ride vehicle is moving uphill at a constant speed. The slope of the hill is 10°. The mass of the vehicle plus the riders is 2000 kg. The vehicle is powered by a wheeled linear induction motor. The pole pitch of the motor is 0.5 m. The friction coefficient of the surface is 0.01. The primary circuit resistance R_1 is 0.5 Ω, and the secondary resistance referred to the primary circuit R_2' is 1.0 Ω. The equivalent inductance of the primary and secondary circuits is 0.01 H. The line-to-line voltage of the primary windings of the motor is 480 V, and its frequency is 15 Hz. Assume the drag force is 500 N and compute the speed of the vehicle.

Solution

The system can be described by Figure 12.28. The weight of the vehicle (mg) is resolved into two components, one normal to the road surface F_{normal} and the other, F_g, parallel to the road surface pulling the vehicle down due to gravity.

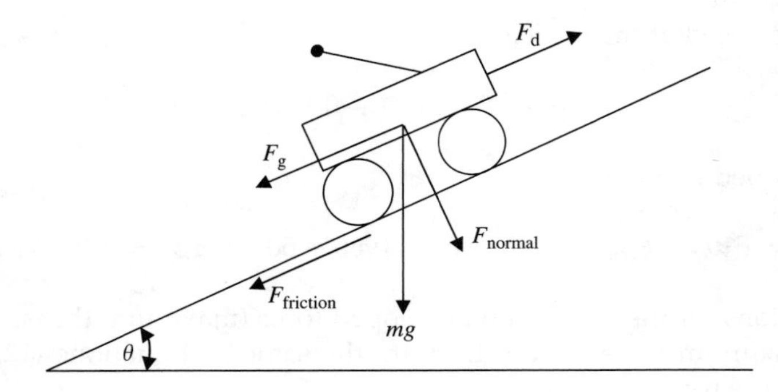

FIGURE 12.28
Force vectors of a vehicle moving uphill.

F_{normal} produces the frictional force $F_{friction}$.

$$F_d = F_g + F_{friction} + F_{air}$$

$$F_{normal} = mg \cos\theta = 2000 \times 9.8 \times \cos 10 = 1.93 \times 10^4 \text{ N}$$

$$F_{friction} = \mu F_{normal} = 0.01 \times 1.93 \times 10^4 = 193 \text{ N}$$

$$F_g = mg \sin\theta = 2000 \times 9.8 \sin 10 = 3.403 \times 10^3 \text{ N}$$

$$F_d = F_g + F_{friction} + F_{air} = 4.097 \text{ kN}$$

The synchronous speed of the motor is

$$v_s = 2\tau_p f = 2 \times 0.5 \times 15 = 15 \text{ m/s}$$

The motor speed can be computed using the relationship in Equation (12.68). But first, let us compute the maximum force and the slip at maximum force.

$$S_{max} = \frac{R_2'}{\sqrt{R_1^2 + X_{eq}^2}} = \frac{1.0}{\sqrt{0.5^2 + (2\pi \times 15 \times 0.01)^2}} = 0.9373$$

$$F_{max} = \frac{3V^2}{2v_s\left[R_1 + \sqrt{R_1^2 + X_{eq}^2}\right]}$$

$$= \frac{480^2}{2 \times 15\left[0.5 + \sqrt{0.5^2 + (2\pi \times 15 \times 0.01)^2}\right]} = 4.901 \text{ kN}$$

Now we have the information needed to compute the slip of the motor.

$$S = \frac{F_d}{F_{max}} S_{max} = \frac{4.097}{4.901} 0.9373 = 0.784$$

Hence, the speed of the vehicle is

$$v = v_S(1 - S) = 15(1 - 0.784) = 3.24 \text{ m/s or } 11.66 \text{ km/h}$$

12.3.2 Magnetically Levitated (Maglev) Induction Motor

One of the main limitations of the wheeled linear induction motor is the friction between the wheels and the track. Besides the extra power consumption, the friction limits the maximum speed of the vehicle as well as its maximum acceleration and deceleration.

The Maglev technology allows the vehicle to magnetically levitate with virtually no friction between the bogie of the vehicle (undercarriage) and the track. The technology is based on having magnetic poles on the track that are similar to the magnetic poles on the bogie of the vehicle (both are North or South). These poles repel each other and the bogie then levitates. Hence, a Maglev train can achieve very high speeds and very smooth rides because the vehicle essentially flies. In addition, the Maglev trains can climb steeper hills that regular trains cannot reach. With the current technology, a Maglev train can reach speeds up to 500 km/h, which is much faster than any other ground transportation system. The travel time of a medium trip (1000 km) on a Maglev train is essentially equal to the travel time of the airplane when you include the check-in time at the airport as well as the taxing time of the airplane.

There are two types of Maglev systems: *levitation* and *propulsion*. In both systems, magnetic force is developed along the track to repel the bogie of the vehicle. This is done in various ways; one of them is to install magnetic coils along the track as shown in Figure 12.29. These coils are excited by a separate source when the vehicle approaches, thus keeping the vehicle levitated along the track. This system, however, is suitable for short tracks as the excitation of long tracks is expensive and difficult to implement. Another alternative is to

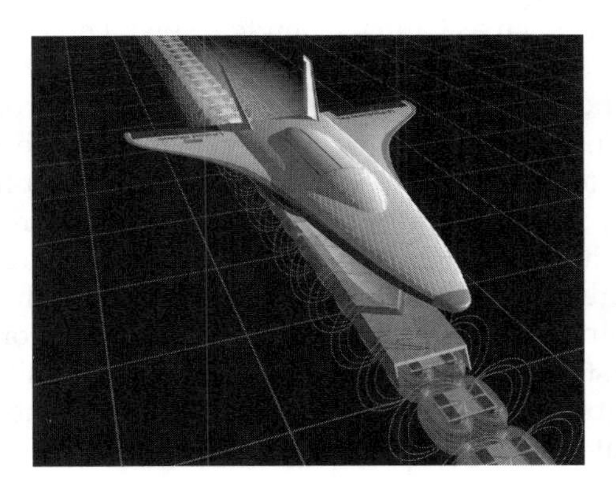

FIGURE 12.29
(see color insert following Page 208) Concept of Maglev for a space launcher. (Courtesy of NASA Marshall Space Flight Center.)

induce a voltage in the coils of the track by the vehicle itself. When the vehicle moves over a coil in the track, the electric magnets installed in the bogie are switched on to induce a voltage on the coil of the track. If the windings of the coil are wrapped in the direction that creates poles with similar polarity to the magnetic poles at the bottom of the bogie's magnets, a repulsive force is created that levitates the vehicle.

The propulsion Maglev is similar to the levitated Maglev, but the magnetic force of the track coils propels the vehicle along the track in addition to lifting the bogie. If you place two magnets on top of each other with their similar poles facing each other, you feel a lifting force. If you fix the bottom magnet to a table and move the top magnet away while maintaining its vertical distance, the repulsion force moves the top magnet horizontally as well as vertically (i.e., propulsion and levitation). The Maglev propulsion is created by a similar method where each magnet on the bogie is switched on as it leaves one of the track coils.

Besides ground transportation, the U.S. Sandia National Laboratories developed the Maglev technology for high-power, high-thrust military applications. The Maglev is also developed for satellite launchers where the spacecraft accelerates to 1000 km/h at the end of the track causing it to take off like an airplane. While airborne, the spacecraft uses conventional rocket engines to reach its orbit.

12.4 Synchronous Generator

The synchronous generator is the machine used most in generating electricity. All power plants use synchronous generators to convert the mechanical power of the turbine into electrical power. The synchronous generators can have enormous capacity where a single generator can produce over 2 GW of electric power. The large ones are used in nuclear power plants and major hydroelectric plants such as China's Yangtze River Three-Gorges Hydraulic Power Plant.

Figure 12.30 shows the stator of one of the synchronous generators of the Grand Coulee Hydroelectric Power Plant on the Columbia River in Washington State. To put the size into perspective, the photograph on the right side shows part of the stator during the construction of the generator. The area in this photo is almost the same as the boxed area in the left side photograph. Figure 12.31 shows the rotor of the same synchronous generator.

The main components of the synchronous generator are shown in Figure 12.32. The stator of the synchronous machine is similar to the stator of the induction motor; it consists of three-phase windings mounted symmetrically inside the stator core. The stator windings are also known as the *armature windings*. The winding of the rotor of the synchronous machine is excited by an external dc source through a slip-ring system similar to the one shown in

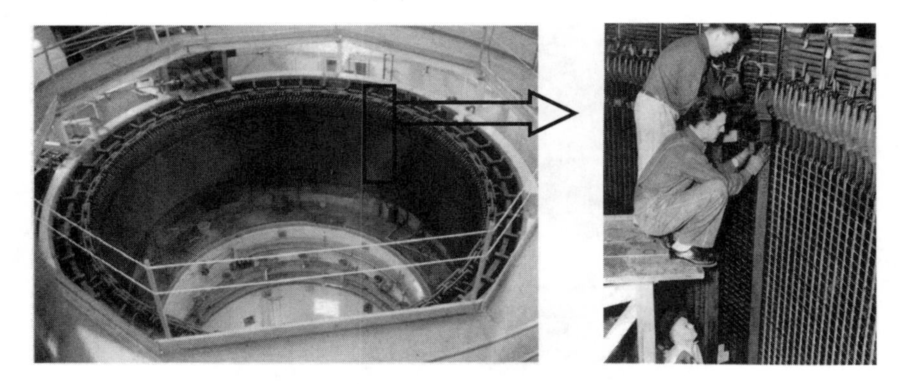

FIGURE 12.30
Stator of a synchronous generator in the Grand Coulee Dam (courtesy of the U.S. Bureau of Reclamation).

FIGURE 12.31
Rotor of a synchronous generator in the Grand Coulee Dam (courtesy of the U.S. Bureau of Reclamation).

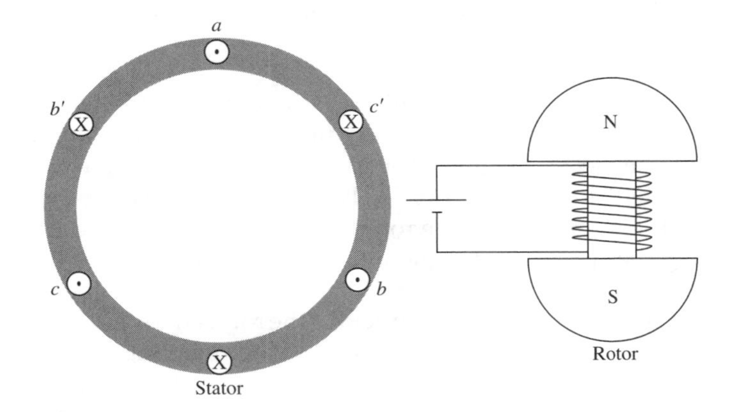

FIGURE 12.32
Basic components of a synchronous generator.

FIGURE 12.33
(see color insert following Page 208) Machine model showing slip-ring arrangement.

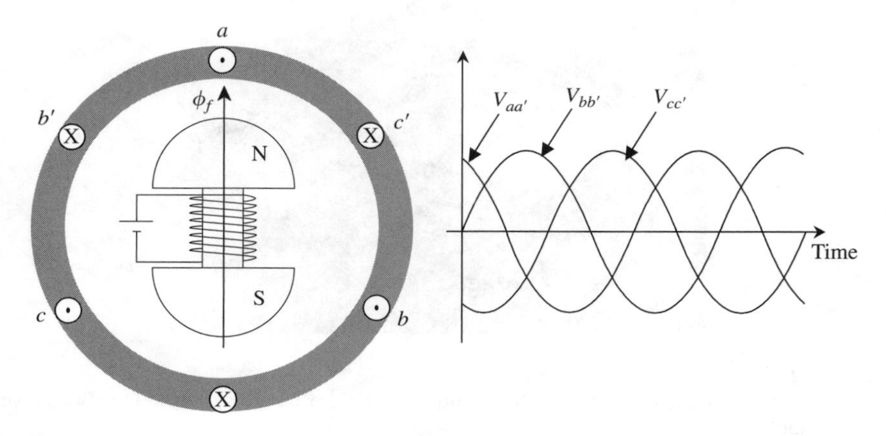

FIGURE 12.34
Three-phase voltage waveforms due to the clockwise rotation of the rotor magnet.

Figure 12.33. The rotor winding, which is also known as the *field winding* or *excitation winding*, produces a stationary flux with respect to the rotor; that is, the rotor is an electric magnet. The rotor is assembled inside the stator as shown on the left side of Figure 12.34. Two sets of ball bearings are used on both ends of the rotor to allow the rotor to spin freely inside the stator.

The rotor of the synchronous generator is connected to a prime mover such as a hydroelectric or thermal turbine. When the turbine spins the rotor of the synchronous generator, the magnetic field of the rotor cuts the stator windings thus inducing a sinusoidal voltage across the stator windings. Since the three stator windings are equally spaced from each other, the induced voltages across the phase windings are shifted by 120° from each other as shown on the right side of Figure 12.34. The frequency of the induced voltage is dependent on the speed of the rotor. The relationship between the speed of the magnetic field and the frequency of the induced voltage is given in Equation (12.9). Hence, to maintain the frequency of the induced voltage at 60 Hz, the rotor speed must be

$$n_s = \frac{7200}{p} \tag{12.69}$$

Any slight variation in this synchronous speed will result in a frequency variation for the induced voltage. This can cause a stability problem if the machine is connected to the fixed frequency power grid. Therefore, the speed of the turbine is delicately controlled so that the rotor is always spinning at the synchronous speed.

When the field circuit of the synchronous generator is excited and the rotor is spinning, a balanced three-phase voltage is induced in the stator windings. If an external three-phase load is connected across the stator windings as shown in Figure 12.35, three-phase currents will flow into the load. The current of the load is known as the *armature current*. If the load is balanced, the armature currents are balanced and a single-phase representation is all we need to model the generator.

The airgap of the synchronous generator has two fields as shown in Figure 12.36:

1. The excitation field ϕ_f, which is stationary with respect to the rotor. When a turbine spins the rotor at the synchronous speed, ϕ_f rotates with respect to the stator at the synchronous speed.

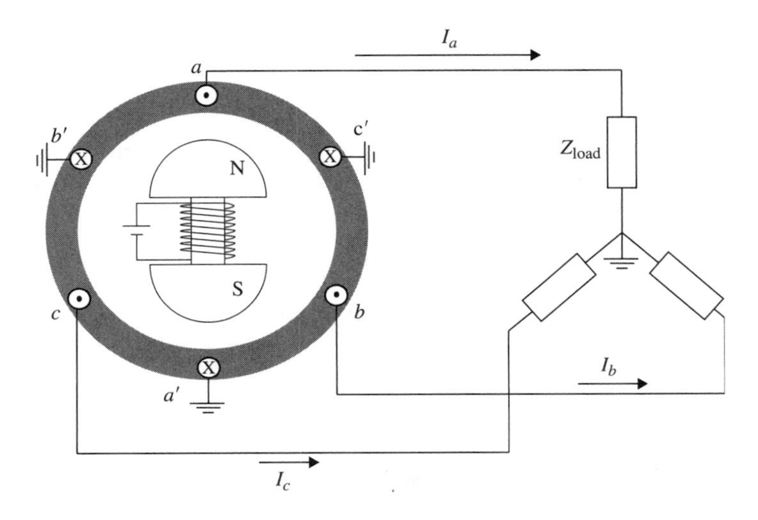

FIGURE 12.35
Loaded synchronous generator.

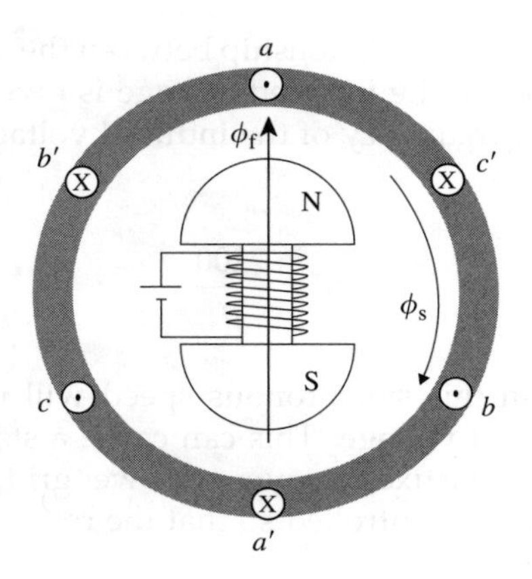

FIGURE 12.36
Total airgap flux inside the synchronous generator.

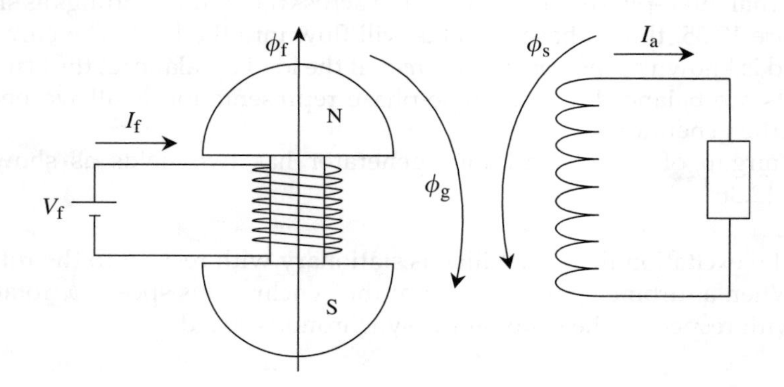

FIGURE 12.37
Representation of the synchronous generator.

2. The stator flux ϕ_s, due to the three phase armature currents. This flux, which is only present when the generator is loaded, rotates at the synchronous speed inside the airgap similar to the rotating field inside the induction motor.

Hence, the net magnetic field in the airgap ϕ_g is the phasor sum of both fields.

$$\overline{\phi}_g = \overline{\phi}_f + \overline{\phi}_s \tag{12.70}$$

The equivalent circuit of the synchronous generator can be developed by using the schematic of the single-phase representation of the generator in Figure 12.37. The figure shows the rotor circuit excited by a dc source V_f that produces an excitation current I_f. The excitation current produces the field ϕ_f. Moreover, the armature currents produce the rotating field ϕ_s, which is known as the *armature reaction*.

Assume first that the armature is unloaded, that is, $I_a = 0$ and $\phi_s = 0$. In this case, the induced voltage across the armature winding e_f is a function only of ϕ_f.

$$e_f = N\frac{d\phi_f}{dt} \tag{12.71}$$

where N is the number of turns in the armature winding. e_f is directly proportional to the excitation current I_f. The rms value of e_f is E_f, which is known as the *equivalent excitation voltage*.

Now assume that the electric load is connected across the armature windings. In this case, the induced voltage across the winding, e_g, is a function of the total airgap flux ϕ_g.

$$e_g = N\frac{d\phi_g}{dt} \tag{12.72}$$

The rms component of e_g is E_g. This induced voltage is not equal to the voltage across the load V_t because of the reactance X and resistance R of the armature winding. Hence,

$$\bar{E}_g = \bar{V}_t + \bar{I}_a(R + jX) \tag{12.73}$$

where V_t is the phase terminal voltage across the load and I_a is the armature current. The equivalent circuit representing Equation (12.73) is shown in Figure 12.38.

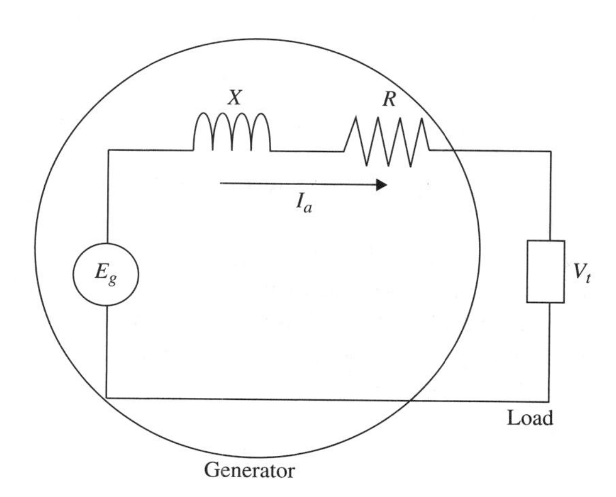

FIGURE 12.38
Equivalent circuit of loaded synchronous generator.

Note that the magnitude of E_g is a function of ϕ_f and ϕ_s. Hence, E_g can changes its value with any change in the load current. Therefore, it is difficult to analyze the synchronous generator at any loading condition by the above model. However, we can modify the equivalent circuit by representing E_g in terms of E_f, which is independent of the load current. This can be done by representing the instantaneous voltage e_g in Equation (12.72) as

$$\bar{e}_g = N\frac{d\bar{\phi}_g}{dt} = N\frac{d\bar{\phi}_f}{dt} + N\frac{d\bar{\phi}_s}{dt} \tag{12.74}$$

The magnitude of the term $N(d\bar{\phi}_f/dt)$ is equal to e_f, which is the *induced voltage* across the stator winding due to flux ϕ_f. The magnitude of the second term $N(d\bar{\phi}_s/dt)$ is the voltage across the stator windings due to the armature currents that produce the flux ϕ_s. Hence, the second term represents a *voltage drop* across the stator windings. Since Faraday's law gives a positive sign for the induced voltage and a negative sign for the drop voltage, Equation (12.74) can be rewritten as

$$\bar{e}_g = N\frac{d\bar{\phi}_f}{dt} + N\frac{d\bar{\phi}_s}{dt} = \bar{e}_f - \bar{e}_s \tag{12.75}$$

where e_s is the voltage drop across the armature winding due to the armature current. The magnitude of e_s is

$$e_s = N\frac{d\phi_s}{dt} = L_a\frac{di_a}{dt} \tag{12.76}$$

where i_a is the instantaneous current in the armature winding (the current that produces ϕ_s) and L_a is an equivalent inductance. Based on Equation (12.76), we can conclude that e_s is a voltage drop across an inductor L_a. In rms terms, Equation (12.74) can be rewritten as

$$\bar{E}_g = \bar{E}_f - \bar{E}_s = \bar{E}_f - \bar{I}_a\bar{X}_a \tag{12.77}$$

where $X_a = \omega L_a$. Using Equation (12.77), we can modify the equivalent circuit in Figure 12.38 as shown in Figure 12.39. X_s in the figure is known as the *synchronous reactance*, and is equal to $X + X_a$.

The model for the synchronous generator can be further simplified by ignoring the resistance of the armature windings. This is justified for large machines with windings made of large cross-section wires. In this case, the synchronous generator model can be represented by

$$\bar{E}_f = \bar{V}_t + \bar{I}_a\bar{X}_s \tag{12.78}$$

The phasor diagram of Equation (12.78) is shown in Figure 12.40. Note that E_f leads V_t.

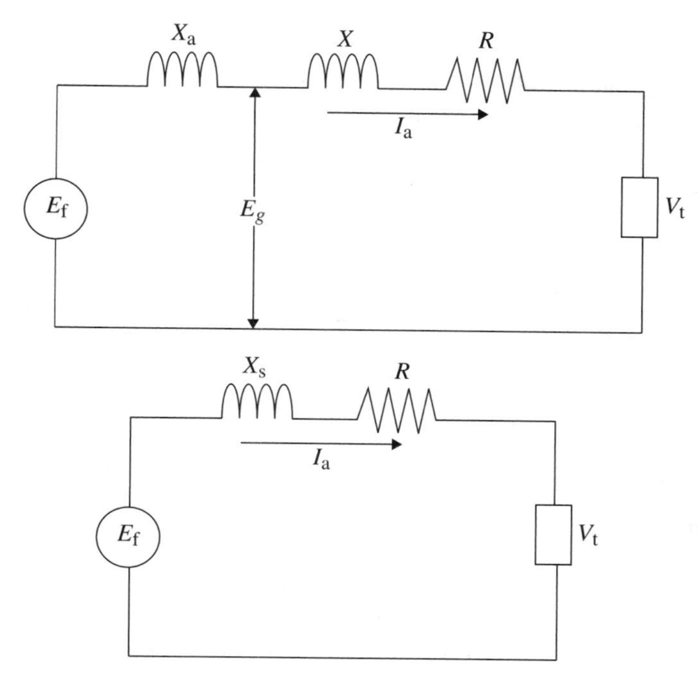

FIGURE 12.39
Equivalent circuit of the synchronous generator.

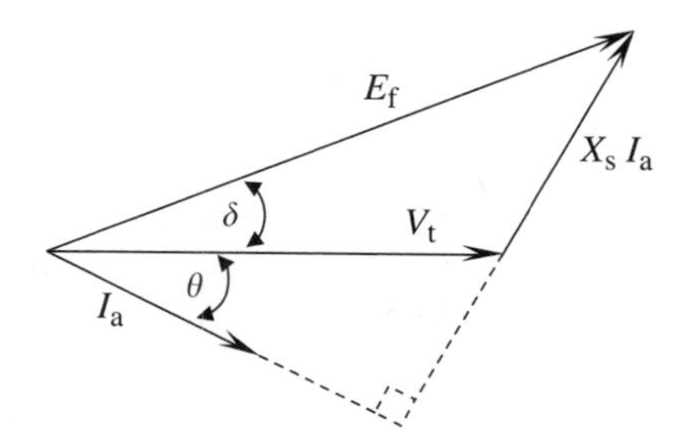

FIGURE 12.40
A phasor diagram of the synchronous generator.

EXAMPLE 12.13 A 60 Hz, four-pole synchronous generator of a thermal power plant has a synchronous reactance of 5 Ω. The stator windings are connected in wye, and its line-to-line voltage is 15 kV. The line current of the generator is 1 kA at 0.9 power factor lagging. Compute the following:

a. The equivalent field voltage E_f.
b. The real power delivered to the power grid.
c. The reactive power delivered to the power grid.

Solution

a. The equivalent field voltage can be computed by Equation (12.78) assuming the reference to be the phase terminal voltage.

$$\overline{E}_{\text{f}} = \overline{V}_{\text{t}} + \overline{I}_{\text{a}}\overline{X}_{\text{s}} = \frac{15}{\sqrt{3}}\angle 0° + (1\angle -\cos^{-1} 0.9)5\angle 90° = 11.74\angle 22.54°\ \text{kV}$$

b. $P = \sqrt{3}V_{\text{t-ll}}I_a \cos\theta = \sqrt{3} \times 15 \times 1 \times 0.9 = 23.38$ MW
where $V_{\text{t-ll}}$ is the line-to-line value of the terminal voltage

c. $Q = \sqrt{3}V_{\text{t-ll}}I_a \sin\theta = \sqrt{3} \times 15 \times 1 \times \sin(\cos^{-1} 0.9) = 11.32$ MVAr

---

12.4.1 Synchronous Generator Connected to an Infinite Bus

The infinite bus, by definition, is a large power grid consisting of a number of large generators. The voltage and frequency of the infinite bus are constant and cannot be changed regardless of any action made by any one generator. The infinite bus can absorb or deliver any amount of active or reactive powers from any single generator without any change in its voltage or frequency.

When a synchronous generator is connected to an infinite bus as shown in Figure 12.41, the voltage and frequency at the infinite bus are constant and cannot be changed by any change in the excitation current I_{f} or the speed of the generator. This is much like having two objects tied together. If the mass of one object is much larger than the mass of the other, any attempted movement of the small object may not affect the large object. The infinite bus in this scenario is the large object and the synchronous generator is the small one.

When the generator is connected directly to an infinite bus, the terminal voltage of the machine V_{t} in Figure 12.39 is equal to the infinite bus voltage V_0. Since V_0 is constant, V_{t} is also constant. Keep in mind that E_{f} is a function of the field current I_{f}. Hence, E_{f} is controlled by the operator of the generator. The phasor diagram of the synchronous machine connected to an infinite bus is shown in Figure 12.42. Note that the only difference between this phasor diagram and the one in Figure 12.40 is that the terminal voltage of the generator is equal to the infinite bus voltage.

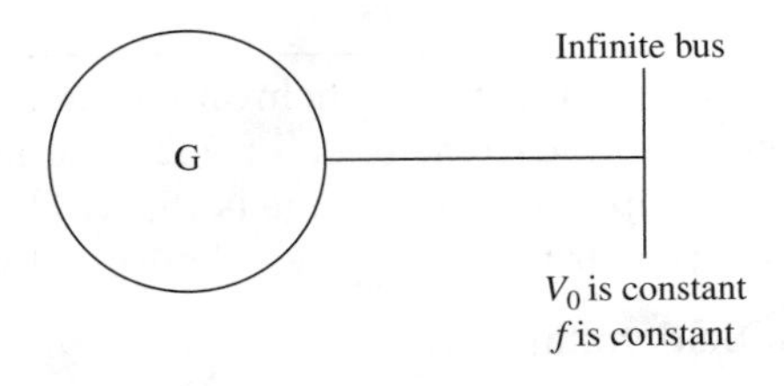

FIGURE 12.41
Synchronous generator connected to an infinite bus.

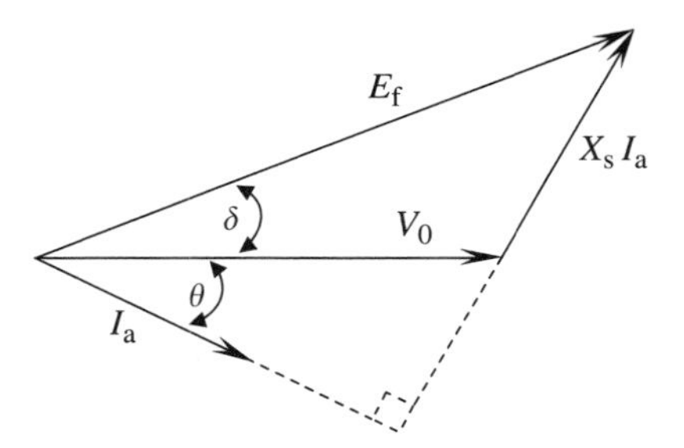

FIGURE 12.42
Phasor diagram of a synchronous generator connected to an infinite bus.

12.4.1.1 Real Power of a Synchronous Generator

The real power delivered to the infinite bus by the synchronous generator is

$$P = 3V_t I_a \cos\theta = 3V_0 I_a \cos\theta = 3E_f I_a \cos(\theta + \delta) \tag{12.79}$$

where θ is the phase angle of the current and δ is the angle between the infinite bus voltage and the equivalent field voltage E_f. δ is known as the *power angle*. In Equation (12.79), V_t, V_0, and E_f are phase quantities. Examining the phasor diagram in Figure 12.42 leads to the following relationship:

$$I_a X_s \cos\theta = E_f \sin\delta \tag{12.80}$$

Hence,

$$I_a \cos\theta = \frac{E_f \sin\delta}{X_s} \tag{12.81}$$

Substituting Equation (12.81) into Equation (12.79) yields the equation for the real power.

$$P = \frac{3V_0 E_f}{X_s} \sin\delta \tag{12.82}$$

Equation (12.82) is called the *power–angle equation*; it relates the output power of the generator to the power angle δ as well as other parameters and variables. The equation shows that the maximum power that can be generated by the machine P_{max}, which is known as the *pullover* power, occurs when $\delta = 90°$.

$$P_{max} = \frac{3V_0 E_f}{X_s} \tag{12.83}$$

Figure 12.43 shows the power–angle curve representing Equation (12.82). Let us assume that the internal losses of the generator are insignificant. In this

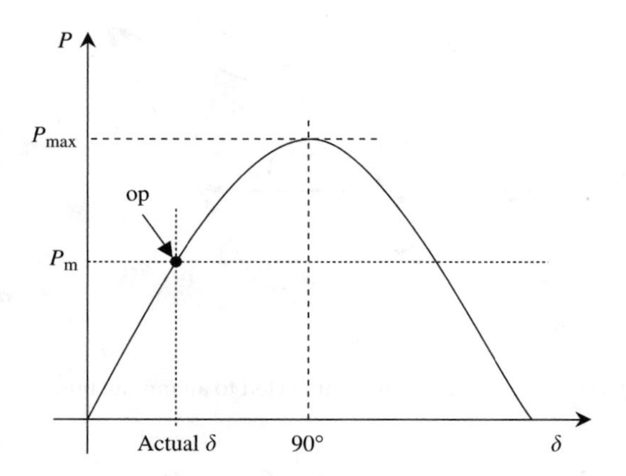

FIGURE 12.43
Real power as a function of the power angle.

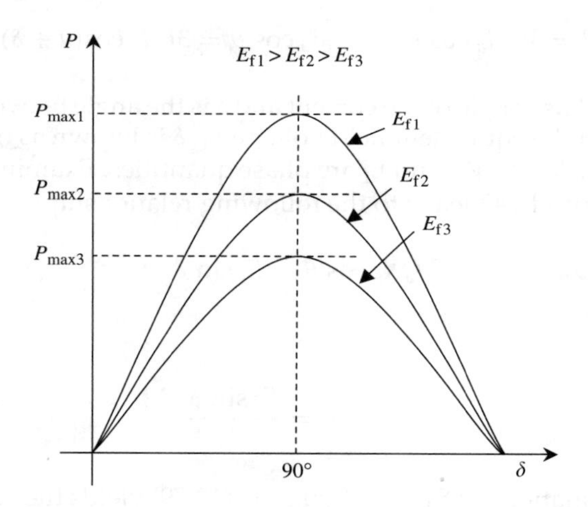

FIGURE 12.44
Pullover power due to changes in excitation current.

case, the input mechanical power P_m to the generator is equal to the electrical power P delivered to the infinite bus. The mechanical power P_m is controlled by the operator at the power plant. Hence, the point at which the mechanical power equals the electrical power is the operating point op of the generator. The power angle at this point is the actual angle between the infinite bus voltage V_0 and the equivalent field voltage E_f.

As seen in Equation (12.83), the maximum power P_{max} of the system increases when E_f is increased. However, the power delivered to the infinite bus will not increase unless the input mechanical power to the generator increases. So increasing P_{max} increases the capacity of the generator to deliver more power. Changing E_f is done by changing the excitation current I_f, which is continuously controlled at the power plant. Figure 12.44 shows various values of the pullover powers due to different E_f.

EXAMPLE 12.14 A synchronous generator is connected directly to an infinite bus. The voltage of the infinite bus is 23 kV. The excitation of the generator is adjusted until the equivalent field voltage E_f is 25 kV. The synchronous reactance of the machine is 2 Ω.

1. If the output mechanical power of the turbine is 100 MW, compute the power angle.
2. Compute the pullover power.
3. If the excitation of the generator is decreased by 10%, compute the power angle and the pullover power.

Solution

1. The electric power of the generator is equal to the mechanical power of the turbine. By direct substitution in Equation (12.82), we can compute the power angle. But keep in mind that all voltages in the equation are phase quantities and all powers are for the three phases. Also, the synchronous generators are almost always connected in wye.

$$P = \frac{3V_0 E_f}{X_s} \sin \delta$$

$$100 = \frac{3(23/\sqrt{3})(25/\sqrt{3})}{2} \sin \delta$$

$$\delta = 20.35^\circ$$

2. The equation for the pullover power is given in (12.83).

$$P_{max} = \frac{3V_0 E_f}{X_s} = \frac{23 \times 25}{2} = 287.5 \text{ MW}$$

3. Assume a linear relationship between the excitation current and the equivalent field voltage. When the excitation is decreased by 10%, the equivalent field voltage decreases by 10%.

$$P = \frac{3V_0 E_f}{X_s} \sin \delta$$

$$100 = \frac{23 \times (0.9 \times 25)}{2} \sin \delta$$

$$\delta = 22.74^\circ$$

$$P_{max} = \frac{3V_0 E_f}{X_s} = \frac{23 \times (0.9 \times 25)}{2} = 258.75 \text{ MW}$$

12.4.1.2 Reactive Power of Synchronous Generator

The reactive power delivered by the synchronous generator to the infinite bus Q_t can be computed from the phasor diagram in Figure 12.42.

$$Q_t = 3V_0 I_a \sin\theta \tag{12.84}$$

Using trigonometric relations, the following relationship can be obtained:

$$\begin{aligned} I_a X_s \sin\theta &= E_f \cos\delta - V_0 \\ I_a \sin\theta &= \frac{E_f \cos\delta - V_0}{X_s} \end{aligned} \tag{12.85}$$

Substituting Equation (12.85) into (12.84) yields

$$Q_t = 3V_0 I_a \sin\theta = \frac{3V_0}{X_s}(E_f \cos\delta - V_0) \tag{12.86}$$

Equation (12.86) shows that the reactive power delivered to the infinite bus is dependent on the magnitude of E_f. The sign of the reactive power can be positive, negative, or zero depending on the level of excitation of the synchronous generator.

1. *Over-excited generator*: This is when the excitation is adjusted so that $E_f \cos\delta > V_0$. The reactive power in this case is positive, meaning that the generator delivers reactive power to the infinite bus. The phasor diagram in this case is shown in Figure 12.42. Note that the current in this case lags the voltage of the infinite bus.
2. *Under-excited generator*: If the excitation is adjusted so that $E_f \cos\delta < V_0$, the reactive power is negative; that is, the generator is receiving reactive power from the infinite bus. The phasor diagram in this case is shown in Figure 12.45 where the current is leading the infinite bus voltage.

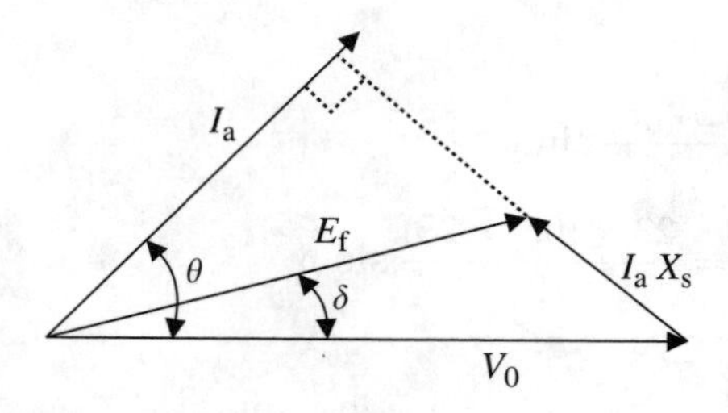

FIGURE 12.45
Under-excited synchronous machine.

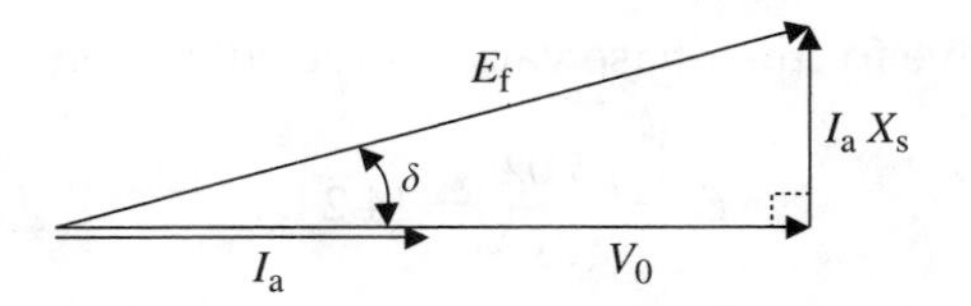

FIGURE 12.46
Synchronous machine with exact excitation.

3. *Exactly-excited generator*: If the excitation is adjusted so that $E_f \cos\delta = V_0$, there is no reactive power at the terminals of the generator. The phasor diagram in this case is shown in Figure 12.46. The current in this case is in phase with the infinite bus voltage.

EXAMPLE 12.15 For the synchronous machine in Example 12.14, compute the equivalent excitation voltage E_f required to deliver 100 MW of real power and no reactive power to the infinite bus.

Solution

The terminal voltage of the machine is unchanged since it is connected to an infinite bus. However, we cannot assume that the power angle is equal to the one computed in Example 12.14 because the equivalent excitation voltage must change to modify the reactive power. As seen in Figure 12.44, the change in the excitation alone leads to a change in the power angle.

Our problem has two unknowns: the equivalent excitation voltage and the power angle. To solve for these two unknowns, we need two equations, namely, the real and reactive power equations.

$$P = \frac{3V_0E_f}{X_s}\sin\delta$$

$$100 = \frac{3(23/\sqrt{3})E_f}{2}\sin\delta$$

$$E_f \sin\delta = 5.02\ \text{kV}$$

For zero reactive power at the generator terminals, the generator must be exactly excited. Hence,

$$E_f\cos\delta = V_0 = \frac{23}{\sqrt{3}} = 13.28\ \text{kV}$$

Hence,

$$\frac{E_f\sin\delta}{E_f\cos\delta} = \frac{5.02}{13.28} = 0.38$$

$$\tan\delta = 0.38$$

$$\delta = 20.7^\circ$$

Now we can solve for the phase value of the equivalent excitation voltage.

$$E_{\mathrm{f}} = \frac{5.02}{\sin\delta} = 14.2\ \mathrm{kV}$$

The line-to-line equivalent excitation voltage is $14.2\sqrt{3} = 24.59$ kV.

12.4.2 Synchronous Generator Connected to an Infinite Bus through a Transmission Line

Most power plants are located in remote areas where the energy resources are available or easily transportable, and the possible pollutions from the power plants are kept away from the population centers. The energy generated by these power plants must then be transmitted to the load centers (cities, factories, and the like) by a system of transmission lines. Figure 12.47 shows a one-line diagram of a simple system where the generator is connected to an infinite bus through a transmission line. The terminal voltage of the generator is V_{t} and the voltage at the infinite bus is V_0. The transmission line is a high-voltage wire that has a resistance and an inductive reactance X_l. However, the resistance is much smaller than the inductive reactance and is often ignored.

The equivalent circuit of the system is shown in Figure 12.48. The power angle δ in this case is defined as the angle between the voltage of the infinite bus V_0 and the equivalent field voltage E_{f}. The basic equations of the system are

$$\begin{aligned}\overline{E}_{\mathrm{f}} &= \overline{V}_{\mathrm{t}} + \overline{I}_{\mathrm{a}}\overline{X}_{\mathrm{s}} \\ \overline{V}_{\mathrm{t}} &= \overline{V}_0 + \overline{I}_{\mathrm{a}}\overline{X}_l \\ \overline{E}_{\mathrm{f}} &= \overline{V}_0 + \overline{I}_{\mathrm{a}}\overline{X}_l + \overline{I}_{\mathrm{a}}\overline{X}_{\mathrm{s}}\end{aligned} \tag{12.87}$$

These equations can be interpreted by the phasor diagram in Figure 12.49, where the current is assumed to be lagging the infinite bus voltage. Equation (12.87) can also be simplified as

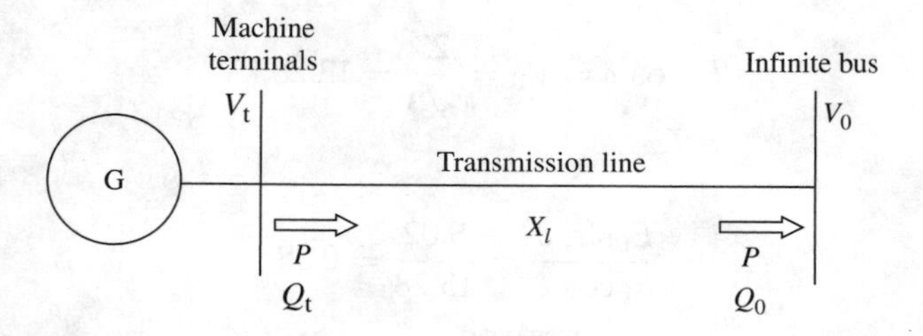

FIGURE 12.47
A synchronous generator connected to infinite bus through a transmission line.

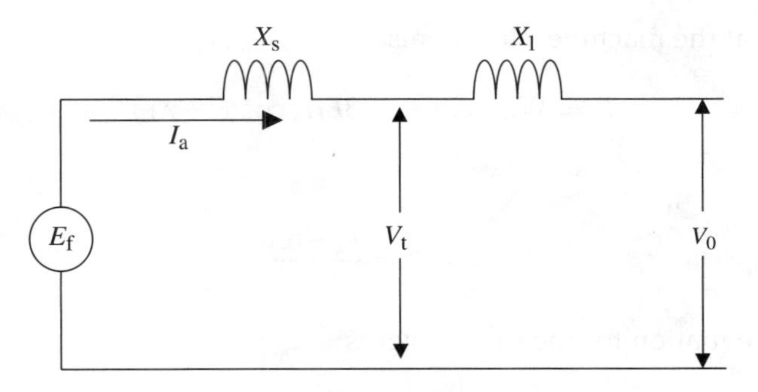

FIGURE 12.48
Equivalent circuit of the system in Figure 12.47.

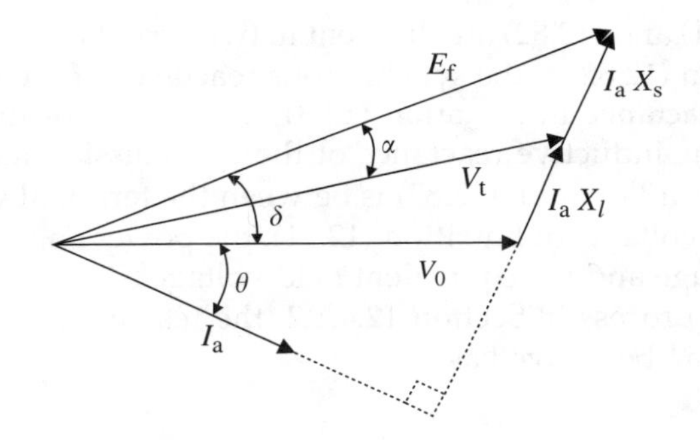

FIGURE 12.49
Phasor diagram of Equation (12.87).

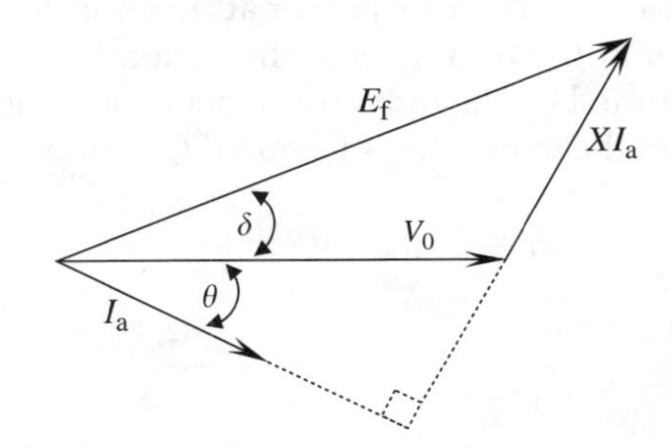

FIGURE 12.50
Simplified phasor diagram of Equation (12.87).

$$\overline{E}_f = \overline{V}_0 + \overline{I}_a(\overline{X}_s + \overline{X}_l) = \overline{V}_0 + \overline{I}_a\overline{X} \tag{12.88}$$

where $\overline{X} = \overline{X}_s + \overline{X}_l$. The simplified phasor diagram is shown in Figure 12.50.

The real and reactive power equations can be developed similar to the process in Sections 12.4.1.1 and 12.4.1.2. Because we are ignoring the resistance of the transmission line, the real power at the infinite bus is the same as the real power at the machine's terminals.

$$P = 3V_0 I_a \cos\theta = 3E_f I_a \cos(\theta + \delta) \tag{12.89}$$

where

$$I_a \cos\theta = \frac{E_f \sin\delta}{X} \tag{12.90}$$

Hence, the equation for the real power is

$$P = \frac{3V_0 E_f}{X} \sin\delta \tag{12.91}$$

Equations (12.91) and (12.82) are different in two aspects: The inductive reactance in Equation (12.82) is the synchronous reactance of the machine, while the inductive reactance in Equation (12.91) is the sum of the synchronous reactance and the inductive reactance of the transmission line. In addition, the power angle in Equation (12.82) is between the terminal voltage and the equivalent field voltage. In Equation (12.91), the power angle is between the infinite bus voltage and the equivalent field voltage.

Similar to the process in Section 12.4.1.2, the reactive power delivered to infinite bus Q_0 can be derived as

$$Q_0 = 3V_0 I_a \sin\theta = \frac{3V_0}{X}(E_f \cos\delta - V_0) \tag{12.92}$$

As seen in Figure 12.47, the reactive power at the infinite bus is not the same as the reactive power at the terminals of the generator. The difference is the reactive power consumed by the inductive reactance of the transmission line. The reactive power at the generator's terminal Q_t can be computed as

$$Q_t = 3V_t I_a \sin(\theta + \delta - \alpha) \tag{12.93}$$

where

$$I_a X_s \sin(\theta + \delta - \alpha) = E_f \cos\alpha - V_t \tag{12.94}$$

Hence, the reactive power at the terminals of the generator is

$$Q_t = \frac{3V_t}{X_s}(E_f \cos\alpha - V_t) \tag{12.95}$$

EXAMPLE 12.16 A synchronous generator is connected to an infinite bus through a transmission line. The infinite bus voltage is 23 kV, the inductive reactance of the transmission line is 1 Ω, and the synchronous reactance of the machine is 2 Ω. When the excitation of the generator is adjusted so that the line-to-line equivalent field voltage is 28 kV, the generator delivers 100 MW of real power to the infinite bus. Compute the following:

a. The terminal voltage of the machine.
b. The reactive power delivered to the infinite bus.
c. The reactive power consumed by the transmission line.
d. The reactive power at the terminals of the generator.

Solution

a. The terminal voltage of the generator can be computed if we know the current in the transmission line. The current is a function of the phasor difference between the infinite bus voltage and the equivalent field voltage. From the power equation we can compute the angle of the equivalent field voltage.

$$P = \frac{3V_0 E_f}{X} \sin \delta$$

$$100 = \frac{23 \times 28}{(2+1)} \sin \delta$$

$$\delta = 27.76^\circ$$

The current in the transmission line can be computed by using the model in Figure 12.48.

$$\bar{I}_a = \frac{\bar{E}_f - \bar{V}_0}{\bar{X}} = \frac{(28/\sqrt{3})\angle 27.76^\circ - (23/\sqrt{3})\angle 0^\circ}{3\angle 90^\circ} = 2.53\angle -7.75^\circ \text{ kA}$$

$$\bar{V}_t = \bar{V}_0 + \bar{I}_a \bar{X}_l = \frac{23}{\sqrt{3}}\angle 0^\circ + (2.53\angle -7.75^\circ)1\angle 90^\circ = 13.85\angle 10.44^\circ \text{ kV}$$

The line-to-line terminal voltage is $\sqrt{3} \times 13.85 = 23.99$ kV

b. $$Q_0 = \frac{3V_0}{X}(E_f \cos\delta - V_0) = \frac{\sqrt{3} \times 23}{3}\left(\frac{28}{\sqrt{3}}\cos(27.76^\circ) - \frac{23}{\sqrt{3}}\right)$$

$$= 13.62 \text{ MVAr}$$

c. The reactive power consumed by the transmission line is

$$Q_l = 3I_a^2 X_l = 3 \times (2.53)^2 \times 1 = 19.25 \text{ MVAr}$$

d. The reactive power at the terminals of the generator can be computed by two methods. The first is the sum of the reactive powers at the infinite bus and the reactive power consumed by the load.

$$Q_t = Q_0 + Q_l = 32.87 \text{ MVAr}$$

The other method is to use Equation (12.95). But first, we need to calculate the angle α using the phasor diagram in Figure 12.49.

$$\alpha = \delta - \angle \overline{V}_t = 27.76^\circ - 10.44^\circ = 17.32^\circ$$

$$Q_t = \frac{3V_t}{X_s}(E_f \cos\alpha - V_t) = \frac{3 \times 13.85}{2}\left(\frac{28}{\sqrt{3}}\cos 17.32^\circ - 13.85\right)$$

$$= 32.87 \text{ MVAr}$$

EXAMPLE 12.17 A synchronous generator of a power plant is connected to an infinite bus through a transmission line. The infinite bus voltage is 23 kV, the inductive reactance of the transmission line is 1 Ω, and the synchronous reactance of the machine is 2 Ω. The reactive power measured at the machine's terminals is zero, and the machine delivers 1 kA to the infinite bus. Compute the following:

a. The equivalent field voltage.
b. The real power delivered to the infinite bus.
c. The real power at the terminals of the machine.

Solution

a. Since there is no reactive power at the terminals of the generator, the current I_a is in phase with the terminal voltage V_t. Using the equivalent circuit in Figure 12.48, the phasor diagram can be constructed as shown in Figure 12.51.

The equivalent field voltage can be computed using Equation (12.87).

$$\overline{E}_f = \overline{V}_t + \overline{I}_a \overline{X}_l$$

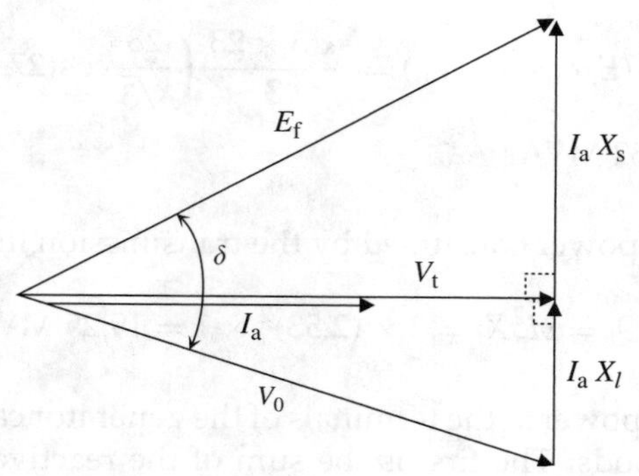

FIGURE 12.51
Phasor diagram of system in Example 12.17.

Since I_a is in phase with V_t, the voltage drop $I_a X_l$ is leading V_t by 90°. Hence we can use the Pythagorean theorem to compute the terminal voltage.

$$V_t = \sqrt{V_0^2 - (I_a X_l)^2} = \sqrt{\left(\frac{23}{\sqrt{3}}\right)^2 - (1 \times 1)^2} = 13.24 \text{ kV}$$

Similarly,

$$E_f = \sqrt{V_t^2 + (I_a X_s)^2} = \sqrt{(13.24)^2 + (1 \times 2)^2} = 13.4 \text{ kV}$$

The line-to-line equivalent excitation voltage is $\sqrt{3} \times 13.4 = 23.2$ kV.

b. The real power delivered to the infinite bus is

$$P = \frac{3V_0 E_f}{X} \sin \delta$$

where

$$\delta = \sin^{-1}\left(\frac{I_a X_l}{V_0}\right) + \sin^{-1}\left(\frac{I_a X_s}{E_f}\right)$$

$$= \sin^{-1}\left(\frac{1 \times 1}{23/\sqrt{3}}\right) + \sin^{-1}\left(\frac{1 \times 2}{13.4}\right) = 12.91°$$

$$P = \frac{3V_0 E_f}{X} \sin \delta = \frac{23 \times 23.2}{3} \sin 12.91° = 39.7 \text{ MW}$$

c. Since we ignore the resistance of the transmission line, the real power at the terminals of the machine is the same as the real power at the infinite bus.

12.4.3 Increasing Transmission Capacity

If we wish to increase the generated electrical power of the synchronous machine, we must increase the mechanical power of the turbine. In addition, the transmission system must be able to transmit the extra power. The power system operates securely when the capacity of the transmission system is higher than the generated electric power. The capacity of the power system can be viewed as the maximum power that can be transmitted by the system. In Equation (12.91), the system capacity is the maximum power.

$$P_{max} = \frac{3V_0 E_f}{X} \tag{12.96}$$

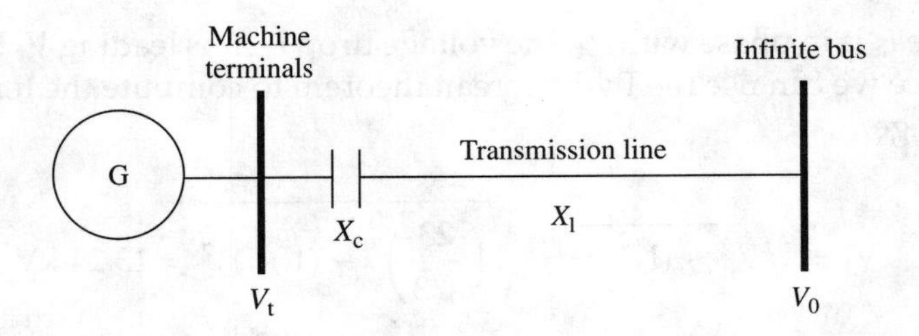

FIGURE 12.52
Capacitor in series with the transmission line.

The power plant cannot generate more than the transmission capacity P_{max}. However, it is possible to increase the transmission capacity if we increase E_f or reduce the total inductive reactance X. There are two common methods to reduce X: (1) inserting a capacitor in series with the transmission line and (2) using parallel transmission lines.

12.4.3.1 Increasing Transmission Capacity by Series Capacitor

A capacitor can be connected in series with the transmission line, as shown in Figure 12.52, to reduce the total inductive reactance of the system. Recall that without the capacitor, $X = X_s + X_l$. Now, with the capacitor, the total reactance is

$$\begin{aligned}\overline{X} &= \overline{X}_s + \overline{X}_l + \overline{X}_c = jX_s + jX_l - jX_c = j(X_s + X_l - X_c)\\ X &= X_s + X_l - X_c\end{aligned} \tag{12.97}$$

where X_c is the capacitive reactance of the capacitor.

EXAMPLE 12.18 A synchronous generator is connected to an infinite bus through a transmission line. The infinite bus voltage is 230 kV and the equivalent field voltage of the machine is 210 kV. The transmission line inductive reactance is 10 Ω, and the synchronous reactance of the machine is 2 Ω.

1. Compute the capacity of the system.
2. If a capacitor is connected in series with the transmission line to increase the transmission capacity by 25%, compute its reactance.

Solution

1. The system capacity is

$$P_{max} = \frac{3V_0E_f}{X} = \frac{230 \times 210}{(10+2)} = 4.025 \text{ GW}$$

2. The system capacity increases by using a capacitor.

$$1.25 \times 4025 = \frac{3V_0E_f}{X_{new}} = \frac{230 \times 210}{(10 + 2 - X_c)}$$

Hence,

$$X_c = 2.4\ \Omega$$

This is a large value of capacitive reactance, almost 1.1 mF. Since it carries large currents at high voltages, the capacitor is very large in size and can only be installed in major substations.

12.4.3.2 Increasing Transmitted Capacity by Using Parallel Lines

Parallel lines are often used to increase the transmission capacity of the system. It is intuitive that two lines can transmit more power than any one line.

Figure 12.53 shows a schematic of a system with two parallel lines (TL1 and TL2), and Figure 12.54 shows a photo of a transmission tower with two parallel circuits (each circuit consists of three-phase lines).

The equivalent circuit of the system is shown in Figure 12.55. In this system, the total reactance between the infinite bus and the equivalent field voltage is

$$X = X_s + \frac{X_{l1}X_{l2}}{X_{l1} + X_{l2}} \tag{12.98}$$

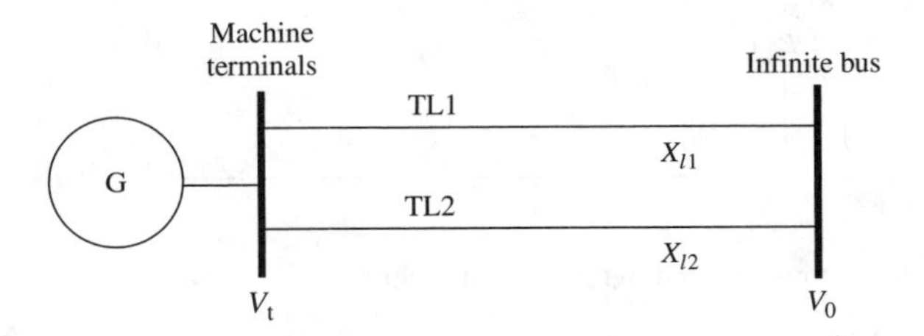

FIGURE 12.53
Two transmission lines in parallel.

EXAMPLE 12.19 A synchronous generator is connected to an infinite bus through a transmission line. The infinite bus voltage is 230 kV and the equivalent field voltage of the machine is 210 kV. The transmission line inductive reactance is 10 Ω and the synchronous reactance of the machine is 2 Ω.

1. Compute the capacity of the system.
2. Compute the capacity of the system if a second transmission line with a 10 Ω inductive reactance is connected in parallel with the first line.

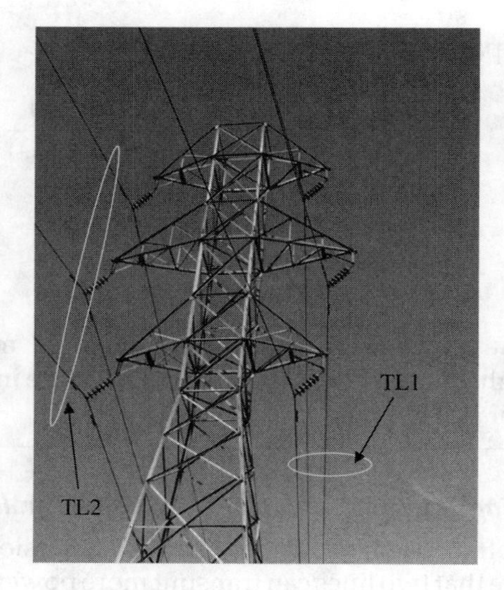

FIGURE 12.54
A tower with two transmission lines (the line is a three-phase circuit).

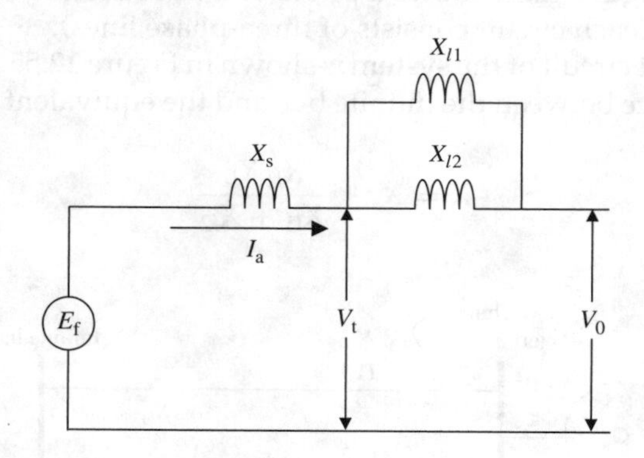

FIGURE 12.55
Equivalent circuit of power system with two parallel lines.

Solution

1. The system capacity is

$$P_{\max} = \frac{3V_0 E_f}{X} = \frac{230 \times 210}{(10 + 2)} = 4.025 \text{ GW}$$

2. After the second line is added, the system capacity is

$$P_{\max} = \frac{3V_0 E_f}{X_{\text{new}}} = \frac{3V_0 E_f}{(X_s + X_{l1} X_{l2} / (X_{l1} + X_{l2}))} = \frac{230 \times 210}{2 + \frac{10}{2}} = 6.9 \text{ GW}$$

The capacity of the system increases by about 71% when two parallel lines are used. Can you tell why the capacity is not doubled? Also, repeat the example for three parallel lines.

12.5 Synchronous Motor

The synchronous machine can be used as a generator or motor. As a motor, the synchronous machine is used in applications that demand constant and precise speeds, such as electric clocks, movie cameras, traction wheels, uniform actuation, gate and governor control, constant feed industrial processes, and many others. The synchronous motor is also used to compensate the reactive power of heavy industrial loads.

For small synchronous motors, the rotor is often made of a ferrite permanent magnet. In newer synchronous motors, the rotor field is generated by a strong permanent magnet made out of rare earth material such as samarium-cobalt. These permanent magnet motors can be as powerful as 100 hp. One great advantage of these newer motors is its high power/volume ratio, which make them among the smallest machines. Furthermore, the rare earth permanent magnet cannot be easily demagnetized like the ferrite material, so it can last for a long time. Because of these advantages, the rare earth permanent magnet synchronous motor is used to actuate newer commercial airplanes using fly-by-wire designs. The motor is also used in ground transportation; the new hybrid electric vehicles in Figure 12.56 use this type of motor, which is normally in the range of 10 to 70 hp.

For large machines, especially the ones used to regulate the reactive power, the rotor is made of an electric magnet that is excited externally by a separate dc source. This is exactly the same rotor used for the synchronous generator; a photo of its slip ring arrangement is shown in Figure 12.33.

The three-phase currents of the stator produce rotating magnetic field at synchronous speed n_s exactly like the induction machine. The magnetic field

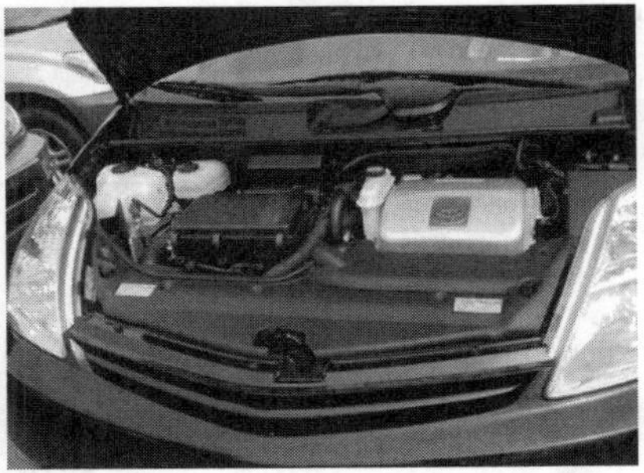

FIGURE 12.56
(see color insert following Page 208) Hybrid electric vehicle uses permanent magnet synchronous motor.

of the rotor, which is stationary with respect to the rotor, aligns itself with the rotating field thus spinning the rotor at the synchronous speed. This is similar to two magnets: when one is moving the other is following its motion at the same speed. The equivalent circuit of the synchronous generator, shown in Figure 12.39, is also applicable to the synchronous motor as shown in Figure 12.57. The only differences are: (1) the stator is excited by a three-phase source and (2) the armature current of the motor is flowing in the reverse direction when compared with the armature current of the generator. The equivalent circuit of the synchronous motor can be further simplified by ignoring the resistance of the armature winding. This is justified for large machines where the stator windings are made of large cross-section wires. The simplified equivalent circuit of the synchronous motor is shown in Figure 12.58.

The equation of the synchronous motor can be written as

$$\bar{V}_t = \bar{E}_f + \bar{I}_a X_s \tag{12.99}$$

The phasor diagram of Equation (12.99) is shown in Figure 12.59. Note that for the synchronous motor, E_f lags V_t; this is opposite to the synchronous generator. Note that the power factor angle θ is the angle between the armature current and the terminal voltage of the motor.

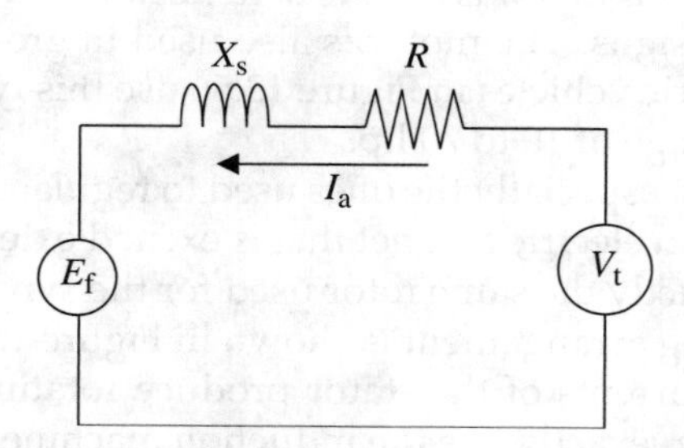

FIGURE 12.57
Equivalent circuit of a synchronous motor.

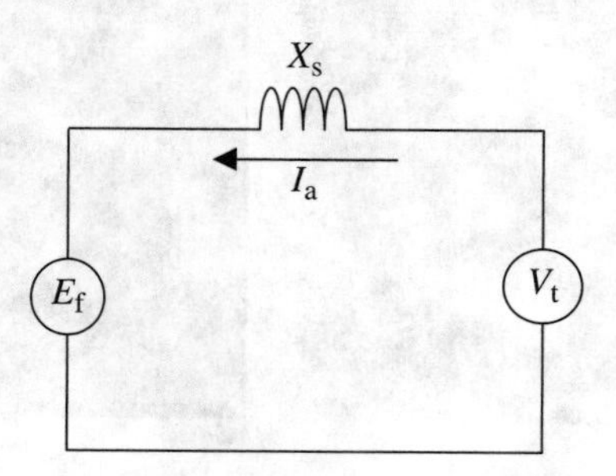

FIGURE 12.58
Simplified equivalent circuit of a synchronous motor.

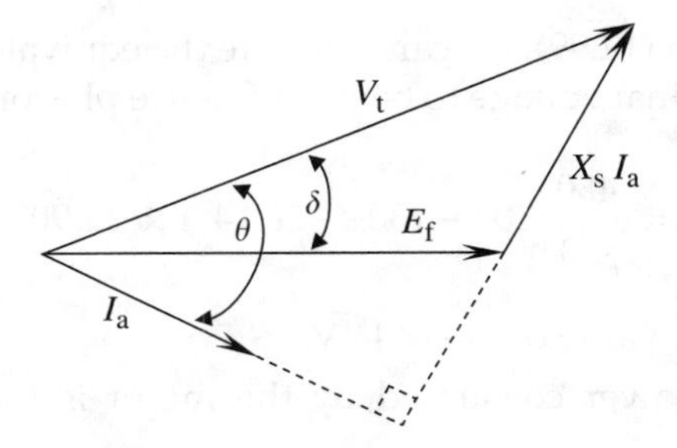

FIGURE 12.59
Phasor diagram of a synchronous motor.

EXAMPLE 12.20 A 30 hp, 4-pole, 480 V, 60 Hz, three-phase, wye-connected synchronous motor has a synchronous reactance of 2 Ω. At full load, the armature current of the motor is 30 A and is lagging the terminal voltage. Ignore all losses and compute the following:

a. The torque of the motor.
b. The equivalent field voltage.
c. The reactive power consumed by the motor.

Solution

a. We can compute the torque of the motor if we know the output power and the speed of the motor. The output power at full load is 30 hp.

$$P_{out} = \frac{30}{1.34} = 22.39 \text{ kW}$$

$$n_s = 120\frac{f}{p} = 120\frac{60}{4} = 1800 \text{ rpm}$$

$$T = \frac{P_{out}}{\omega_s} = \frac{22390}{(2\pi/60) \times 1800} = 118.78 \text{ Nm}$$

b. Using the circuit in Figure 12.58, we can compute the equivalent field voltage. But first, we need to compute the phase angle of the current. Since the machine losses are ignored, the output power is equal to the input power, hence

$$P_{in} = P_{out} = \sqrt{3} V I_a \cos\theta$$

$$22390 = \sqrt{3}\, 480 \times 30 \times \cos\theta$$

$$\theta = 26.14°$$

Using Equation (12.99), we can compute the equivalent field voltage. Select the terminal voltage to be the reference phasor of the equation.

$$\overline{E}_f = \overline{V}_t - \overline{I}_a \overline{X}_s = \frac{480}{\sqrt{3}} \angle 0^\circ - (30\angle -26.14^\circ) \times 2\angle 90^\circ = 256.4\angle -12.13^\circ \text{ V}$$

The line-to-line value of E_f is 444 V.

c. The reactive power consumed by the motor is the input reactive power.

$$Q_{in} = \sqrt{3} V I_a \sin\theta = \sqrt{3}\, 480 \times 30 \times \sin(26.14) = 10.988 \text{ kVAr}$$

12.5.1 Reactive Power Control

The main relationship of the synchronous motor is given by Equation (12.99). The magnitude of V_t is often fixed, but E_f is controlled by adjusting the dc field current in the rotor circuit. Hence, the armature current is dependent on the value of E_f.

There are three possible scenarios for the current in Equation (12.99): lagging, leading, and in-phase with the terminal voltage. The three possible phasor diagrams are shown in Figure 12.59 and Figure 12.60. In Figure 12.59, E_f is adjusted so that $E_f \cos\delta < V_t$. In this case, the current I_a lags V_t, and the power factor measured at the terminals of the motor is lagging.

In Figure 12.60(a), E_f is increased so that $E_f \cos\delta > V_t$. In this case, the current leads the terminal voltage and the power factor measured at the terminals of the motor is leading.

In Figure 12.60(b), E_f is adjusted so that $E_f \cos\delta = V_t$. In this case, the current is in-phase with the terminal voltage and the power factor measured at the terminals of the motor is unity.

The reactive power Q at the terminals of the motor can be computed as

$$Q = 3 V_t I_a \sin\theta \tag{12.100}$$

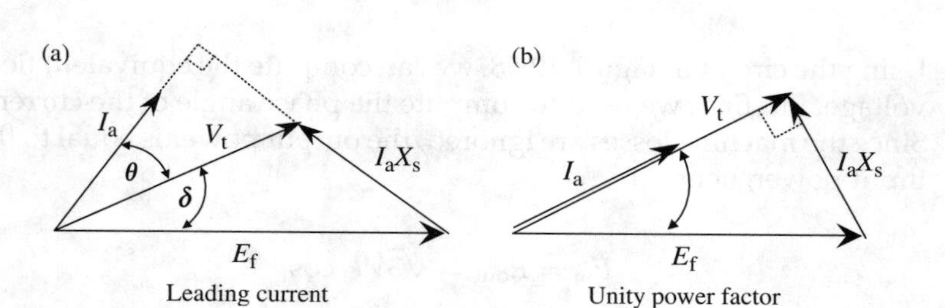

FIGURE 12.60
Phasor diagram of a synchronous motor.

where V_t is a phase quantity. By examining the phasor diagrams of the synchronous motor, we can show that

$$I_a X_s \sin\theta = E_f \cos\delta - V_t \quad (12.101)$$

Substituting the current in Equation (12.101) into (12.100) yields

$$Q = \frac{3V_t}{X_s}[E_f \cos\delta - V_t] \quad (12.102)$$

The reactive power at the terminals of the motor is leading when the magnitude of Q in Equation (12.102) is positive. This means that the motor is delivering reactive power to the ac source. When Q is negative, the reactive power is lagging and the motor consumes reactive power.

EXAMPLE 12.21 A 5 kV, 60 Hz synchronous motor has a synchronous reactance of 4 Ω. The motor runs unloaded and its equivalent field voltage is adjusted to 6 kV. Ignore all losses and compute the reactive power at the terminals of the motor.

Solution

Since the motor is unloaded, the real power is zero. Hence, the current I_a is either leading or lagging V_t by 90°. However, since the power angle is zero and $E_f > V_t$, the current must be leading the terminal voltage as shown in the phasor diagram in Figure 12.61.

Using Equation (12.102), we can compute the reactive power.

$$Q = \frac{3V_t}{X_s}[E_f \cos\delta - V_t] = \frac{3 \times (5000/\sqrt{3})}{4}\left[\frac{6000}{\sqrt{3}} - \frac{5000}{\sqrt{3}}\right] = 1.25 \text{ MVAr}$$

The positive value of the reactive power means the synchronous motor is delivering reactive power to the ac source.

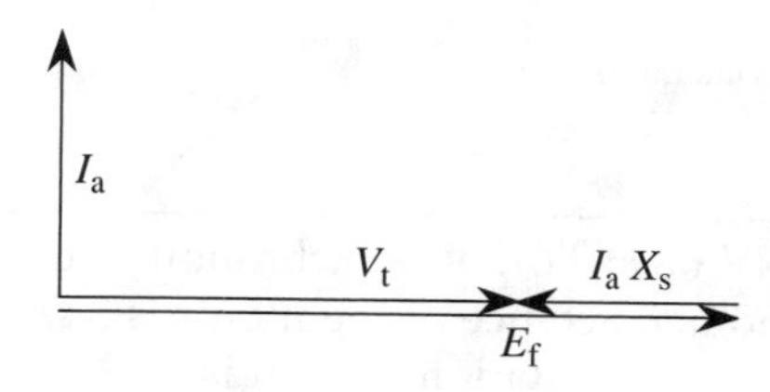

FIGURE 12.61
Phasor diagram of a synchronous motor running at no load.

12.5.2 Real Power

The input power to the synchronous motor from the ac source is

$$P = 3V_t I_a \cos\theta \qquad (12.103)$$

where V_t is a phase quantity. By examining any phasor diagram in Figures 12.59 and 12.60, we can show that

$$I_a X_s \cos\theta = E_f \sin\delta \qquad (12.104)$$

Substituting I_a of Equation (12.104) into Equation (12.103) yields

$$P = 3\frac{V_t E_f}{X_s}\sin\delta \qquad (12.105)$$

Since the synchronous machine rotates at a synchronous speed, we can write the developed torque equation as

$$T = \frac{P}{\omega_s} = \frac{3V_t E_f}{\omega_s X_s}\sin\delta \qquad (12.106)$$

Figure 12.62 shows the torque curve representing Equation (12.106). The maximum torque of the motor occurs at $\delta = 90°$.

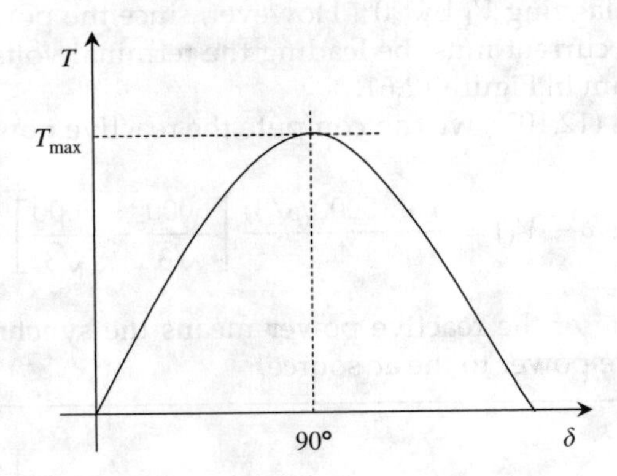

FIGURE 12.62
Torque curve of a synchronous motor.

EXAMPLE 12.22 A 3 kV, 60 Hz, 6-pole synchronous motor is driving a load of 6 kNm. The synchronous reactance of the motor is 5 Ω. The equivalent field voltage as a function of the excitation current is

$$E_f = 40 I_f$$

Compute the following:

a. The power consumed by the load.
b. The minimum excitation current of the machine.
c. The reactive power at the motor terminals.

Solution

a. The power of the motor is

$$P = T\omega_s$$

where

$$\omega_s = \frac{2\pi}{60}\left(120\frac{60}{6}\right) = 125.66 \text{ rad/s}$$

$$P = T\omega_s = 6000 \times 125.66 = 754 \text{ kW}$$

b. The minimum excitation current is the value that makes the maximum torque of the motor in Equation (12.106) equal the load torque. Any excitation current less than this value will produce torque less than that needed by the load.

$$T_l = T_{max} = \frac{3V_t E_{f\,min}}{\omega_s X_s}$$

$$6000 = \frac{3(3000/\sqrt{3})E_{f\,min}}{125.66 \times 5}$$

$$E_{f\,min} = 725.5 \text{ V}$$

Hence, the minimum excitation current is

$$E_{f\,min} = 40 I_{f\,min}$$

$$I_{f\,min} = \frac{E_{f\,min}}{40} = 18.14 \text{ A}$$

c. The reactive power at the terminals of the motor is

$$Q = \frac{3V_t}{X_s}[E_f \cos\delta - V_t]$$

Since the machine is running under minimum excitation, the power angle is 90°.

$$Q = \frac{-3V_t^2}{X_s} = -1.8 \text{ MVAr}$$

The negative sign of the reactive power means the machine is consuming reactive power from the source.

12.6 Stepper Motor

The position control devices have two main features: (1) the rotor position is controlled and (2) the position is held to the desired angle for as long as needed. These features are essential in several applications that demand fine movements such as hard disk drives, printers, plotters, fax machines, medical equipment, laser guiding systems, robots, and actuators. One major drawback for the motors discussed so far is their lack of position control. However, the stepper motors can achieve fine position control with a step resolution of one degree.

Stepper motors come in three basic types: (1) variable reluctance, (2) permanent magnet, and (3) hybrid. The stator of these three types is similar to that for the induction or synchronous machine. However, for a stepper motor, any number of phases could be used depending on the design preference and required resolution. Figure 12.63 shows a stator with four phases. The stator windings are not imbedded inside the cylindrical stator, but are wrapped around salient poles extended inside the airgap.

When any of the stator windings (a–a', b–b', c–c', or d–d') is excited, the flux of the winding is moving along the axis between the opposing pole of the same phase (e.g., a and a') as shown on the right side in Figure 12.63. The direction of the flux is dependent on the polarity of the applied voltage; that is, the flow of the current determines the direction of the flux in the airgap.

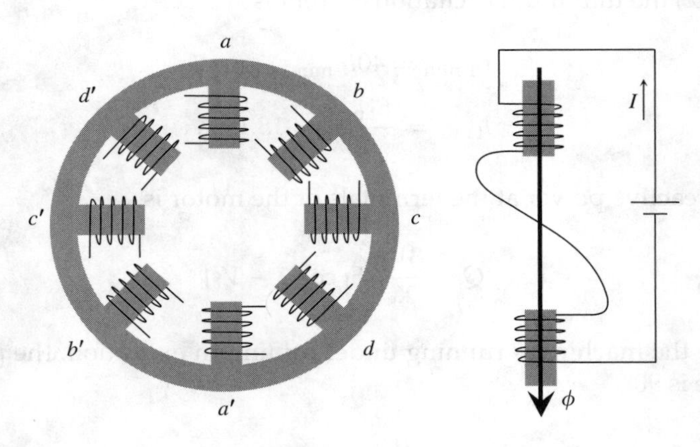

FIGURE 12.63
Stator of the stepper motor.

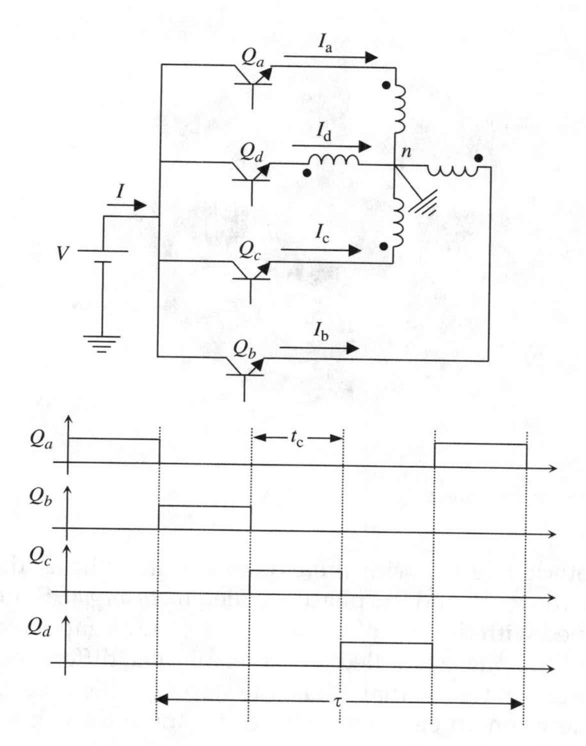

FIGURE 12.64
Switching of stepper motor winding.

The phases of the stator are switched by a power electronic converter; the most common one is the dc/ac converter such as the one shown in Figure 12.64. Each transistor switches one phase, and all phases are connected in a star configuration with the common neutral point providing the return path for all currents. The switching pattern of the transistors is a design parameter depending on the resolution of the step angle. In case one phase is switched at the time, the typical switching pattern is given at the bottom of Figure 12.64. The period of the switching cycle τ is the sum of the conduction period t_c of all phases.

12.6.1 Variable Reluctance Stepper Motor (VRSM)

This is the oldest type of stepper motor. Unlike the synchronous or induction motors, the rotor of the VRSM is neither a magnet nor has any windings, but is made out of a multi-tooth iron material. The stepper motors are described by their number of phases in the stator windings and the number of poles (or the number of teeth) in its rotor. Figure 12.65 shows a 4-phase, 6-tooth stepper motor.

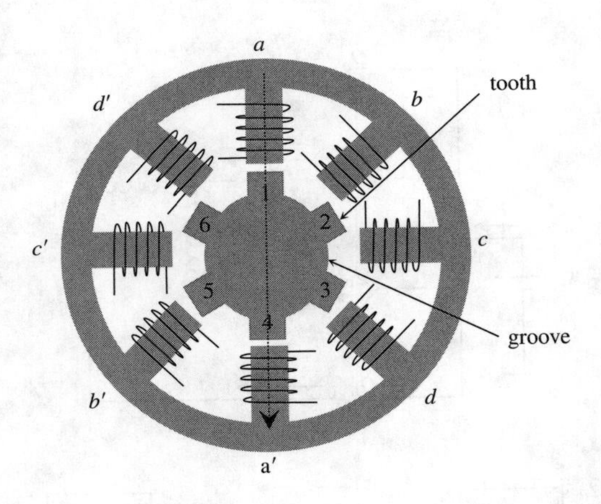

FIGURE 12.65
Variable reluctance stepper motor.

When a tooth is aligned with the axis of a stator phase, the airgap distance between the tooth and the phase is called *tooth airgap*. Similarly, when a groove is aligned with the axis of a stator phase, the airgap distance between the groove and the phase is called *groove airgap*. The difference between the two airgaps causes a torque that aligns the stepper. This is because the force of attraction between an energized phase and the rotor is governed by the electromagnetic force equation.

$$F = \frac{\mu_o}{4\pi d} I^2 \tag{12.107}$$

where μ_o is the absolute permeability (1.257×10^{-6} H/m), d is the airgap distance between the energized phase of the stator and the rotor, and I is the current of the winding. Since the tooth airgap is smaller than the groove airgap, the force of attraction of the tooth is higher than that for the groove. Hence, the tooth is always aligned with the energized pole. (in the figure, phase a–a' is energized and teeth 1 and 4 are aligned along the axis of phase a.)

If we disconnect phase a and energize phase b, teeth 2 and 5 move under poles b–b', thus rotating one step in the counterclockwise direction. The step angle δ is a function of the number of phases in the stator and the number of teeth in the rotor.

$$\delta = \frac{360}{N_{\text{ph}} N_{\text{t}}} \tag{12.108}$$

where N_{ph} is the number of stator phases and N_{t} is the number of teeth in the rotor. For a variable reluctance stepper, the rotor pole pair pp consists of one tooth and one groove. Thus, the relationship between the number of rotor poles p and number of rotor teeth is

$$\begin{aligned} \text{pp} &= N_\text{t} \\ p &= 2N_\text{t} \end{aligned} \tag{12.109}$$

Equation (12.108) can be written as a function of the rotor poles.

$$\delta = \frac{720}{N_{\text{ph}}p} \tag{12.110}$$

In the 4-phase, 6-tooth stepper motor in Figure 12.65, the number of rotor poles is 12 and the step angle is 15°. A finer step angle can be achieved if we increase the number of stator phases, increase the number of teeth, or energize more than one phase at a time.

The stepper motors are often rated by their number of steps per revolution s, which can be as high as 400 steps per revolution (spr).

$$s = \frac{360}{\delta} = \frac{N_{\text{ph}}p}{2} \tag{12.111}$$

To rotate the motor continuously, the switching of the phases must also be continuous. The motor speed n can be calculated in a similar process to that in Equation (12.9).

$$n = \frac{120f}{p} = \frac{120}{\tau p} \tag{12.112}$$

where τ is the switching period in seconds. The unit of n in Equation (12.112) is rpm.

The rotating stepper motor can also be rated by its step rate s_r, which is the number of steps per second (sps). Hence,

$$s_r = \frac{sn}{60} = \frac{N_{\text{ph}}}{\tau} \tag{12.113}$$

EXAMPLE 12.23 A 6-phase variable reluctance stepper motor has a step angle of 10°. Compute the following:

a. The number of teeth in the rotor.
b. If the conduction period of each transistor is 2 ms, compute the continuous speed of the motor.

Solution

a. $\delta = \dfrac{360}{N_{\text{ph}}N_t}$

$N_t = \dfrac{360}{10N_{\text{ph}}} = \dfrac{360}{60} = 6$ teeth. Hence, the motor is a 12-pole machine.

b. $\tau = t_c N_{\text{ph}} = 2 \times 6 = 12$ ms

$$n = \frac{120}{\tau p} = \frac{120}{0.012 \times 12} = 833.33 \text{ rpm}$$

12.6.2 Permanent Magnet Stepper Motor (PMSM)

Because its rotor has no magnet, the variable reluctance stepper develops little torque. Therefore, it is only used in applications such as disk drives, printers, or low-torque actuations. For higher torque applications, the permanent magnet stepper is often selected. The rotor of the PMSM is a cylinder with several permanent magnets as shown in Figure 12.66. The magnet provides better holding force under the energized stator as compared to the VRSM.

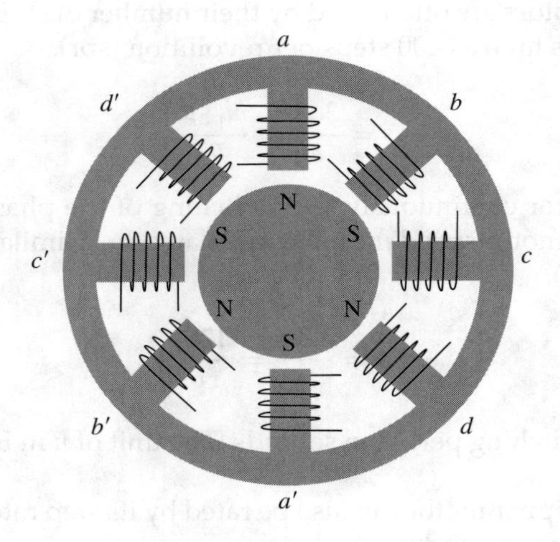

FIGURE 12.66
Permanent magnet stepper motor.

One common characteristic of the permanent magnet motors is that they tend to 'cog' as you turn the rotor manually without energizing the stator. This is because the rotor magnetic poles are attracted to the salient iron poles of the stator. This force of attraction is known as the *detent force*.

Equations (12.110) to (12.113) for the VRPM are also applicable to the PMSM. The only difference is that the PMSM has poles instead of teeth.

EXAMPLE 12.24 A 2-phase, 4-pole permanent magnet stepper motor is driven by a 100 Hz dc/ac converter. Compute the following:

a. The speed of the motor.
b. The conduction period of each transistor in the converter.

Solution

a. $n = \frac{120f}{p} = \frac{120 \times 100}{4} = 3000$ rpm

b. $\tau = t_c N_{ph}$

$$t_c = \frac{\tau}{N_{ph}} = \frac{1}{fN_{ph}} = \frac{1}{100 \times 2} = 5 \text{ ms}$$

12.6.3 Hybrid Stepper Motor (HSM)

The hybrid stepper motor provides the best resolution and the highest torque among all other types of stepper motors. The HSM can achieve up to 400 steps per revolution. The rotor of the HSM is a multi-tooth rotor like the VRSM, but the teeth are also magnets. This structure, which is called salient-pole rotor, provides a better path for the flux between the rotor and stator poles as compared to the PMSM. Thus, stronger attraction force is developed.

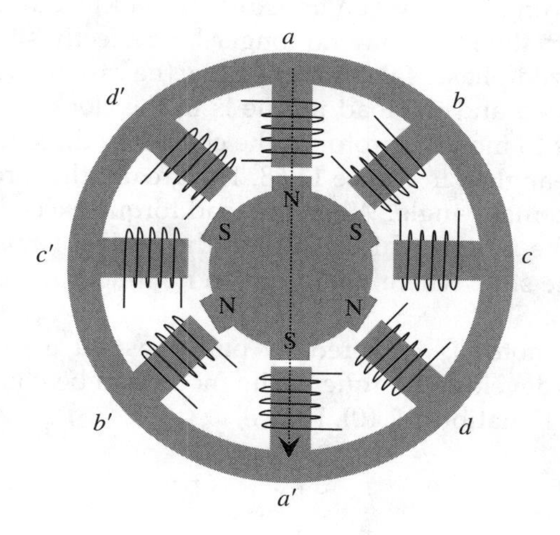

FIGURE 12.67
Hybrid stepper motor.

12.6.4 Holding State of the Stepper Motor

When the stepper motor is unloaded, the rotor is perfectly aligned with the energized phase as shown in Figures 12.65 to 12.67. In this case, there is no angle between the axes of the energized phase and the attracted poles of the rotor. If no other switching action is made, the rotor will maintain its holding position.

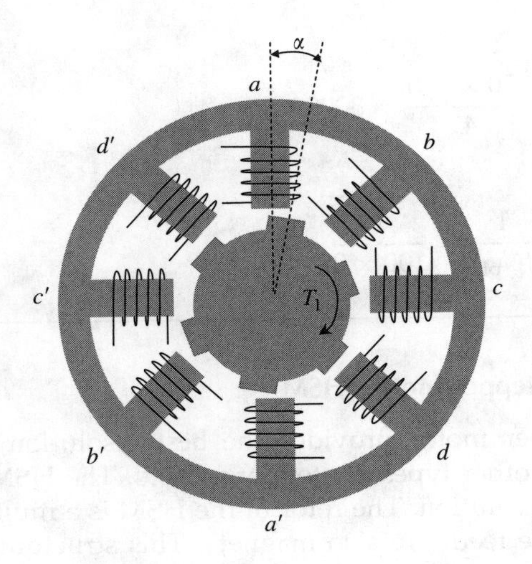

FIGURE 12.68
Displacement angle of the stepper motor.

During the holding state, when an external load torque T_l is applied on the stepper motor, the rotor may no longer be perfectly situated under the axis of the energized phase. Take, for example, the case in Figure 12.68 when phase *a–a′* is excited and the load torque is in the clockwise direction. The load torque tends to move the rotor in the clockwise direction, thus causing the displacement angle α in Figure 12.68. The greater the torque applied, the larger the displacement angle. When the load torque exceeds the maximum static torque of the motor, which is called the *holding torque* T_{h}, the motor enters an unstable state where the rotor can no longer maintain its holding state.

If the stepper motor is powered by one phase at a time, as given in Figure 12.64, the developed torque of the motor can be computed by using Faraday's rule in Equation (12.10). Hence,

$$\begin{aligned} T_{\mathrm{d}} &\sim \text{force} \sim I \\ T_{\mathrm{d}} &= KI \end{aligned} \tag{12.114}$$

where K is the torque constant that depends upon the motor geometry and the property of its windings. The unit of K is volt second. I is the current of the winding. When the axes of the stator windings and the attracted poles are not perfectly aligned, the torque equation can be generalized as

$$T_{\mathrm{d}} = KI \sin\theta \tag{12.115}$$

where θ is the electrical displacement angle. Equation (12.115) is plotted in Figure 12.69, and it shows that the maximum holding torque T_{h} occurs at

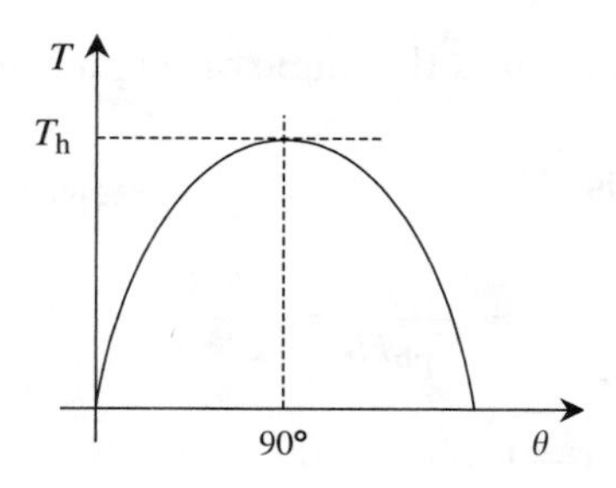

FIGURE 12.69
Torque curve of the stepper motor.

$\theta = 90°$. Hence,

$$T_h = KI \tag{12.116}$$

The mechanical displacement angle of the rotor shaft α is equal to θ only if the motor is a 2-pole machine. For any other number of poles, the relationship between the mechanical and electrical displacement angles is

$$\alpha = \frac{\theta}{p/2} = \frac{\theta}{N_t} \tag{12.117}$$

EXAMPLE 12.25 A 10-teeth, 2-phase variable reluctance stepper motor is loaded by 10 mNm torque. The torque constant of the motor is 3 Vs and the holding torque is 30 mNm. Compute the following:

a. The current of the stator windings.
b. The mechanical displacement angle.
c. The displacement error.
d. The current that reduces the displacement angle by 50%.

Solution

a. The holding torque is the maximum torque in Equation (12.115).

$$T_h = KI; \qquad I = \frac{T_h}{K} = \frac{30}{3} = 10 \text{ mA}$$

b. The mechanical angle of the rotor can be computed by Equation (12.115).

$$T_d = KI\sin(N_t\alpha)$$
$$10 = 30\sin(10\alpha)$$

Hence, $\alpha = 2°$.

c. The displacement error is the angle of displacement with respect to the step angle.

The step angle is

$$\delta = \frac{360}{N_{\text{ph}} N_{\text{t}}} = \frac{360}{2 \times 10} = 18^\circ$$

Displacement error $= (2/18) \times 100 = 11.11\%$.

d. For $\alpha = 1^\circ$,

$$T_d = K I \sin(N_{\text{t}} \alpha)$$
$$10 = 3I \sin(10)$$

Hence, $I = 19.2$ mA.

12.6.5 Rotating of Stepper Motor

During the holding state, the stepper motor must develop enough torque to compensate for the load torque. When the motor rotates, the developed torque of the motor must compensate for three components: (1) load torque T_{l}, (2) friction and drag torque T_{f}, and (3) acceleration or deceleration torque T_j.

$$T_{\text{d}} = T_{\text{l}} + T_{\text{f}} + T_j \tag{12.118}$$

The acceleration and deceleration torque is also known as the inertia torque.

$$T_j = J\frac{\text{d}\omega}{\text{d}t} = \frac{2\pi}{60} J \frac{\text{d}n}{\text{d}t} \tag{12.119}$$

where J is the inertia of the rotor and load in Kg m^2.

EXAMPLE 12.26 A 6-phase variable reluctance stepper motor with a step angle of 7.5° and a step rate of 200 sps is driving an inertia load. The inertia of the load plus the rotor is 1 mg m^2, and the friction torque is 15 mNm. Compute the following:

a. The number of teeth of the rotor.
b. The switching period τ.
c. The speed of the motor.
d. The input power at steady state (ignore electrical losses).
e. The inertia torque to accelerate the motor at 40 rad/s^2.

Solution

a. $\delta = \dfrac{720}{N_{ph}p}$

$$p = \frac{720}{\delta N_{ph}} = \frac{720}{7.5 \times 6} = 16 \text{ poles}$$

Hence, the number of teeth is 8.

b. $s_r = \dfrac{N_{ph}}{\tau}$; $\quad \tau = \dfrac{6}{200} = 30$ ms

c. $n = \dfrac{120}{\tau p} = \dfrac{120}{30 \times 16} = 250$ rpm

d. Since the load is just inertia, the developed torque of the motor at the steady state compensates for only the friction torque. Hence, the input power is

$$P = T_f\omega = 15 \times \frac{2\pi}{60} 250 = 392.7 \text{ mW}$$

e. $T_j = J\dfrac{d\omega}{dt} = 1 \times 40 = 0.04$ mNm

Exercise

12.1. The output power of a 4-pole, three-phase induction motor is 100 hp. The motor operates at a slip of 0.02 and efficiency of 90%. Compute the following:

a. The input power of the motor in kilowatts.
b. The shaft torque (output torque).

12.2. The synchronous speed of a 60 Hz, three-phase induction motor is 1200 rpm. What is the synchronous speed of the motor if it is used in a 50 Hz system?

12.3. A 100 hp, 60 Hz, three-phase induction motor has a slip of 0.03 at full load. Compute the efficiency of the motor at full load when the friction and windage losses are 900 W, the stator core loss is 4200 W, and the stator copper loss is 2700 W.

12.4. A three-phase induction motor is rated at 50 hp, 480 V, 60 Hz, and 1150 rpm. If the motor operates at full load, compute the following:

a. The developed torque.
b. The frequency of the rotor current.

12.5. A three-phase, 480 V, 60 Hz, 12-pole induction motor has the following parameters:

$$R_1 = 1.0\ \Omega; \quad R_2' = 0.5\ \Omega; \quad X_{eq} = 10\ \Omega; \quad X_c = 100\ \Omega; \quad R_c = 800\ \Omega$$

Calculate the following:

a. The slip at maximum torque.
b. The current at maximum torque.
c. The speed at maximum torque.
d. The maximum torque.

12.6. A 480 V, 3-phase, 6-pole, slip-ring, Y-connected induction motor has $R_1 = R_2' = 0.1\ \Omega$ and $X_{eq} = X_1 + X_2' = 0.5\ \Omega$. The motor slip at full load is 3%, and its efficiency is 90%. Calculate the following:

a. The starting current. (You may ignore the magnetizing current.)
b. The starting torque.
c. The maximum torque.
d. The value of the resistance that should be added to the rotor circuit to reduce the starting current by 50%.
e. What is the starting torque of case (d)?
f. Calculate the value of the resistance that should be added to the rotor circuit to increase the starting torque to the maximum value.
g. What is the starting current of case (f)?

12.7. A three-phase, 4-pole, 480 V, 60 Hz, Y-connected slip-ring induction motor has the following parameters:

$$R_1 = R_2' = 0.5\ \Omega; \qquad X_1 + X_2' = 5\ \Omega$$

Computer the following:

a. The starting torque.
b. The inserted rotor resistance that increases the starting torque by a factor of 4.

12.8. A three-phase, 2.2 kV, 60 Hz, Y-connected slip-ring induction motor has the following parameters:

$$R_1 = R_2' = 0.2\ \Omega; \qquad X_1 + X_2' = 1.5\ \Omega$$

The motor is running at 570 rpm. Ignore the rotational losses and calculate the full load torque.

12.9. A 480 V, 60 Hz, three-phase induction motor has the following parameters:

$$R_1 = R_2' = 0.3\ \Omega; \qquad X_{eq} = 1.0\ \Omega; \qquad X_c = 600\ \Omega$$

At full load, the motor speed is 1120 rpm, the rotational loss is 400 W, and the core loss is 1 kW. Calculate the following:

a. The motor slip.
b. The developed torque at full load.
c. The developed power in horsepower.

d. The rotor current.

e. The copper losses.

f. The input power.

g. The reactive power consumed by the motor.

h. The power factor of the motor.

12.10. The speed of a 12-pole, 480 V three-phase induction motor at full load is 560 rpm. The motor has the following parameters:

$$R_1 = 0.1\ \Omega; \qquad R_2' = 0.5\ \Omega; \qquad X_1 + X_2' = 5\ \Omega$$

a. Compute the developed torque.

b. While the torque is unchanged, the voltage is changed to reduce the speed of the motor to 520 rpm. Compute the new voltage.

12.11. A train is driven by a linear induction motor at 80 km/h when the frequency of the primary windings is 15 Hz. The frontal area of the train is 25 m^2 and its coefficient of drag is 0.9. The friction coefficient between the wheels of the train and the track is 0.1, and the weight of the train is 500,000 kg. The pole pitch of the vehicle is 3 m. Compute the following:

a. The speed of the thrust force.

b. The slip of the motor.

c. The developed force of the motor at steady state.

d. The developed power in horsepower.

12.12. A train with a linear induction motor is traveling uphill at 80 km/h when the frequency of its primary windings is 15 Hz. The inclination of the hill is 5°. The frontal area of the train is 25 m^2 and its coefficient of drag is 0.9. The friction coefficient between the wheels of the train and the track is 0.1, and the weight of the train is 500,000 kg. The pole pitch of the vehicle is 3 m. Compute the following:

a. The developed force of the motor at steady state.

b. The developed power in horsepower.

12.13. A train with a linear induction motor is traveling downhill at 80 km/h when the frequency of its primary windings is 15 Hz. The inclination of the hill is 5°. The frontal area of the train is 25 m^2 and its coefficient of drag is 0.9. The friction coefficient between the wheels of the train and the track is 0.1, and the weight of the train is 500,000 kg. The pole pitch of the vehicle is 3 m. Compute the following:

a. The developed force of the motor at steady state.

b. The developed power in horsepower.

12.14. A vehicle is powered by a linear induction motor with a pole pitch of 2 m. The primary circuit resistance R_1 is 0.5 Ω, and the secondary resistance referred to the primary circuit R_2' is 1.0 Ω. The equivalent inductance of the primary and secondary circuits is 0.02 H. The frequency of the primary windings is 10 Hz. Assume that the drag force is 500 N and the friction force is 2 kN.

Compute the voltage across the primary windings when the motor travels at 100 km/h.

12.15. A synchronous generator connected directly to an infinite bus is generating at its maximum real power. The line-to-line voltage of the infinite bus is 480 V and the synchronous reactance of the generator is 5 Ω. Compute the reactive power of the generator.

12.16. A 1 GW synchronous generator is connected to an infinite bus through a transmission line. The synchronous reactance of the generator is 9 Ω and the inductive reactance of the transmission line is 3 Ω. The infinite bus voltage is 110 kV. If the generator delivers its rated power at unity power factor to the infinite bus, compute the following:

a. The terminal voltage of the generator.
b. The equivalent field voltage.
c. The real power output at the generator terminals.
d. The reactive power output at the generator terminals.

12.17. A synchronous generator is connected directly to an infinite bus. The voltage of the infinite bus is 15 kV. The excitation of the generator is adjusted until the equivalent field voltage E_f is 14 kV. The synchronous reactance of the machine is 5 Ω. Compute the following:

a. The pullover power.
b. The equivalent excitation voltage that increases the pullover power by 20%.

12.18. A 100 MVA synchronous generator is connected to a 25 kV infinite bus through two parallel transmission lines. The synchronous reactance of the generator is 2.5 Ω and the inductive reactance of each transmission line is 2 Ω. The generator delivers 100 MVA to the infinite bus at 0.8 power factor lagging. Suppose a lightning strike causes one of the transmission lines to open. Assume that the mechanical power and excitation of the generator are unchanged. Can the generator still deliver the same amount of power to the infinite bus?

12.19. A synchronous generator is connected to an infinite bus through a transmission line. The infinite bus voltage is 15 kV and the equivalent field voltage of the machine is 14 kV. The transmission line inductive reactance is 4 Ω and the synchronous reactance of the machine is 5 Ω.

a. Compute the capacity of the system.
b. If a 2 Ω capacitor is connected in series with the transmission line, compute the new capacity of the system.

12.20. A synchronous motor excited by a 5 kV source has a synchronous reactance of 5 Ω. The motor is running without any mechanical load (i.e., the output real power is zero). Ignore all losses and calculate the equivalent field voltage E_f that delivers 3 MVAr to the source. Also, draw the phasor diagram.

12.21. A 6-pole, 60 Hz synchronous motor is connected to an infinite bus of 15 kV through a transmission line. The synchronous reactance of the motor is 5 Ω and the inductive reactance of the transmission line is 2 Ω. When E_f

is adjusted to 16 kV (line-to-line), the reactive power at the motor's terminals is zero.

a. Draw the phasor diagram of the system.
b. Calculate the terminal voltage of the motor.
c. Calculate the developed torque of the motor.

12.22. A 12-pole, 480 V, 60 Hz synchronous motor has a synchronous reactance of 5 Ω. The field current is adjusted so that the equivalent field voltage E_f is 520 V (line-to-line). Calculate the following:

a. The maximum torque.
b. The power factor at the maximum torque.

12.23. An industrial load is connected across a three-phase, Y-connected source of 4.5 KV (line-to-line). The real power of the load is 160 kW, and its power factor is 0.8 lagging.

a. Compute the reactive power of the load.
b. A synchronous motor is connected across the load to improve the total power factor to unity. The synchronous motor is running unloaded (no mechanical load). If the synchronous reactance of the motor is 10 Ω, compute the equivalent field voltage of the motor.

12.24. A 10-tooth variable reluctance stepper motor is loaded by 10 mNm torque. The phase voltage is adjusted so that the winding current is 20 mA. The torque constant K of the motor is 3 Vs. Compute the following:

a. The holding torque of the motor.
b. The mechanical displacement angle.

12.25. An 8-pole, three-phase permanent magnet stepper motor has a torque constant K of 2 Vs. When the load torque is 20 mNm, the mechanical displacement angle is 10°. Compute the following:

a. The holding torque.
b. The current of the stator windings.
c. The stator current that reduces the mechanical displacement angle to 4°.

12.26. An 8-phase variable reluctance stepper motor with a step angle of 4.5° and a step rate of 300 steps per second is driving an inertia load. The inertia of the load plus the rotor is 4.4 mgm^2, and the friction torque is 13 mNm. Compute the following:

a. The number of teeth of the rotor.
b. The switching period τ.
c. The speed of the motor.
d. The input power at the steady state (ignore electric losses).
e. The inertia torque to accelerate the motor at 20 rad/s^2.

13

Power Distribution and Blackouts

The electric power systems in North America and Europe are probably the most complex systems ever built by man. In the United States, the power system contains several thousands of major generating units, tens of thousands of transmission lines, millions of transformers, and hundreds of millions of protection and control devices. Figure 13.1 shows the electric energy produced in the United States and worldwide. The power systems in 2003 produced over 16 PetaWh (16×10^{15} Wh) of electric energy worldwide and over 4 PetaWh in the United States. This staggering amount of energy translates into over 13 MWh per person during 2003 in the United States. The worldwide consumption, although overwhelming, is expected to increase even more primarily because of the rapid industrial development in Asia. To match the ever-growing demand for electrical energy, the capacity of the power system must continuously increase. In 2002, the capacity of the U.S. generation exceeded 900 GW as shown in Figure 13.2. If continuously operated, the U.S. generators could have produced about 8 PetaWh in 2002.

It is remarkable that power engineers maintain the stable operation of these immense and complex power systems. Given the size and dynamic nature of the power system, it is incredible that a fewer outages occur every year.

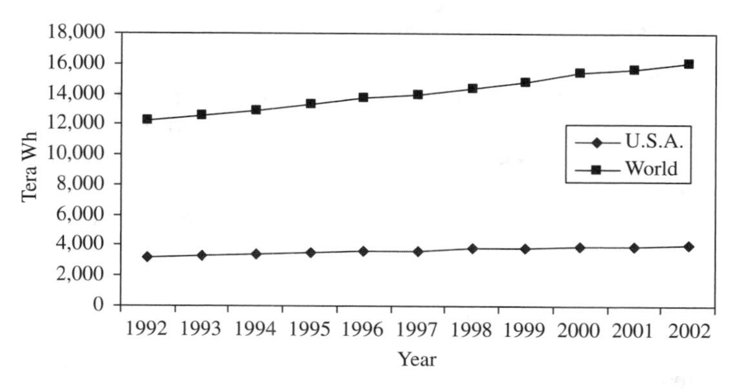

FIGURE 13.1
Net generation of electricity.

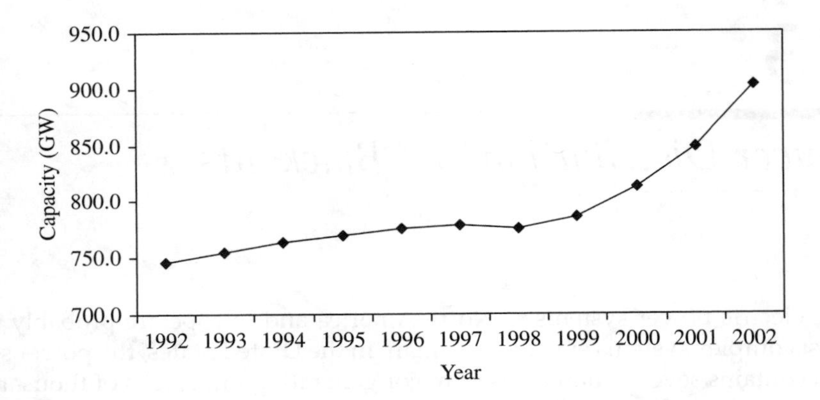

FIGURE 13.2
Generation capacity in the United States.

Power outages are often described as *blackouts* and *brownouts*. The *blackout* refers to a major outage where most or all loads in a given region are left without power. The term *brownout* is used when a portion of the system in a given region is left without power. The borderline between the two situations is fuzzy, and sometimes these two terms are used interchangeably.

Blackouts and brownouts, in general, occur due to the lack of generation and/or the lack of transmission lines. The questions most people ask are whether we can predict the blackout ahead of time, and whether we can prevent it from happening. These difficult questions have no general answers for all blackout scenarios. Although it is technically possible to build a very robust power system by constructing a large number of distributed generations and multiple redundant transmission lines, the technical solution may not be cost effective and is often unacceptable to the public. The public views toward building new generations and new transmission lines are generally negative. Utilities often face a great deal of difficulties when expansions are proposed to meet new growth. The public, in general, prefers using renewable energy and conservations instead of building new fossil fuel or nuclear power plants near their towns. Furthermore, the utilities often face tremendous resistance when proposing new transmission lines, especially when the right of way is near residential areas.

Renewable energy and conservations are often stated as ways to reduce the frequency of blackouts and lessen their impacts. Given the current status of the technology, these ideas cannot really save the power system from blackouts as per the following discussions:

- Besides hydrogenation, renewable energy is not yet a viable option to replace the large generators we have today. However, in 10 to 20 years, the technology may provide us with high-energy density renewable systems that can replace the generators we use today.

- Conservation will help tremendously if it is implemented on a large scale where conservation circuits are imbedded in the designs of the electrical equipment. Relying on individuals to reduce their energy consumption has been proven ineffective because most people buy comfort rather than electricity. Most of us do not think in terms of how many kilowalts we consume to make the home cool or warm. Rather, we set the controllers to our comfort level, and we let electricity flow until we reach this level.
- Besides renewable energy and conservation, blackouts can have lesser impact if distributed generations are used. Small generating units can be installed to provide electricity to an individual building or a group of buildings either continuously or as backup systems. These generating units have different design configurations ranging from a small natural gas power plant the size of a household water heater (power plant in a box), to large gas or renewable systems. The main obstacle facing these systems is their high cost, which can only be justified for sensitive loads such as hospitals.

Another question that is often asked after every blackout is why the power system cannot be controlled as we do for phone and internet networks? After all, they are similarly large and complex. Actually, there are several reasons that make the control of power systems unique and different from all other large and complex networks; among them are:

- The power grid is not a programmable network like phone or internet systems. Electricity cannot be sent from point to point in packages. The power system is rather a giant reservoir, where all generators deliver their energy to the reservoir and any load can tap from that reservoir. It is impossible to know which generator produces the energy you consume in your home.
- Since there is no effective way to store large amounts of electric energy, the generated power must immediately be consumed by all loads and the various losses in the system. A deficiency or surplus in power may lead to a blackout if not corrected within seconds (milliseconds in some cases). For the internet network, when the packets sent are at a higher rate than what is being processed by the user's node, the extra packets are often stored temporarily until the node is available. Even if the packets are dropped, the network can still function.
- The amount of energy that must be controlled at all times in the power system is immense. The internet and phone networks process an infinitesimal amount of energy compared to the power grid.
- The immense masses of the generators, turbines, and motors in the power systems create relatively large delays in the control actions imposed on the system. Hence, instant correction is impossible. This problem is virtually nonexistent in phone and internet networks.

- Overloaded equipment in the power systems are often tripped to protect the equipment from being damaged. This could initiate outages. In phone or internet networks, overloading the equipment often leads to just a delay in transmission.

Based on the above discussions and the status of current technologies, our best option to reduce blackouts is to enhance the reliability of the existing power systems by building more generating plants, more transmission lines, and improving the monitor and control systems. These options, which are discussed in the following sections, are costly solutions and must always be justified socially, environmentally, and economically. The installation of a new generating plant or new transmission line is often resisted by nearby communities because of several reasons; some of them are (1) the transmission lines have unattractive sights; (2) the perception that the magnetic field of the power lines may have negative biological effects; (3) fossil fuel power plants produce pollutions and nuclear power plants are perceived to be radiation hazards; (4) power plants occupy a large size of real estate; and (5) substations are perceived to be incompatible with city views. Some of these problems can be lessened by attractive new designs of transmission towers and substations, and by reducing the pollution levels of power plants. If and when renewable energy becomes reliable and viable, existing fossil fuel generations or transmission lines can be dismantled.

13.1 Topology of Power Systems

Early power systems were simple in design but highly unreliable. They consisted mainly of generating units connected to load centers in *radial* fashion as shown in Figure 13.3. The generator is connected to a transformer (xfm_g) to raise the voltage to the transmission voltage level. The transmission lines

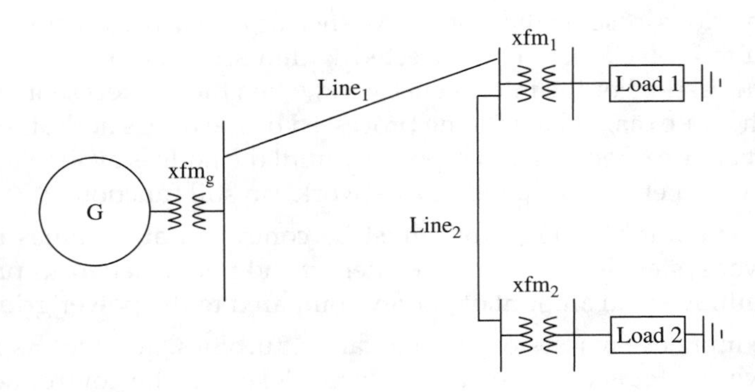

FIGURE 13.3
Radial power system.

connect the generating stations to the load centers (two of them are shown in the figure). At the load centers, transformers (xfm_1 and xfm_2) are used to step down the voltage to the users' level. Another form of the radial configuration is shown in Figure 13.4.

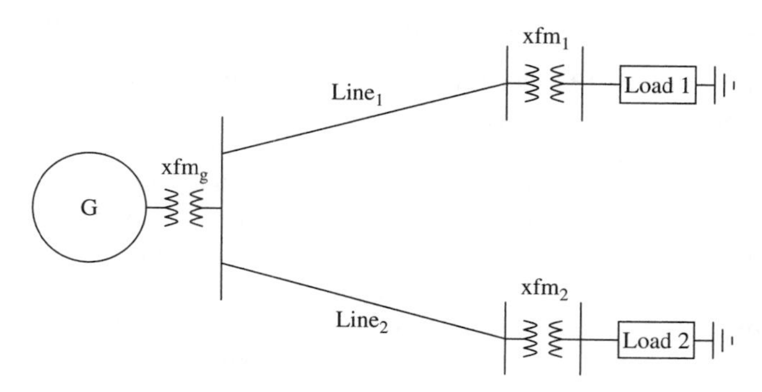

FIGURE 13.4
Another radial power system.

Both systems have reliability problems. In the first system, if $line_1$ is tripped (opened) for any reason, both loads are left without power; that is, the system is in a *blackout* state. If instead $line_2$ is tripped, only load 2 is disconnected and the system is in the *brownout* state. The second system is better designed, because when any of the two lines is tripped, only the load being fed by that line is left without power. Hence, single line tripping causes a brownout but not a blackout.

13.1.1 Enhancing Power System Reliability by Adding Transmission Lines

For the radial systems in Figure 13.3 and Figure 13.4, a single line outage can cause at least one load to be disconnected. To correct this problem a third line can be added as shown in Figure 13.5. In this new system, tripping any one

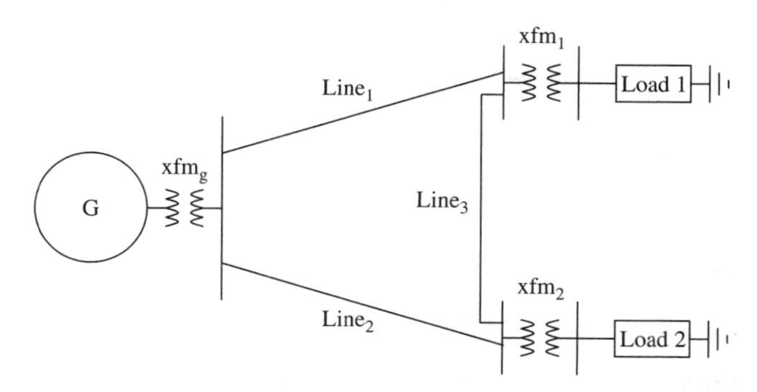

FIGURE 13.5
Network power system.

line will not result in a power outage. The transmission system in this case is connected in a *network* configuration, where each load is fed by multiple lines. Thus, the network connection is much more reliable than the radial connections.

13.1.2 Enhancing Power System Reliability by Adding Generation

The system in Figure 13.5 is unreliable in terms of generation — if the generator is tripped due to any reason, all loads will be left without power. The obvious solution to this problem is to add more generators to the system at different locations. An example is shown in Figure 13.6 for a two-generator, five-transmission line system. Tripping out any one line, or one generator, may not result in loss of service to any load. However, there is no guarantee that the system is secure under any tripping scenario as discussed in the following sections. You may find scenarios where the system is insecure, especially when the system is heavily loaded and the line with most current is tripped.

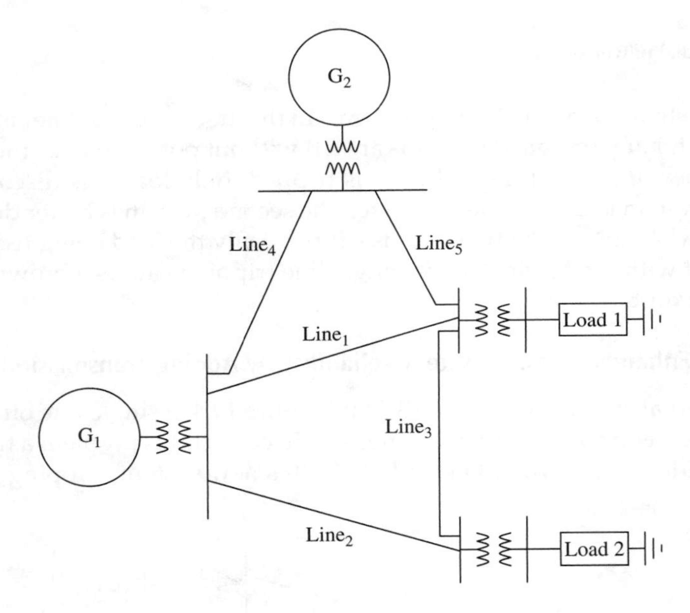

FIGURE 13.6
Network power system with multiple generators.

13.2 Analysis of Power Networks

To detect situations that may lead to blackouts (e.g., over-current, over-voltage, or under-voltage), a method known as load flow analysis is used to study the power network. The load flow is a circuit analysis tool that relates the structure of the network and the load demands to the currents and voltages throughout the network. After performing the load flow analysis, the status of the power system can be determined. The power system is in a *secure state* if all currents and voltages of the major equipment (transmission lines, transformers, etc.) are within their design range. The system is *insecure* if any major equipment (transmission lines, transformers, etc.) is operating below its normal voltage or higher than its rated current. If the load flow analysis identified overloaded lines, the system is insecure and the operator must shift some of the loads to other lines thus saving the system from potential outage. If the voltage at any load center is lower than its normal range, the load is served by an insufficient number of lines and the operator must shift the flow of power through additional lines.

Because the power system is not fully monitored, the load flow analysis uses existing measurements of powers and various voltages to solve for the currents and voltages everywhere else in the network. The method, which is often numerical, uses a set of nonlinear equations relating the measurements to the unknowns. However, if we assume that all system impedances as well as the voltages of all generators are known, the load flow analysis turns into a solution of a set of linear equations. The procedure starts by developing the impedance diagram of the power system; the diagram for the system in Figure 13.5 is shown in Figure 13.7. Z_{11} is the impedance of load 1 plus its own transformer (xfm_1), Z_{22} is the impedance of load 2 including its transformer, and Z_{12} is the impedance of transmission line between bus 1 and 2, and so on. The voltage at each bus is labeled according to the bus number, and the direction of the current can be chosen arbitrarily. The final solution of the current determines its direction; if the current is positive, the chosen direction is correct, and if it is negative the actual direction is opposite to the chosen one.

A convenient choice for the initial directions of the transmission line currents is to assume that the flow is from the larger numbered bus to the smaller

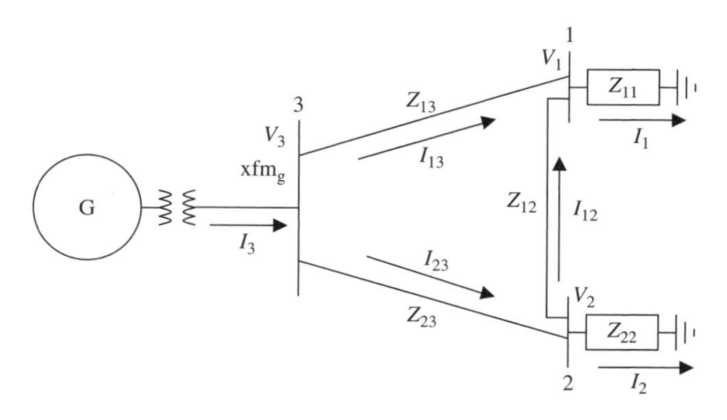

FIGURE 13.7
Current distribution of the system in Figure 13.5.

numbered bus. In this case, the node equations at buses 1, 2, and 3 are

$$\begin{aligned} \bar{I}_1 &= \bar{I}_{12} + \bar{I}_{13} \\ \bar{I}_2 &= \bar{I}_{23} - \bar{I}_{12} \\ \bar{I}_3 &= \bar{I}_{13} + \bar{I}_{23} \end{aligned} \tag{13.1}$$

The load current can be written as a function of the bus voltage and the load admittance as follows:

$$\begin{aligned} \bar{I}_1 &= \overline{V}_1 \overline{Y}_{11} \\ \bar{I}_2 &= \overline{V}_2 \overline{Y}_{22} \end{aligned} \tag{13.2}$$

where $\overline{Y}_{ij} = 1/\overline{Z}_{ij}$. The currents of the transmission lines can also be written as

$$\begin{aligned} \bar{I}_{12} &= (\overline{V}_2 - \overline{V}_1)\overline{Y}_{12} \\ \bar{I}_{13} &= (\overline{V}_3 - \overline{V}_1)\overline{Y}_{13} \\ \bar{I}_{23} &= (\overline{V}_3 - \overline{V}_2)\overline{Y}_{23} \end{aligned} \tag{13.3}$$

Substituting the currents in Equation (13.3) into Equation (13.1) yields

$$\begin{bmatrix} \bar{I}_1 \\ \bar{I}_2 \\ -\bar{I}_3 \end{bmatrix} = \begin{bmatrix} -\left(\overline{Y}_{12} + \overline{Y}_{13}\right) & \overline{Y}_{12} & \overline{Y}_{13} \\ \overline{Y}_{12} & -\left(\overline{Y}_{12} + \overline{Y}_{23}\right) & \overline{Y}_{23} \\ \overline{Y}_{13} & \overline{Y}_{23} & -\left(\overline{Y}_{13} + \overline{Y}_{23}\right) \end{bmatrix} \begin{bmatrix} \overline{V}_1 \\ \overline{V}_2 \\ \overline{V}_3 \end{bmatrix} \tag{13.4}$$

Substituting the load currents in Equation (13.2) into Equation (13.4) yields

$$\begin{bmatrix} \bar{I}_1 \\ \bar{I}_2 \\ -\bar{I}_3 \end{bmatrix} = \begin{bmatrix} -\left(\overline{Y}_{12} + \overline{Y}_{13}\right) & \overline{Y}_{12} & \overline{Y}_{13} \\ \overline{Y}_{12} & -\left(\overline{Y}_{12} + \overline{Y}_{23}\right) & \overline{Y}_{23} \\ \overline{Y}_{13} & \overline{Y}_{23} & -\left(\overline{Y}_{13} + \overline{Y}_{23}\right) \end{bmatrix} \begin{bmatrix} \overline{Z}_{11}\bar{I}_1 \\ \overline{Z}_{22}\bar{I}_2 \\ \overline{V}_3 \end{bmatrix} \tag{13.5}$$

Equation (13.5) can be written in the matrices form.

$$\begin{aligned} [I] &= [Y]\big([V] + [Z][I]\big) \\ [I] &= ([1] - [Y][Z])^{-1}[Y][V] \end{aligned} \tag{13.6}$$

EXAMPLE 13.1 The power system in Figure 13.7 has the following data:

Impedance of load 1 plus its transformer is $\overline{Z}_{11} = 20 + j10\ \Omega$

Impedance of load 2 plus its transformer is $\overline{Z}_{22} = 30 + j5\ \Omega$

Impedance of line$_1$ is $\overline{Z}_{13} = j0.5\ \Omega$

Impedance of line$_2$ is $\overline{Z}_{23} = j0.6\ \Omega$

Impedance of line$_3$ is $\overline{Z}_{12} = j0.4\ \Omega$

All impedances are referred to the high voltage side of the transformers. The voltage at the high voltage side of xfm$_g$ is fixed at 230 kV (line-to-line). The capacity of any line (maximum current the line can carry) is 6 kA. Assume that all connections are in wye.

1. Compute the power delivered to each load, and the power produced by the generator.
2. Assume that transmission line$_3$ is tripped; repeat part 1.
3. Is the system in part 2 secure?
4. Assume that line$_1$ is tripped while line$_2$ and line$_3$ are in service; repeat part 1.
5. Is the system in part 4 secure?

Solution

1. To compute the powers, we need to compute the currents of each load. We have five unknowns, I_1, I_2, I_3, V_1, and V_2. Equation (13.5) can be used to solve for the currents.

$$\begin{bmatrix} \bar{I}_1 \\ \bar{I}_2 \\ -\bar{I}_3 \end{bmatrix} = \begin{bmatrix} -(\overline{Y}_{12} + \overline{Y}_{13}) & \overline{Y}_{12} & \overline{Y}_{13} \\ \overline{Y}_{12} & -\left(\overline{Y}_{12} + \overline{Y}_{23}\right) & \overline{Y}_{23} \\ \overline{Y}_{13} & \overline{Y}_{23} & -\left(\overline{Y}_{13} + \overline{Y}_{23}\right) \end{bmatrix} \begin{bmatrix} \overline{Z}_{11}\bar{I}_1 \\ \overline{Z}_{22}\bar{I}_2 \\ \overline{V}_3 \end{bmatrix} \tag{13.7}$$

Equation (13.7) can be further rearranged as follows:

$$\begin{bmatrix} 1 + (\overline{Y}_{12} + \overline{Y}_{13})\overline{Z}_{11} & -\overline{Y}_{12}\overline{Z}_{22} & 0 \\ -\overline{Y}_{12}\overline{Z}_{11} & 1 + \left(\overline{Y}_{12} + \overline{Y}_{23}\right)\overline{Z}_{22} & 0 \\ -\overline{Y}_{13}\overline{Z}_{11} & -\overline{Y}_{23}\overline{Z}_{22} & 1 \end{bmatrix} \begin{bmatrix} \bar{I}_1 \\ \bar{I}_2 \\ -\bar{I}_3 \end{bmatrix} = \begin{bmatrix} \overline{Y}_{13} \\ \overline{Y}_{23} \\ -\left(\overline{Y}_{13} + \overline{Y}_{23}\right) \end{bmatrix} \overline{V}_3 \tag{13.8}$$

By directly substituting the variables in this example into Equation (13.8), the currents of the system can be computed. Keep in mind that the phase voltage of V_3 is $230/\sqrt{3}$.

$$\begin{bmatrix} \bar{I}_1 \\ \bar{I}_2 \\ \bar{I}_3 \end{bmatrix} = \begin{bmatrix} 5.82\,\angle{-29.4^\circ} \\ 4.3\,\angle{-11.6^\circ} \\ 10.0\,\angle{-21.8^\circ} \end{bmatrix} \text{kA}$$

The power consumed by load 1 is

$$P_1 = 3I_1^2\,R_{11} = 3 \times 5.82^2 \times 20 = 2.032\ \text{GW}$$

The power consumed by load 2 is

$$P_2 = 3I_2^2 R_{22} = 3 \times 4.3^2 \times 30 = 1.664 \text{ GW}$$

The power produced by the generator is the sum of P_1 and P_2 since the lines have no resistance.

$$P_3 = P_1 + P_2 = 3.696 \text{ GW}$$

2. If line$_3$ is tripped, $Y_{12} = 0$, and the current can be computed similar to part 1.

$$\begin{bmatrix} \bar{I}_1 \\ \bar{I}_2 \\ \bar{I}_3 \end{bmatrix} = \begin{bmatrix} 5.31 \angle{-37°} \\ 4.35 \angle{-10.6°} \\ 9.41 \angle{-25°} \end{bmatrix} \text{kA}$$

The power consumed by load 1 is

$$P_1 = 3I_1^2 R_{11} = 3 \times 5.31^2 \times 20 = 1.692 \text{ GW}$$

The power consumed by load 2 is

$$P_2 = 3I_2^2 R_{22} = 3 \times 4.35^2 \times 30 = 1.703 \text{ GW}$$

The power produced by the generator is the sum of P_1 and P_2 since the lines have no resistance.

$$P_3 = P_1 + P_2 = 3.395 \text{ GW}$$

3. To check the security of the system, we need to compute the currents in the transmission lines and compare them with their capacities. But first, we need to compute the bus voltages

$$\overline{V}_1 = \overline{Z}_{11} \times \bar{I}_1 = 116.86 - \text{j}\,21.25 \text{ kV}$$
$$\overline{V}_2 = \overline{Z}_{22} \times \bar{I}_2 = 132.3 - \text{j}\,2.25 \text{ kV}$$

The magnitudes of the line currents are

$$I_{12} = |(\overline{V}_2 - \overline{V}_1)\overline{Y}_{12}| = 0 \text{ kA}$$
$$I_{13} = |(\overline{V}_3 - \overline{V}_1)\overline{Y}_{13}| = 5.31 \text{ kA}$$
$$I_{23} = |(\overline{V}_3 - \overline{V}_2)\overline{Y}_{23}| = 4.35 \text{ kA}$$

All the currents are lower than the line capacities, which is 6 kA. The system is secure and can continue to provide power to the customers.

4. If $line_1$ is tripped, $Y_{13} = 0$, and the currents are

$$\begin{bmatrix} \bar{I}_1 \\ \bar{I}_2 \\ \bar{I}_3 \end{bmatrix} = \begin{bmatrix} 5.77 \angle -30.34^\circ \\ 4.29 \angle -11.88^\circ \\ 9.94 \angle -22.47^\circ \end{bmatrix} \text{kA}$$

The power consumed by load 1 is

$$P_1 = 3I_1^2 R_{11} = 3 \times 5.77^2 \times 20 = 1.998 \text{ GW}$$

The power consumed by load 2 is

$$P_2 = 3I_2^2 R_{22} = 3 \times 4.29^2 \times 30 = 1.656 \text{ GW}$$

The power produced by the generator is the sum of P_1 and P_2 since the lines have no resistance.

$$P_3 = P_1 + P_2 = 3.654 \text{ GW}$$

5. To check the security of the system, we need to compute the currents in the remaining lines.

$$\overline{V}_1 = \overline{Z}_{11} \times \bar{I}_1 = 128.8 - j\,8.5 \text{ kV}$$
$$\overline{V}_2 = \overline{Z}_{22} \times \bar{I}_2 = 130.5 - j\,5.5 \text{ kV}$$

The magnitudes of the line currents are

$$I_{12} = |(\overline{V}_2 - \overline{V}_1)\overline{Y}_{12}| = 5.77 \text{ kA}$$
$$I_{13} = |(\overline{V}_3 - \overline{V}_1)\overline{Y}_{13}| = 0 \text{ kA}$$
$$I_{23} = |(\overline{V}_3 - \overline{V}_2)\overline{Y}_{23}| = 9.94 \text{ kA}$$

Note that $line_2$ carries 9.94 kA, which is higher than the capacity of the line (6 kA). In this case, the system is insecure and $line_2$ will eventually trip to prevent it from overheating and sagging too deeply and touch trees or structures. When this line is tripped, no load will be served.

13.3 Electric Energy Demand

The decision to construct new generation or transmission facilities is mainly based on reliability concerns as well as existing and predicted future demands for energy. However, the demand varies on an hourly basis, and the maximum demand occurs probably a few times a year during heat waves or cold spells. Therefore, it is hard to justify the cost of new facilities based on the maximum demand alone. Instead, utilities regularly depend on neighboring utilities to provide support during the high demand periods.

During a typical day, the energy demand fluctuates as shown in the typical load curve in Figure 13.8. In most utilities, the peak energy consumptions occur twice in a typical day: the first is around 9:00 A.M. when factories, shops, and offices are at their peak consumption; the second is around 6:00 P.M. when people are at home preparing meals, turning on lights, and watching television.

To provide reliable service to its customers, every utility must be able to meet the daily peak demands. Since the high energy demand occurs only for a few hours every day, it is uneconomical to build generating plants to be used only during the peak loads. Instead, it would be more profitable for the utility to build its generating capacity to meet the average daily demand. But what should be done when the load is higher than the available generation? Consider, for example, the load profile in Figure 13.9. The dark areas represent the periods when the generating capacity is below the demand, and the dashed areas represent the times when the generating capacity is higher than the demand. The obvious economical solution to the deficit and surplus powers is to trade electricity with neighboring utilities. When a utility has surplus, it sells the excess power to another utility. When it has a deficit, the utility buys the extra power from another utility with surplus. This arrangement makes sense economically, but requires all utilities to be interconnected through a mesh of transmission lines.

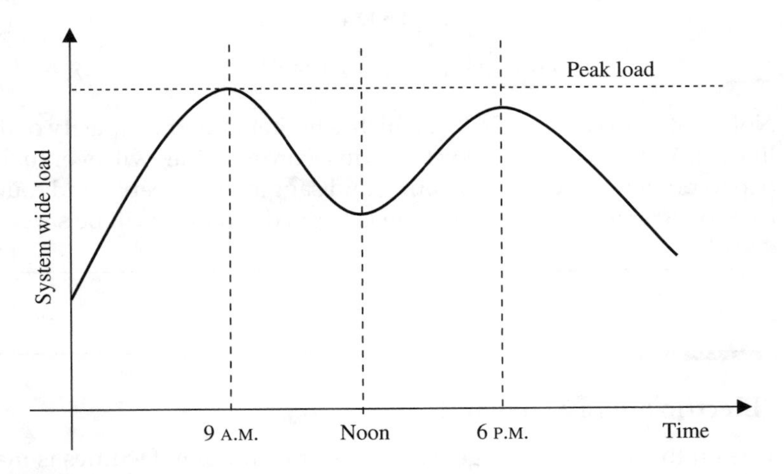

FIGURE 13.8
Daily system load.

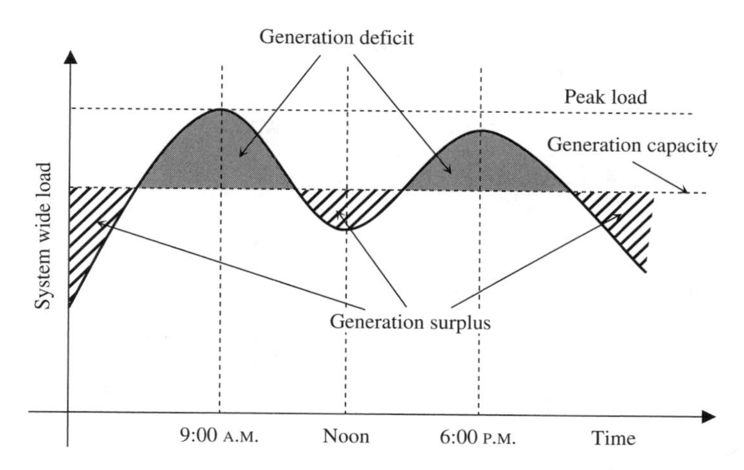

FIGURE 13.9
Setting the generation capacity.

EXAMPLE 13.2 The system load demand for a given day can be approximated by

$$P = 2 - 0.3\cos(0.5417t - 1)\mathrm{e}^{-0.01t} \text{ GW}$$

where t is the time of the day in hours using the 24-h clock. Compute the following:

1. The times of the peak demands.
2. The peak demands.
3. The average demand.
4. If the generation capacity of the utility is 2 GW, compute the power to be imported during the first peak.

Solution

1. To compute the times of the peaks, we need to set the derivative of the power equation with respect to time to zero.

$$\frac{\mathrm{d}P}{\mathrm{d}t} = 0$$

$$0.01 \ \cos(0.5417t - 1) + 0.5417\sin(0.5417t - 1) = 0$$

$$\tan(0.5417t - 1) = -0.01846$$

$$t = \frac{1 + \tan^{-1}(-0.01846)}{0.5417}$$

Solving the above equation using the four quadrants of the function, the first peak is at

$$t_1 = 7.6115 = 7{:}37 \text{ A.M.}$$

And the second peak is at

$$t_2 = 19.2 = 7{:}12 \text{ P.M.}$$

2. The first and second peaks are

$$P_{\text{peak1}} = 2 - 0.3\cos(0.5417t_1 - 1)\mathrm{e}^{-0.01t_1} = 2.278 \text{ GW}$$

$$P_{\text{peak2}} = 2 - 0.3\cos(0.5417t_2 - 1)\mathrm{e}^{-0.01t_2} = 2.2475 \text{ GW}$$

3. The average power during the 24-h period is

$$P_{\text{ave}} = \frac{1}{24}\int_0^{24} P\,\mathrm{d}t$$

The power equation can be rewritten in the following form:

$$P = 2 - 0.162\mathrm{e}^{-0.01t} \times \cos(0.5417t) - 0.2525\mathrm{e}^{-0.01t} \times \sin(0.5417t)$$

where

$$\int e^{ax}[c\,\cos(bx) + k\,\sin(bx)]\,\mathrm{d}x$$

$$= \frac{\mathrm{e}^{ax}}{a^2 + b^2}[(ka + cb)\sin(bx) + (ca - kb)\cos(bx)]$$

Hence,

$$P_{\text{ave}} = \frac{1}{24}\int_0^{24} (2 - 0.162\mathrm{e}^{-0.01t} \times \cos(0.5417t)$$

$$- 0.2525\mathrm{e}^{-0.01t} \times \sin(0.5417t))\mathrm{d}t$$

$$P_{\text{ave}} = 2 - \left[\frac{e^{-0.01t}}{0.01^2 + 0.5417^2}[(0.08523)\sin(0.5417t)\right.$$

$$\left. - (0.1384)\cos(0.5417t)]\right]_0^{24}$$

$$P_{\text{ave}} = 1.7688 \text{ GW}$$

4. The power to be imported during the first peak is

$$P_{\text{import}} = P_{\text{peak1}} - P_{\text{ave}} = 2.278 - 1.7688 = 509.2 \text{ MW}$$

13.4 Electric Energy Trade

As mentioned earlier, the trading of energy between utilities makes great economic sense. Take, for example, the two utilities in Figure 13.10 that are connected by a transmission line often called a *tie line*. Assume that the two utilities are located in two different longitudes (two different time zones). Assume that the time zone of the western utility lags the eastern utility by 3 h. At 9:00 A.M. Eastern Time (ET), the eastern utility is at its peak demand while the western utility has surplus generation since it is 6:00 A.M. in the West. Hence, the extra power can be delivered to the eastern utility by the western utility. At noon ET, the western utility is at its peak demand while the eastern utility has surplus. So the power can flow from east to west.

Another example of power exchange occurs between utilities at different latitudes. Take the case in Figure 13.11 where the northern area is assumed to

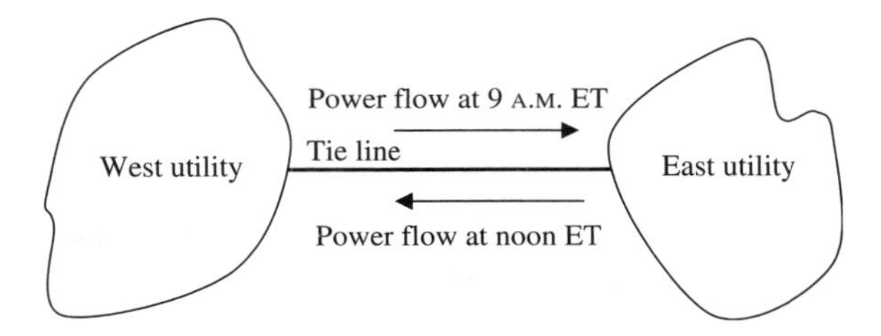

FIGURE 13.10
Two utilities at different time zones.

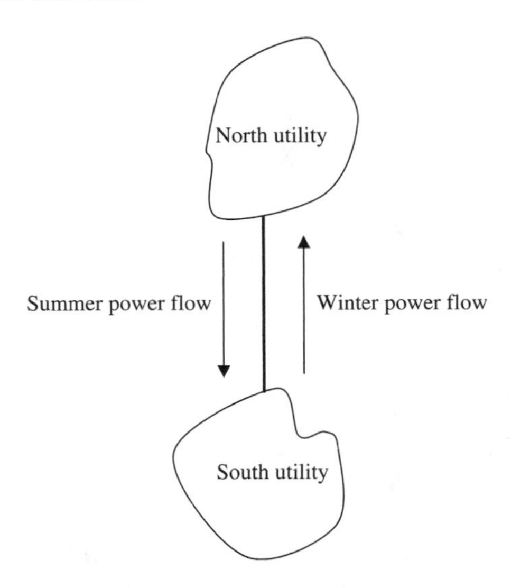

FIGURE 13.11
Trade between utilities at different latitudes.

be generally cooler than the southern areas; an example in the United States is the Northwest and California. Heating loads are dominant in the Northwest while cooling loads are prevalent in California. In the summer, California has high energy demand because of the use of the air conditioning equipment, while the temperature in the Northwest is mild. Hence, the Northwest has surplus energy while California has a deficit and the power moves from north to south. In the winter, the scenario is reversed.

EXAMPLE 13.3 The load demand for a given utility can be approximated by

$$P = 4 - \cos(0.5t - 1.2) \text{ GW}$$

where t is the time of the day in hours using the 24-h clock. The generating capacity of the utility is 4 GW. Compute the following:

1. The energy available for export.
2. The imported energy during the time of energy deficit.
3. The net energy trade.

Solution

1. The power available for trading is the difference between the actual demand and the generation capacity.

$$\Delta P = P_{\text{ave}} - P = \cos(0.5t - 1.2)$$

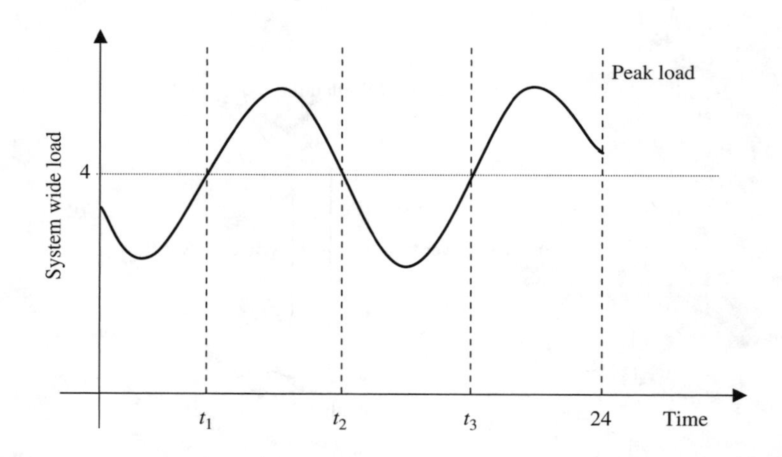

Before we compute the traded energy, we need to identify the times in the figure when the demand is equal to the capacity.

$$\Delta P = \cos(0.5t - 1.2) = 0$$

Hence,

$$t_1 = 5.542$$
$$t_2 = 11.825$$
$$t_3 = 18.108$$

The energy available for export is

$$E_{\text{export}} = \int_0^{t_1} \Delta P \,\mathrm{d}t + \int_{t_2}^{t_3} \Delta P \,\mathrm{d}t$$

$$E_{\text{export}} = 0.5\sin(0.5t - 1.2)\big|_0^{5.542} + 0.5\sin(0.5t - 1.2)\big|_{11.825}^{18.108} = 1.966 \text{ GWh}$$

2. The imported energy is

$$E_{\text{import}} = \int_{t_1}^{t_2} \Delta P \,\mathrm{d}t + \int_{t_3}^{24} \Delta P \,\mathrm{d}t$$

$$E_{\text{import}} = 0.5\sin(0.5t - 1.2)\big|_{5.542}^{11.825} + 0.5\sin(0.5t - 1.2)\big|_{18.108}^{24} = -1.99 \text{ GWh}$$

3. The net trade is

$$E_{\text{net}} = E_{\text{export}} + E_{\text{import}} = 1.966 - 1.99 = -24 \text{ MWh}$$

13.5 World Wide Web of Power

In addition to the economic benefits, interconnecting utilities makes great reliability sense. If some generators in certain areas are out of service, the generation deficit can be compensated by importing extra power from neighboring utilities. Also, if a key transmission line is tripped, other transmission lines in the grid can reroute the power to the customers. Because of these economic and reliability benefits, power systems are continuously interconnected with little regard to political or geographical borders.

Although interconnected, the control and power exchange policies of the power system are left to local pools that are composed of several utilities in a geographical region. In the United States, Canada, and a small part of Mexico, the power grid is divided into ten power pools as shown in Figure 13.12; the pools are called *reliability councils*. An example of a reliability council is the *Western Electricity Coordinating Council* (WECC) which provides electricity to 71 million people in 14 Western states, 2 Canadian provinces, and portions of a Mexican state. The WECC system consists of over one thousand generating units and tens of thousands of transmission lines. It has at least 600 generating units rated at 500 MW or higher. The full WECC system is very complex and

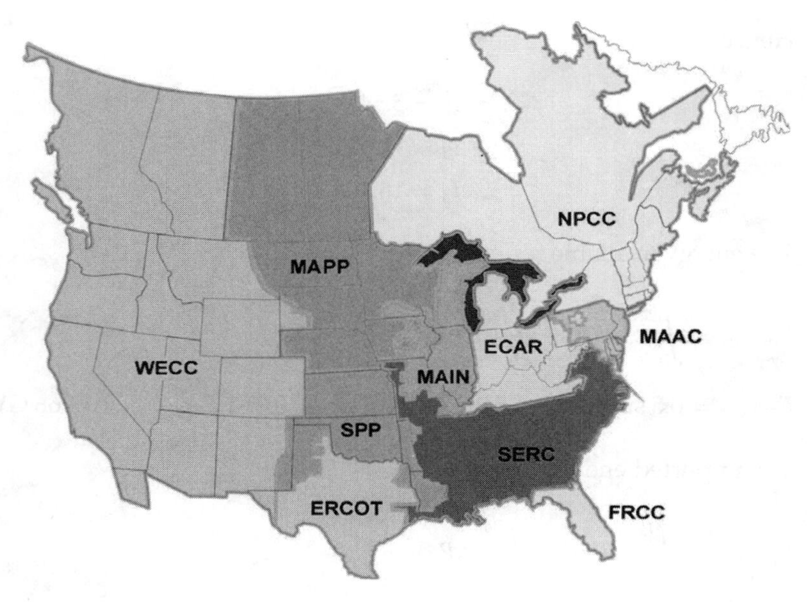

FIGURE 13.12
(see color insert following Page 208) U.S. regional reliability councils.

is hard to display in a single page. However, a small part of the WECC is shown in Figure 13.13.

Another example of a pool is the *Mid-Atlantic Area Council* (MAAC), which encompasses an area of nearly 130,000 km^2 and provides electricity to more than 23 million people in the northeastern region of the United States and western Canada. The capacity of the MAAC is approximately 60 GW and it has over 13,000 km of bulk power transmission lines.

All ten coordinating councils form a nonprofit corporation called the *North American Electric Reliability Council* (NERC). The NERC's mission is to ensure that the bulk electric system in North America is reliable, adequate, and secure by setting, monitoring, and enforcing standards for the reliable operation and planning of the electric grid.

The power systems are also fully interconnected in Western Europe, and their equivalent to NERC is called the *Union for the Co-ordination of Transmission of Electricity* (UCTE). UCTE covered 20 European countries as of 2004, while more countries from Eastern Europe expected to be connected to the UCTE grid. Their service area in 2004 included 400 million people and its total annual consumption is approximately 2100 TWh. North Africa is also being connected with Europe through submarine cable systems under the Mediterranean Sea. In other regions of the world, the power systems are linked together whenever possible. Indeed, many consider that power engineers have succeeded in creating a borderless power grid that truly unifies the world!

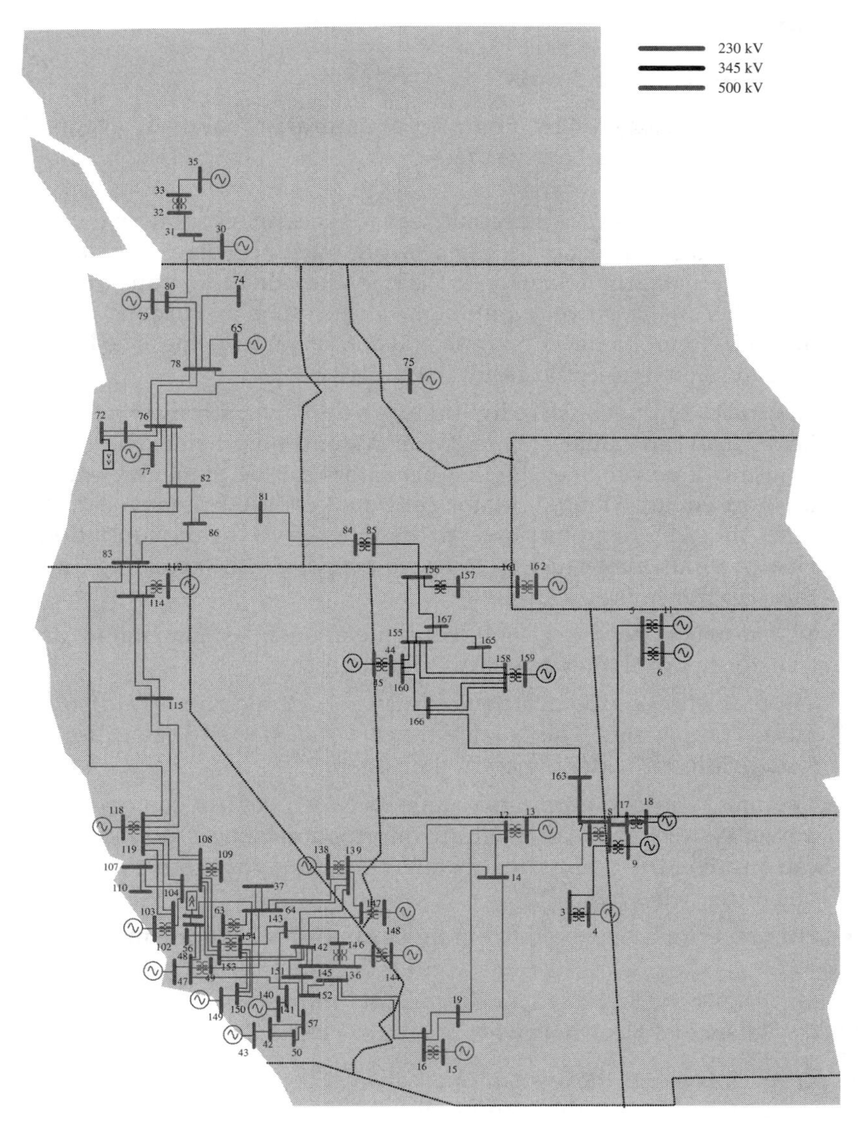

FIGURE 13.13
Major components of the WECC.

Although the interconnection of the power networks has many great advantages, it has two major drawbacks:

1. The power grid becomes incredibly complex, creating enormous challenges for monitoring, operation, and control of the system.
2. Major failures in one area could affect other areas, thus creating wider blackouts. This was evident in the major blackouts in the United States in 1965, 1976, and 2003.

13.6 Anatomy of Blackouts

Blackouts can be triggered by a number of natural or manmade events. The list is long, but the typical events are:

- Faults in transmission lines could lead to excessive currents in the system. The faulted section of the network must be isolated quickly to prevent any thermal damage to the transmission lines, transformers, and other major power equipment. This is done by tripping (opening) the circuit breakers on both ends of the faulted line. The loss of transmission lines could result in outages.
- Natural calamities such as lightning, earthquakes, strong winds, and heavy frost can damage major power system equipment. When lightning hits a power line, the line insulators can be damaged leading to short circuits (faults). Major earthquakes could damage substations, thus, interrupting power to the areas served by the substations. Heavy winds may cause trees to fall on power lines creating faults that trip transmission lines.
- When major power system equipment such as generators and transformers fail, they may lead to outages.
- The protection and control devices may not operate properly to isolate faulty components. This is known as *hidden failure* and can cause the fault to affect a wider region in the system.
- The breaks in communication links between control centers in the power system may lead to wrong information being processed, thus causing control centers to operate asynchronously, and probably negating each other.
- Human errors can lead to tripping of important equipment.

13.6.1 Balance of Electric Powers

The power balance of the system must always be maintained to ensure its stable operation; that is, the sum of all generations must equal the sum of all loads plus all losses. This simple relationship must be preserved at the system-wide level, and at the power pool level. Take, for example, the power pool in Figure 13.14 that represents a geographical area or a service territory of a utility or a group of utilities. The power pool has a total power generation P_g from all its power plants, and a total system load P_{load} representing customer demand plus the losses in the system. The pool imports power P_{import} from neighboring utilities and exports power P_{export} to other neighboring utilities. The balanced power equation of the pool is

$$P_g + P_{import} = P_{load} + P_{export} \tag{13.9}$$

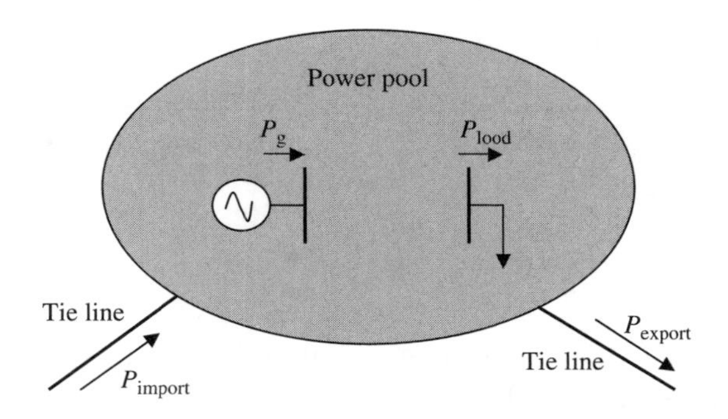

FIGURE 13.14
Power balance of a pool.

Now let us assume that the tie line transmitting P_{import} is lost for some reason. Then, the power balance is not maintained, and

$$P_g < (P_{\text{load}} + P_{\text{export}}) \tag{13.10}$$

The fundamental question is how to balance the power in the pool. In this simple example, we have four options:

1. Reduce P_{export}, which may result in a power deficit in the neighboring utility.
2. If the generated power is below the capacity of the pool, increase the generation P_g.
3. Reduce the demand by disconnecting some loads. This is called *rolling blackouts*.
4. Find another utility that can transmit the needed power through other transmission routes.

All these solutions must be implemented within a short time (milliseconds in some cases), otherwise the system may collapse. Why? The answer is in the following section.

13.6.2 Balance of Electrical and Mechanical Powers

The synchronous generator is an electro-mechanical converter as depicted in Figure 13.15. Its input power is mechanical P_m and its output is electrical power P_g. If you ignore the internal losses, the balanced operation of the generator is achieved when

$$P_m = P_g \tag{13.11}$$

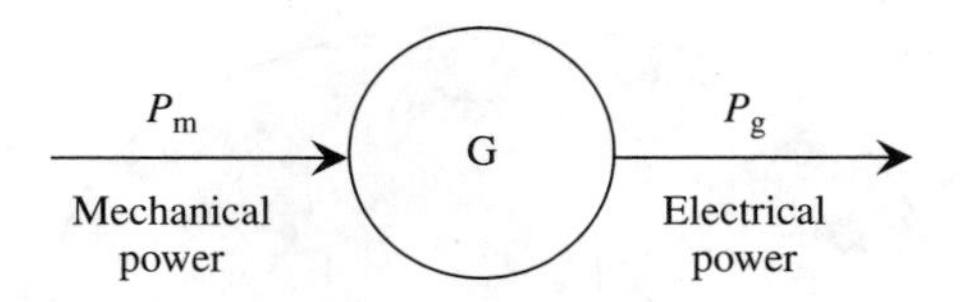

FIGURE 13.15
Power balance of synchronous generator.

Keep in mind that the electrical power P_g is determined by the customers loads of the system and any import or export transaction as given in Equation (13.9). Although the utilities do not often know in advance their customers' intention, the utility must continually adjust the mechanical power into its generators so that Equation (13.11) is satisfied at all times. As long as the balance is maintained, the system is stable and all loads are served.

Among the problems that lead to blackouts is the uncompensated imbalance between the mechanical and electrical powers of the generators. When the input power to the generator P_m is greater than the output power P_g, the excess power is stored in the rotating mass of the generator–turbine unit in the form of kinetic energy. If $P_m < P_g$, the electric power shortage is recovered from the kinetic energy stored in the rotating mass. The dynamic equation of the electro-mechanical powers of the generator can be written as

$$M_{eq}\frac{d\omega}{dt} = P_m - P_g \tag{13.12}$$

where M_{eq} is the equivalent inertia of the rotating mass of the generator–turbine unit, ω is the angular speed of the rotor, $d\omega/dt$ is the angular acceleration of the rotor, and $M_{eq}d\omega/dt$ is the kinetic power.

As explained in Chapter 12, the speed of the synchronous machine ω_s is constant during the steady state operation (stable operation), and is equal to

$$\begin{aligned} \omega_s &= 2\pi\frac{n_s}{60} \text{ rad/s} \\ n_s &= 120\frac{f}{p} \text{ rpm} \end{aligned} \tag{13.13}$$

where f is the frequency of the output voltage of the generator, p is the number of poles inside the machine, and n_s is the synchronous speed of the magnetic field, which is the same as the rotor speed in the steady state.

Now let us assume that the synchronous machine is connected to an *infinite bus* as shown in Figure 13.16. As discussed in Chapter 12, the infinite bus is a term used to describe a large system that is constant frequency and constant voltage. The equivalent circuit of the synchronous generator, which is also given in Chapter 12, is depicted on the left side of Figure 13.17.

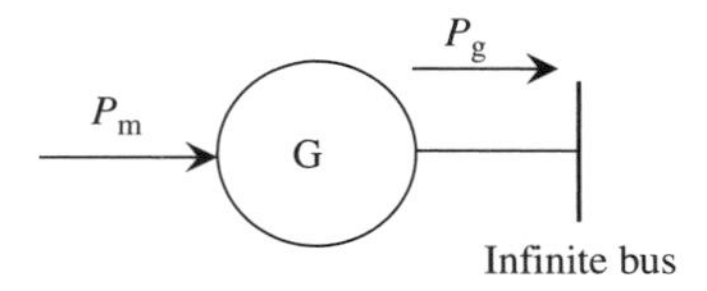

FIGURE 13.16
Phasor diagram of synchronous generator.

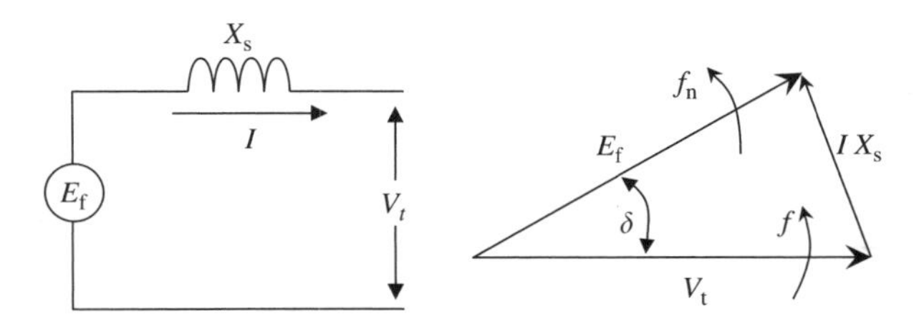

FIGURE 13.17
Model of synchronous generator.

E_f is the equivalent field voltage, V_t is the terminal voltage of the generator (the infinite bus voltage in our case), and X_s is the synchronous reactance of the generator.

As explained in Chapter 7, any phasor has three attributes: magnitude, angle, and frequency. Although we draw the phasor as a stationary vector, it is actually rotating at its own frequency. In our example, and as shown on the right side of Figure 13.17, V_t rotates at the frequency of the infinite bus voltage f, which is constant (60 Hz in the United States or 50 Hz in Europe). E_f rotates at frequency f_n corresponding to the actual rotor speed n. As given in Chapter 12, f_n can be computed by using Equation (13.13).

$$f_n = \frac{p}{120}\, n \tag{13.14}$$

When the rotor of the generator rotates at the synchronous speed, $n = n_s$, the frequency of E_f is equal to the system frequency; that is, $f_n = f$, and the angle δ is constant.

Now let us use Equation (13.12) to discuss the three possible scenarios in Table 13.1. In scenario 1, the mechanical and electrical powers are equal. Hence, the acceleration $d\omega/dt = 0$, and the rotor speed ω is equal to the synchronous speed ω_s. Thus, $f_n = f$. Consequently, as shown in Figure 13.17, the power angle δ is constant. The difference between $\overline{E}_f$ and $\overline{V}_t$, which is $\overline{I}_a\overline{X}_s$, is also constant. Since X_s is fairly constant, the current inside the machine is also constant. This is a balanced (steady state) operation and the generator is *in synchronism* with the rest of the power system.

For scenario 2 where $P_m > P_g$, the generator accelerates and the rotor speed becomes greater than the synchronous speed. Hence, $f_n > f$, and the

TABLE 13.1
Effects of Power Imbalance

Scenario	Power Equation	Rotor Acceleration dω/dt	Rotor Speed ω	Kinetic Energy of Rotating Mass
1	$P_m = P_g$	$\frac{d\omega}{dt} = 0$	$\omega = \omega_s$	Unchanged. Machine is in synchronism. This is the steady state operation.
2	$P_m > P_g$	$\frac{d\omega}{dt} > 0$	$\omega > \omega_s$	Increases. Machine is out of synchronism.
3	$P_m < P_g$	$\frac{d\omega}{dt} < 0$	$\omega < \omega_s$	Decreases. Machine is out of synchronism.

machine in this case is said to be *out of synchronism* with the rest of the power system. Because of the difference in frequencies, δ keeps increasing, thus, the current inside the machine continuously increases. When the current reaches the thermal limit of the machine, the generator is disconnected from the grid to prevent any further damage to the machine. This can happen in just a few milliseconds.

In scenario 3 when $P_m < P_g$, the rotor decelerates and its speed falls below the synchronous speed. Hence, $f_n < f$, and the machine is *out of synchronism* with the rest of the power system. Unless corrected, E_f will eventually lag V_t making the generator act as a motor and reverse the flow of power. During this period, the current inside the machine will increase to a high level that shuts down the generating system.

Based on the above discussions, we can conclude that every generator in the power network must operate under scenario 1 that satisfies Equation (13.11). Thus, the power plant controller must continuously adjust the mechanical power of the turbine to match the changes in the load demand to keep the generators of the plant in synchronism with the rest of the power network.

13.6.3 Control Actions for Decreased Demand

Substituting the value of P_g in Equation (13.9) into Equation (13.11), yields

$$P_m = P_{load} + P_{export} - P_{import} \tag{13.15}$$

Assume that P_{import} and P_{export} are constant. When the load demand P_{load} is reduced, the mechanical power into the generator must be reduced to match the new demand. This is a relatively simple process if the change in demand is slow (minutes or hours). For thermal power plants, the reduction of P_m is done by gradually reducing the steam entering the turbines. For a hydroelectric power plant, the governor reduces the amount of water entering the turbines.

However, if the reduction in demand is sudden and large, the control action is stern and very fast. For thermal power plants, the steam is allowed to escape in to the air, thus rapidly reducing the amount of steam entering the turbine. However, for a hydroelectric power plant, changing the water flow is a slow process because rapid closure of the water valve can damage the penstock due to excessive water hammers. Instead, resistive loads, known as braking resistances, are connected to the system to consume the excess electric energy until the water flow is reduced.

13.6.4 Control Actions for Increased Demand

When the electrical demand P_{load} is greater than the output mechanical power of the turbine P_{m}, the steam of the thermal power plant, or water of the hydroelectric power plant, must increase to match the new demand as shown in Equation (13.15). This is achievable if the change in demand is slow. However, matching a rapid increase in demand is a much more difficult process as the increase in mechanical power is a slow process; for hydroelectric plants, the water time constant is about 7 to 10 s. Until the mechanical power increases, the hope is that enough kinetic energy is stored in the system to match the new demand for a short period. If not, one solution would be to increase the imported power or decrease the exported power. This must be coordinated with the other utilities to ensure that they are able to cope with the new changes. If none of these solutions is possible, the utility may have to shed some of its customers' loads to maintain the power balance in Equation (13.15). This is known as a *rolling blackout*.

To reduce the reliance on rolling blackouts to balance the system powers, utilities install enough spinning reserves in each pool. The spinning reserves are rotating generators in power plants that are on standby to generate electricity very quickly. These spinning reserves are steam power plants with their steam temperature and pressure kept at values necessary for generating electricity on short notice. When the demand is rapidly increasing, the utility can use its own spinning reserve before it reaches out to neighboring utilities. This way, the reliability of the system is greatly enhanced. Figure 13.18 shows the modified power pool with the spinning reserve P_{s}.

With the spinning reserve, the stable operation of the power pool is when

$$(P_{\text{g}} + P_{\text{import}} + P_{\text{s}}) > (P_{\text{load}} + P_{\text{export}}) \tag{13.16}$$

Hence,

$$(P_{\text{g}} + P_{\text{import}} + P_{\text{s}}) = \lambda(P_{\text{load}} + P_{\text{export}}) \tag{13.17}$$

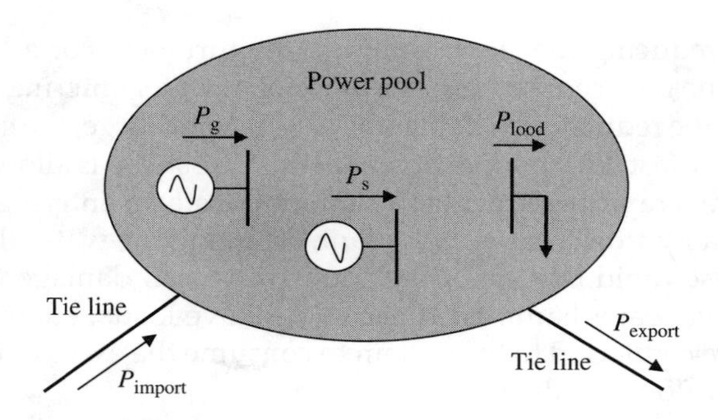

FIGURE 13.18
Power balance of a pool with spinning reserve.

where λ is defined as the *pool power margin*. For a robust system, the margin must be greater than one.

EXAMPLE 13.4 A power pool has a generation capacity of 1.5 GW. The trade commitments and forecasted load at four different times is given in the following table.

Time	P_{load}	P_{export}	P_{import}
t_1	800 MW	200 MW	0
t_2	1 GW	200 MW	0
t_3	1.2 GW	400 MW	100 MW
t_4	1.8 GW	600 MW	200 MW

Compute the spinning reserve at each of the four times to maintain at least a power pool margin of 110%.

Solution

Use Equation (13.17) to compute the spinning reserve.

At t_1,

$$P_s = \lambda(P_{load} + P_{export}) - (P_g + P_{import})$$
$$= 1.1(800 + 200) - (1500 + 0) = -400 \text{ MW}$$

There is no need for a spinning reserve at t_1.

At t_2,

$$P_s = \lambda(P_{load} + P_{export}) - (P_g + P_{import})$$
$$= 1.1(1000 + 200) - (1500 + 0) = -180 \text{ MW}$$

There is no need for a spinning reserve at t_2.
At t_3,

$$
\begin{aligned}
P_s &= \lambda(P_{load} + P_{export}) - (P_g + P_{import}) \\
&= 1.1(1200 + 400) - (1500 + 100) = 160 \text{ MW}
\end{aligned}
$$

At t_2,

$$
\begin{aligned}
P_s &= \lambda(P_{load} + P_{export}) - (P_g + P_{import}) \\
&= 1.1(1800 + 600) - (1500 + 200) = 940 \text{ MW}
\end{aligned}
$$

13.7 Blackout Scenarios

Most major blackouts occur during heavy loading conditions. The worst scenarios occur when the power plants are generating electricity close to their capacities, the transmission lines are heavily loaded, and the spinning reserves are low.

As an example of a blackout, consider the power pools shown in Figure 13.19. The pool in the middle is connected to two neighboring pools (external pool 1 and external pool 2). One tie line connects the power pool with external pool 1, and two tie lines connect the power pool with external pool 2. Assume the power pool is heavily loaded and $P_{load} > P_g$. To compensate for the deficit in power, the power pool imports P_1 from external pool 1, and $P_2 + P_3$ from external pool 2. Thus, the power pool is balanced, where

$$P_g + P_1 + P_2 + P_3 = P_{load} \tag{13.18}$$

If we add the spinning reserve P_s of the pool to the above equation, we get

$$(P_g + P_1 + P_2 + P_3 + P_s) > P_{load} \tag{13.19}$$

So far, the spinning reserve of the power pool is not needed. Assume that all tie lines are heavily loaded (close to their capacities). Now, let us assume that

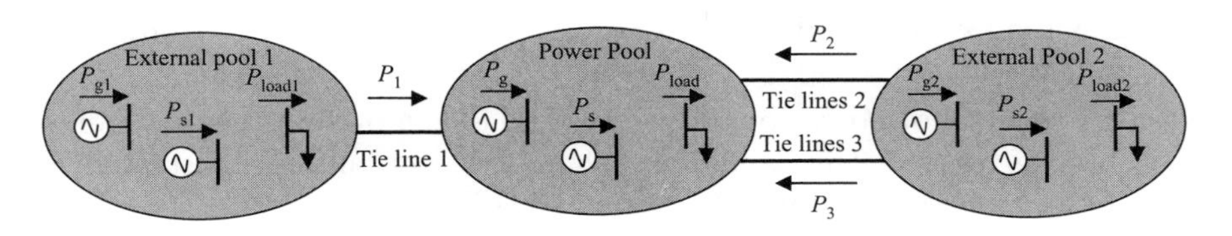

FIGURE 13.19
Blackout scenario.

tie $line_2$ is lost for some reason. In this case, the power pool has a deficit in power since

$$(P_g + P_1 + P_3) < P_{load} \tag{13.20}$$

To compensate for the deficit, the power pool uses its own spinning reserve. But, assume that the spinning reserve is not enough to compensate for the lost power P_2, that is,

$$(P_g + P_1 + P_3 + P_s) < P_{load} \tag{13.21}$$

To correct the problem, tie $line_1$ and $line_3$ need to deliver more power to the pool. This requires that (1) the external pools can provide more power than what is scheduled, and (2) tie $line_1$ and $line_3$ can handle the extra currents. If these two conditions cannot be met, the power pool will be left with a deficit in generation and must quickly trip some of its loads (rolling blackout). Otherwise, its own generating plants will shut down to prevent them from being damaged and the power pool may experience a total blackout.

13.7.1 The Great Northeast Blackout of 1965

The Great Northeast Blackout of 1965 left 30 million people without electricity in New England, New York, and Ontario. Before the blackout, the Ontario hydroelectric plant was exporting about 1.7 GW to the United States through 5 tie lines including the Massena tie line. The blackout started in the following sequence:

1. A relay protecting one of the main transmission lines in the Ontario hydroelectric plant was set too low. The line was heavily loaded and the relay tripped the transmission line. The line was directly connected to a hydroelectric plant on the Niagara River.
2. Because of the line tripping, the flow of power was shifted to the remaining four lines. These lines were heavily loaded before the power shift, and were overloaded after the shift.
3. Overloading the lines resulted in the tripping of the remaining lines successively.
4. The tripping of all lines resulted in a shortage of about 1.5 GW in the Canadian system, and the power reversed its flow to Canada from the United States.
5. The reversal of power overloaded the Massena tie line even further and was tripped. The Canadian system was then completely isolated, and became deficient in generation and collapsed.
6. The U.S. system was also generation deficient since it was dependent on about 1.7 GW from Canada. The New England and New York systems collapsed, and a total blackout occurred in their systems in a matter of seconds. It took 24 h to restore the system.

13.7.2 The Great Blackout of 1977

This blackout affected 8 million people, mainly in New York City. The blackout was the second one for N.Y. City in 12 years and lasted for 25 h.

On the hot evening of July 13, 1977, the electrical demand of New York City was very high, about 6 GW. About 3 GW of the power was imported from neighboring utilities. The area was also experiencing electrical storms (lightning storm). The blackout sequence was as follows:

1. One of the main tie lines serving Manhattan, the Bronx, Brooklyn, Queens, and Staten Island was hit by a stroke of lightning that damaged the insulators of the lines, causing the line to trip.
2. The power was shifted to the other remaining lines, which were heavily loaded.
3. The spinning reserve of the New York City was used to alleviate the stress on the remaining tie lines. The solution was adequate and the system survived the loss of the line.
4. However, the storm was severe and after 20 min another lightning strike hit another major tie line that was tripped.
5. All other lines carried currents beyond their thermal limits and sagged. One of the tie lines sagged too deeply and touched a tree causing a short circuit, and the line was tripped.
6. The New York area had a severe deficit in generation, and the system loads could not be served with the existing generations and transmission systems. The system collapsed in seconds.

13.7.3 The Great Blackout of 2003

This blackout is the worst in history, so far. About 50 million people were affected by the blackout, and the power was out for several days. NASA published interesting satellite photos of the northeastern United States before and after the blackout. These pictures are shown in Figure 13.20.

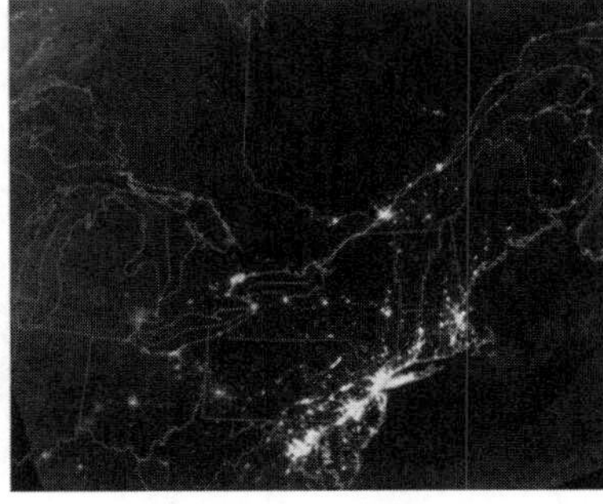

Just before the blackout

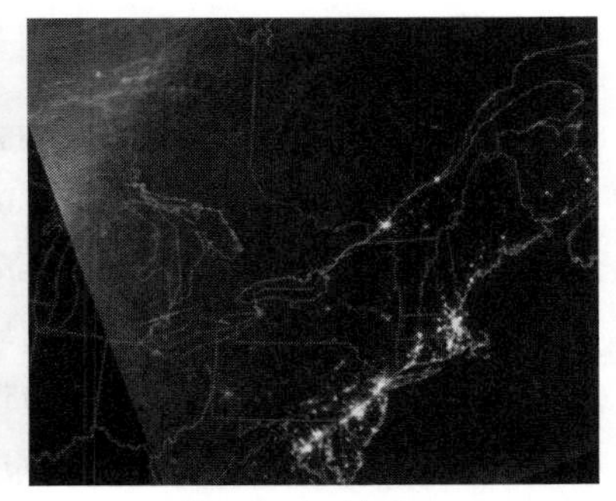

Just after the blackout

FIGURE 13.20
Satellite pictures of the northeast on August 14, 2003 (Courtesy of NASA) U.S.A.

Note how dark Manhattan Island and the Midwest region were during the blackout.

On the hot summer day of August 14, 2003, the systems in the eastern and midwestern United States as well as Ontario, Canada were heavily loaded and the blackout started in the following sequence:

1. A 680 MW coal generation plant in Eastlake, Ohio, tripped.
2. An hour later, a transmission line in northeastern Ohio tripped because it was overloaded.
3. The outage put an extra strain on the other transmission lines, and the lines sagged and touched trees causing short circuits that tripped the lines.
4. Utilities in Canada and the eastern United States experienced wild power swings causing stress on their systems. Power plants and high-voltage electric transmission lines in Ohio, Michigan, New York, New Jersey, and Ontario shut down.
5. In less than 3 min, 21 power plants were shut down including seven nuclear power plants.

Exercise

13.1. What is the spinning reserve?

13.2. State two advantages of interconnecting the power pools.

13.3. State some differences between the power grid and the internet.

13.4. What is the advantage of network grid over radial grid?

13.5. What is the advantage of connecting pools at different time zones?

13.6. Search the web and literatures to find the capacity of the ERCOT power pool in the United States. Find the area it covers and the size of the population it serves.

13.7. Search the web and literatures to find the largest power pool in the United States.

13.8. The power system in Figure 13.5 has the following data:

Impedance of load 1 plus its transformer is $\overline{Z}_{11} = 50 + j5\ \Omega$

Impedance of load 2 plus its transformer is $\overline{Z}_{22} = 40 + j2\ \Omega$

Impedance of line $_1$ is $\overline{Z}_{13} = 0 + j4\ \Omega$

Impedance of line $_2$ is $\overline{Z}_{23} = 0 + j5\ \Omega$

Impedance of line $_3$ is $\overline{Z}_{12} = 0 + j3\ \Omega$

All impedances are referred to the high-voltage side of the transformers. The voltage at the high-voltage side of xfm_g is fixed at 500 kV (line-to-line).

The capacity of any line (maximum current the line can carry) is 15 kA. Assume that all connections are in wye. Compute the power delivered to each load, and the power produced by the generator.

13.9. For the system in the previous problem assume that transmission line 3 is tripped; repeat the solution. Is the system secure?

13.10. The load demand for a given day can be approximated by

$$P = 2 + 2e^{-(t-9)^2/8} \text{ GW}$$

where t is the time of the day in hours using the 24-h clock. Compute the following:

a. The peak demand.

b. The time of the peak demands.

c. The average daily demand.

13.11. For the system in the previous problem, assume that the generation capacity is 1.9 GW. Compute the imported energy to compensate for all demands higher than 1.9 GW.

13.12. A power pool has a generation capacity of 1.5 GW and a demand of 1000 MW. The trade commitments of the pool are: $P_{\text{import}} = 500$ MW and $P_{\text{export}} = 800$ MW. Compute the spinning reserve that maintains the pool margin at 130%.

Appendix: Conversions

Metric Units

Prefix	**Value**
Exa	10^{18}
Peta	10^{15}
Tera	10^{12}
Giga	10^{9}
Mega	10^{6}
Kilo	10^{3}
Hector	10^{2}
Deka	10^{1}
Deci	10^{-1}
Centi	10^{-2}
Milli	10^{-3}
Micro	10^{-6}
Nano	10^{-9}
Pico	10^{-12}
Femto	10^{-15}
Atto	10^{-18}

Distance

1 mi = 1.609 km

1 ft = 0.30480 m

1 yd = 0.91440 m

1 in. = 0.02540 m

Area

$1 \text{ acre} = 4046.9 \text{ m}^2$

$1 \text{ mi}^2 = 2.59 \text{ km}^2$

$1 \text{ km}^2 = 100 \text{ ha}$

Volume

$1 \text{ cu foot} = 0.02832 \text{ m}^3$

$1 \text{ m}^3 = 1.308 \text{ cu yard}$

1 fluid oz = 0.02957 l

1 U.S. gal = 3.78540 l

1 l = 1 kg of water

1 U.S. gal = 3.7854 kg of water

Mass

1 lb = 0.45360 kg

1 oz = 28.35 g

1 ton = 1000 kg

Force

$1 \text{ lb/in}^2 = 7.03 \times 10^2 \text{ kg/m}^2$

$1 \text{ lb/in}^2 = 6.89 \text{ kPa}$

Speed

Speed of light = 299,792,458 m/s

Energy

British thermal unit (BTU) = 252 Cal

British thermal unit (BTU) = 1.0544 kJ (kWh)

1 Nm = 1 Ws

Power

1 hp = 1.34 kW

Key Integrals Used in the Textbook

$$\int \mathrm{e}^{ax}\cos(bx) = \frac{\mathrm{e}^{ax}}{a^2+b^2}[a\cos(bx) + b\sin(bx)]$$

$$\int_0^{24} P_{\max}\mathrm{e}^{-(t-t_o)^2/2\sigma^2}\,\mathrm{d}t \approx P_{\max}\sqrt{2\pi}\sigma$$

$$\int \mathrm{e}^{ax}[c\cos(bx) + k\sin(bx)]\,\mathrm{d}x = \frac{\mathrm{e}^{ax}}{a^2+b^2}[(ka+cb)\sin(bx) + (ca-kb)\cos(bx)]$$

Resistivity

Aluminum	4.0×10^{-08} Ωm
Copper	1.673×10^{-08} Ωm
Nickel	10^{-07} Ωm
Steel	7.2×10^{-07} Ωm

Index